The PDMA ToolBook 1 for New Product Development

Edited by

Paul Belliveau
Paul Belliveau Associates

Abbie Griffin
University of Illinois at Urbana-Champaign

Stephen Somermeyer
Your Encore™

WILEY

John Wiley & Sons, Inc.

This book is printed on acid-free paper. ∞

Copyright © 2002 by John Wiley & Sons, Inc., New York. All rights reserved.

Published simultaneously in Canada.

No part of this publication may be reproduced, stored in a retrieval system or transmitted in any form or by any means, electronic, mechanical, photocopying, recording, scanning or otherwise, except as permitted under Sections 107 or 108 of the 1976 United States Copyright Act, without either the prior written permission of the Publisher, or authorization through payment of the appropriate per-copy fee to the Copyright Clearance Center, 222 Rosewood Drive, Danvers, MA 01923, (978) 750-8400, fax (978) 750-4744. Requests to the Publisher for permission should be addressed to the Permissions Department, John Wiley & Sons, Inc., 605 Third Avenue, New York, NY 10158-0012, (212) 850-6011, fax (212) 850-6008, E-Mail: PERMREQ @ WILEY.COM.

This publication is designed to provide accurate and authoritative information in regard to the subject matter covered. It is sold with the understanding that the publisher is not engaged in rendering professional services. If professional advice or other expert assistance is required, the services of a competent professional person should be sought.

Library of Congress Cataloging-in-Publication Data:

PDMA toolbook for new product development / [edited by] Paul Belliveau, Abbie Griffin, Stephen Somermeyer.
        p.   cm.
"Published simultaneously in Canada."
Includes bibliographical references and index.
   ISBN 0-471-20611-3 (alk. paper)
   1. New products—Management.   2. Product management.   I. Title: Toolbook for new product development.   II. Belliveau, Paul.   III. Griffin, Abbie.   IV. Somermeyer, Stephen.
   HF5415.153 .P355   2002
   658.5'75—dc21   2002003707

Printed in the United States of America.

10 9 8 7 6 5

Contents

Part 4

Contributors

Greg M. Ajamian, *E.I. du Pont de Nemours and Company*, Wilmington, DE
Paul Belliveau, *Paul Belliveau Associates*, Westfield, NJ
Gary Blau, *Purdue University*, West Lafayette, IN
Scott Boyce, *Rohm and Haas Company*, Spring House, PA
Paul Bunch, *Eli Lilly and Company*, Indianapolis, IN
George Castellion, *SSC Associates*, Stamford, CT
Allen Clamen, *ExxonMobil Chemical Company (retired)*, Houston, TX
Robert G. Cooper, *Product Development Institute*, Ancaster, ON, Canada
Mark J. Deck, *Pittiglio Rabin Todd & McGrath*, Waltham, MA
David J. Dunham, *David Dunham & Co.*, Clifton, NJ
Scott J. Edgett, *Product Development Institute*, Ancaster, ON, Canada
Eden Fisher, *Alcoa, Inc.*, Alcoa Center, PA
Stavros Fountoulakis, *Bethlehem Steel Corporation*, Bethlehem, PA
Gregory D. Githens, *Catalyst Management Consulting, LLC*, Columbus, OH
Christine Gorski, *Bank One*, Chicago, IL
Abbie Griffin, *University of Illinois at Urbana-Champaign*, Champaign, IL
Eric J. Heinekamp, *Bank One*, Chicago, IL
Jan Hollander, *Essent Energy*, Den Bosch, The Netherlands
Albert Johnson, *Corning Incorporated*, Corning, NY
Elko J. Kleinschmidt, *Product Development Institute*, Ancaster, ON, Canada
Peter A. Koen, *Stevens Institute of Technology*, Hoboken, NJ
Jan Kratzer, *University of Groningen*, Groningen, The Netherlands
Roger Th. A. J. Leenders, *University of Groningen*, Groningen, The Netherlands
Stephen K. Markham, *North Carolina State University*, Raleigh, NC
Lee Meadows, *Business Genetics, Inc.*, Annapolis, MD
Robert J. Meltzer, *The RJM Consultancy*, Kirkland, WA
Charles Miller, *Insight MAS*, Dublin, OH
Christopher W. Miller, *Innovation Focus Inc.*, Lancaster, PA
James L. Mueller, *J.L. Mueller, Inc.*, Centennial, CO
Pushpinder Puri, *Air Products and Chemicals, Inc.*, Allentown, PA
Rebecca Seibert, *Crompton Corporation*, Middlebury, CT
Stephen Somermeyer, *Your Encore™* Indianapolis, IN
Molly Follette Story, *North Carolina State University*, Raleigh, NC
David C. Swaddling, *Insight MAS*, Dublin, OH
Jo M. L. van Engelen, *University of Groningen*, Groningen, The Netherlands

Chapter 4, "Focusing NPD Research on Customer Perceived Value," describes market research methods and tools that help a firm understand how customers evaluate all of the benefits and costs of an offering and compares them to the benefits and costs of other products or services that they perceive as being alternatives. Although this customer perceived value is the basis upon which customers decide which products and services to purchase, it is difficult to quantify because it is market perceived (not firm imposed), complicated, important only as it is relative to other alternatives, and dynamic because marketplaces are always changing. The chapter overviews techniques for understanding customer wants and needs, identifying value attributes, and understanding market factors, perceived importance, and perceived relative performance. The understanding gained from developing this information can then be applied to specific new product development issues. Firms entering new markets or participating in dynamic markets or business environments will find this tool especially useful in maintaining product success.

1

Fuzzy Front End: Effective Methods, Tools, and Techniques

Peter A. Koen, Greg M. Ajamian, Scott Boyce,
Allen Clamen, Eden Fisher, Stavros Fountoulakis,
Albert Johnson, Pushpinder Puri,
and Rebecca Seibert

The innovation process may be divided into three areas: the fuzzy front end (FFE), the new product development (NPD) process, and commercialization, as indicated in Figure 1-1.[1] The first part—the FFE—is generally regarded as one of the greatest opportunities for improvement of the overall innovation process.[2] Many companies have dramatically improved cycle time and efficiency by implementing a formal Stage-Gate™ (Cooper 1993) or PACE® (McGrath and Akiyama 1996) approach for managing projects in the NPD portion of the innovation process. Attention is increasingly being focused on the front-end activities that precede this formal and structured process in order to increase the value, amount, and success probability of high-profit concepts entering product development and commercialization.

The purpose of this chapter is to provide the reader with the most effective methods, tools, and techniques for managing the FFE.[3] The chapter begins with a brief discussion of the literature and the rationale for developing the new concept development (NCD) model. The next section describes the NCD model. The remaining sections provide a description of the most effective methods, tools, and techniques to be used in each part of the NCD model.

LITERATURE REVIEW AND RATIONALE FOR DEVELOPING THE NCD MODEL

Best practices are well known at the start (Khurana and Rosenthal 1998) and within the NPD portion (Brown and Eisenhardt 1995; Cooper and Klein-schmidt 1987; Griffin and Page 1996) of the innovation process. Similar research on best practices in the FFE is absent. Many of the practices that aid the NPD portion do not apply to the FFE. They fall short, as shown in Table

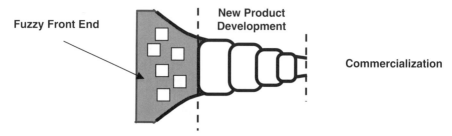

Fuzzy Front End   New Product Development

Commercialization

FIGURE 1-1. The entire innovation process may be divided into three parts: fuzzy front end (FFE), new product development (NPD), and commercialization.
The division between the FFE and the NPD is often less than sharp, since technology development activities may need to be pursued at the intersection.

1-1, because the nature of work, commercialization date, funding level, revenue expectations, activities, and measures of progress are fundamentally different.

Lack of research into best practices made the FFE one of the most promising ways to improve the innovation process. An Industrial Research Institute multicompany project team began studying the FFE in the middle of 1998 to describe and share best practices.[4] However, our work was stymied at first due

TABLE 1-1.
Difference Between the Fuzzy Front End (FFE) and the New Product Development (NPD) Process

	Fuzzy Front End (FFE)	New Product Development (NPD)
Nature of Work	Experimental, often chaotic. "Eureka" moments. Can schedule work—but not invention.	Disciplined and goal-oriented with a project plan.
Commercialization Date	Unpredictable or uncertain.	High degree of certainty.
Funding	Variable—in the beginning phases many projects may be "bootlegged," while others will need funding to proceed.	Budgeted.
Revenue Expectations	Often uncertain, with a great deal of speculation.	Predictable, with increasing certainty, analysis, and documentation as the product release date gets closer.
Activity	Individuals and team conducting research to minimize risk and optimize potential.	Multifunction product and/or process development team.
Measures of Progress	Strengthened concepts.	Milestone achievement.

to the difficulty of comparing FFE practices across companies. The comparison was complicated because there was a lack of common terms and definitions for key elements of the FFE. Without a common language and vocabulary, the ability to create new knowledge and make distinctions between different parts of the process may be impossible (Krough, Ichijo, and Nonaka 2000). Knowledge transfer is ineffective or unlikely if both parties mean different things, even when they are using the same terms. These insights led us to believe that we could improve understanding of the FFE by describing it using terms that mean the same thing to everyone.

To address this shortcoming, we developed a theoretical construct, the NCD model (Koen et al. 2001). It is intended to provide insight and a common terminology for the FFE. Typical representations of the front end consist of a single ideation step (Cooper 1993). However, the actual FFE is more iterative and complex. To create the model, participants provided in-depth reviews of the FFE experience in their companies. Factors common to FFE activities at all companies were identified next. Differences in both terminology and content among FFE activities were then discussed and resolved. We argued with intensity for a long time trying to devise a sequential FFE model similar to the traditional Stage-Gate™ process. All of us had demonstrated success with Stage-Gate™ processes for NPD and assumed that a similar sequential process would work for the FFE. Our argument made us realize that a sequential process model was not appropriate. This important realization allowed us to move from a sequential process model to a nonsequential relationship model.

This chapter presents our understanding of effective tools and techniques in the FFE using the NCD model. The methods, tools, and techniques discussed were determined from the best practices within our companies, an extensive search of the literature, and a review of techniques utilized by consulting firms and our colleagues. In addition, all of the authors have considerable personal experience with the FFE.

The remaining sections start with an overview of the NCD model. Following that, each part of the model is described along with the methods, tools, and techniques that the authors believe are effective.

DEFINITIONS

Opportunity: *A business or technology gap, that a company or individual realizes, that exists between the current situation and an envisioned future in order to capture competitive advantage, respond to a threat, solve a problem, or ameliorate a difficulty.*

Idea: *The most embryonic form of a new product or service. It often consists of a high-level view of the solution envisioned for the problem identified by the opportunity.*

Concept: *Has a well-defined form, including both a written and visual description, that includes its primary features and customer benefits combined with a broad understanding of the technology needed.*

NEW CONCEPT DEVELOPMENT MODEL

The NCD model shown in Figure 1-2 consists of three key parts:

◆ The engine or bull's-eye portion is the leadership, culture, and business strategy of the organization that drives the five key elements that are controllable by the corporation.

◆ The inner spoke area defines the five controllable activity elements (opportunity identification, opportunity analysis, idea generation and enrichment, idea selection, and concept definition) of the FFE.

◆ The influencing factors consist of organizational capabilities, the outside world (distribution channels, law, government policy, customers, competitors, and political and economic climate), and the enabling sciences (internal and external) that may be involved.[5] These factors affect the entire innovation process through to commercialization. These influencing factors are relatively uncontrollable by the corporation.

Several characteristics of the model are worth noting. The inner parts of the NCD are called elements, as opposed to processes. A process implies a struc-

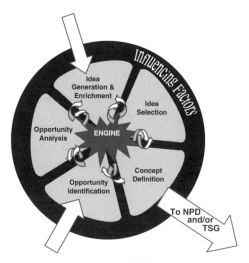

FIGURE 1-2. The new concept development (NCD) construct is a relationship model, not a linear process.
It provides a common language and definition of the key components of the fuzzy front end (FFE). The engine, which represents senior- and executive-level management support, powers the five elements of the NCD model. The engine and the five elements of the NCD model are placed on top of the influencing factors. The circular shape of the NCD model is meant to suggest that ideas and concepts are expected to iterate across the five elements. The arrows pointing into the model represent starting points and indicate that projects begin at either opportunity identification or idea generation and enrichment. The exiting arrow represents how concepts leave the model and enter the new product development (NPD) or technology stage gate (TSG) process.

ture that may not be applicable and could force the use of a set of poorly designed controls to manage FFE activities. In addition, the model has a circular shape, to suggest that ideas are expected to flow, circulate, and iterate between and among all the five elements. The flow may encompass the elements in any order or combination and may use one or more elements more than once. This is in contrast to the sequential NPD or Stage-Gate™ process, in which looping back and redirect or redo activities are associated with significant delays, added costs, and poorly managed projects. Iteration and loop-backs are part of FFE activities. While the inherent looping back may delay the FFE, it typically shortens the total cycle time of product development and commercialization. Clearer definition of market and technical requirements, sources of risk and a well defined business plan for the new product may enable more effective management of the development and commercialization stages with fewer 'redo' or 'redirect' activities. In contrast, the overall project cycle time and costs grow exponentially whenever there is redo activity as the project moves downstream through the NPD or Stage-Gate™ process (Wheelwright and Clark 1992).

An example of looping back and iteration took place when Spence Silver at 3M first identified the strange adhesive that was more tacky than sticky and which later enabled the development of the 3M Post-it notepads. Initially there were no product ideas for this concept—though Silver visited most of the divisions at 3M in order to find one. The initial idea was to develop a bulletin board coated with the tacky adhesive, to which people would attach plain-paper notices. This concept was never realized, and a new concept, which eventually became 3M Post-its, was later proposed by looping back into opportunity identification and opportunity analysis from idea generation and enrichment. Constant iteration and flow within the FFE is a hallmark of activities in this stage of the product development process.

Even though the key elements of the FFE will be discussed in a clockwise progression, they are expected to proceed nonsequentially, as shown by the looping arrows between the elements. Further, the separation between the influencing factors (i.e., environment) and the key elements is not rigid. Interactions and intermingling between the influencing factors, the five key elements, and the engine are expected to occur continuously.

The following sections discuss influencing factors, the engine, and each of the five key elements in more detail. Methods, techniques, and tools utilized will be indicated. Two examples—one market-driven and one technology-driven—highlight the characteristics of each part of the model.

EXAMPLES

The market-driven example is the development of nonfat potato chips using a fat substitute (a substance that provides the same flavor as fat but is not absorbed in the body). The technology-driven example is the development of 3M Post-it notepads (Nayak and Ketteringham 1994).

INFLUENCING FACTORS (THE ENVIRONMENT)

The FFE exists in an environment of influencing factors. The factors are the corporation's organizational capabilities, customer and competitor influences, the outside world's influences, and the depth and strength of enabling sciences and technology. Sustained successful product development can occur only when FFE activities can be accomplished with the company's organizational capabilities. Organizational capabilities determine whether and how opportunities are identified and analyzed, how ideas are selected and generated, and how concepts and technologies are developed. Organizational capabilities can also include organized or structured efforts in acquiring external technology. Electronics and pharmaceutical companies have a long history of augmenting their product development efforts with external licensing, joint development agreements, and the development of testing methodologies and protocols (Slowinski et al. 2000). These capabilities exert influence and give the organization the ability to deal with the influencing factors.

Enabling science and technology is also critical, since technology typically advances by building upon earlier achievements. Science and technology become enabling when they can be used repeatedly in a product or service. "Enabling" is not the same as "mature," which is defined on a technology trend line or penetration curve. It is the point when the technology is developed enough to build it into a manufactured product or regular service offering. Enabling technologies usually provide some degree of enhanced utility, cost avoidance, value, or quality improvement for the customer. Technologies typically become enabling early in their life cycle.

The outside world, government policy, environmental regulations, laws concerning patents, and socioeconomic trends all affect the FFE as well as the new product development or Stage-Gate™ part of the innovation process. Some of these factors are indicated in Porter's "five force" model (1987). Porter's model evaluates the relative power of customers, competitors, new entrants, suppliers, and industry rivalry—a power relationship that determines the intensity of competition and often inspires innovation.

Complementors are companies that are not direct competitors, that serve to help grow one's industry, and should be considered a sixth force (Grove 1999). For instance, complementors to Microsoft are Intel and Dell. Each of these companies complements the others in building an industry. Government law and policy should be considered a seventh force, because of their impact on the use of and profit from a technology.

These factors, constantly influencing people's thoughts and actions, are primary contributors to "serendipitous discovery" of new ideas. Just as a healthy marine environment is essential for a healthy population of aquatic species, so is a supportive climate essential for a productive FFE. These influencing factors are largely uncontrollable by the corporation. However, the response by the engine (corporate culture, leadership, and strategy) greatly affects the NCD's five activity inner elements. The response may also impact the organizational capabilities of the company—internal development as well

---

EXAMPLES

The influencing factors in the nonfat potato chip example would be the increasing consumer desire for nonfat products and cholesterol reduction, the regulatory environment for food, awareness that a competitor was beginning research efforts on fat substitutes that could be used in a nonfat potato chip, and the company's organizational capabilities (from product design, market evaluation, and distribution of potato chip products) in understanding this marketplace.

The influencing factors for 3M Post-it notepads were the organizational capabilities and enabling science in adhesives.

---

as external access through joint development or licensing—although these capabilities usually change much more slowly than the response by the engine.

Effective Methods, Tools, and Techniques

The ability to execute the strategy or plan of action when changes occur is a key tool for addressing influencing factors. For example, Corning enjoyed huge success in developing the successful ceramic substrate for catalytic converters. That success was a direct result of senior executives' early awareness of the Clean Air Act's requirement for reduced emissions and of the huge potential of the business. These factors were so compelling that Corning, in 1970, directed hundreds of scientists and engineers to focus on this single challenge. The resulting product has been used in more than three hundred million automobiles.

New alliances and partnerships may provide the capabilities needed for addressing influencing factors. Examples may be found in the automotive and automotive materials industries. Energy conservation and the drive to improve the quality of life and reduce pollution motivated people in these industries to establish research alliances, industry consortia, and industry-government collaborative R&D ventures. U.S. automakers and their suppliers, government labs, and several universities formed the U.S. Council for Automotive Research (USCAR), an alliance to generate and develop concepts such as a highly fuel-efficient (over eighty miles per gallon) vehicle. This new spirit of collaborative research changed the way the automakers accepted new processes and techniques. Alternative materials such as aluminum, polymers, and composites were able to show their advantages in safety, fuel economy, and vehicle performance.

When the global steel industry sensed a competitive threat, they reacted in turn. Steel industry leaders thought USCAR members could develop new structures and materials that might displace steel. In response to the challenge, more than thirty-five steel producers from around the world formed the Ultra Light Steel Auto Body research consortium. That consortium contracted research to generate and develop new ways to use steel in cars. They developed concept vehicles and built prototypes to show how vehicles and individual components made out of steel can be as much as 40 percent lighter than conventional com-

Most Effective Methods, Tools, and Techniques
Ability to execute the strategy or plan effectively and quickly when the environment changes.

ponents with no cost penalty. They accomplished this through novel architectures, new manufacturing techniques (e.g., hydroforming instead of stamping and welding of parts, tailor-made blanks, laser welding for assembly), and advanced new steel formulations (e.g., complex microstructures to provide for ultrahigh strength combined with light weight and good formability to address engineering and styling demands).

The influencing factors at work on the automakers and their suppliers are inspiring approaches to innovation that bring together the best attributes of multiple materials and organizations' technologies. Overall, the materials innovations are helping produce automobiles that are safer and more fuel-efficient, with longer service lives, adding to customer value.

Ability to execute the strategy or plan depends on quickly and effectively communicating influencing factors throughout the entire organization. Effective communication of the presence and impact of influencing factors and the gathering and organizing of quality information are critical to early foresight. Early foresight in turn provides early warning that gives decision makers time to decide and act. Capacity and time to decide and act are the most valuable resources to have when there are significant shifts in the influencing factors. This is because developing new, enabling technology for new products or services requires a time investment. The Corning, steel, and aluminum industry examples teach us that the impact of influencing factors can be changed favorably by communicating about them in a way that improves foresight and triggers action.

THE ENGINE (LEADERSHIP, CULTURE, AND BUSINESS STRATEGY)

The element of leadership, culture, and business strategy sets the environment for successful innovation. Proficiency in this element distinguishes highly innovative companies from less innovative ones (Koen et al. 2001). Continuous senior management support for innovation has been shown in numerous studies to be critical to new product development success (Cooper and Kleinschmidt 1995; Song and Parry 1996; Swink 2000). In their study of breakthrough projects, Lynn, Morone, and Paulson (1996) indicate that the huge success of Corning's optical fiber, GE's computerized axial tomography scanners, and Motorola's cellular phone—all of which had long gestation periods—were possible only because "senior management persisted because these opportunities made strategic sense. They fit the strategic focus of the business." The entire innovation process (including both FFE and NPD) needs to be aligned with

---

Most Effective Methods, Tools, and Techniques

◆ Create more opportunities by envisioning the future through:
  ➢ Roadmapping
  ➢ Technology trend analysis
  ➢ Customer trend analysis
  ➢ Competitive intelligence analysis
  ➢ Market research
  ➢ Scenario planning

---

to sell GE appliances in its stores without Home Depot carrying the inventory. GE would deliver the appliances directly from its own warehouses. Using this new strategy, GE is on schedule to move 45 percent of its $2.5 billion appliance sales to the Internet, opening whole new segments while decreasing overall transaction costs. Envisioning a new future through the eyes of competition triggered this new strategy.

OPPORTUNITY ANALYSIS

In this element, an opportunity is assessed to confirm that it is worth pursuing. Additional information is needed for translating opportunity identification into specific business and technology opportunities. This involves making early and often uncertain technology and market assessments. Extensive effort may be committed for focus groups, market studies, and/or scientific experiments. However, the effort expended will depend on the value of the information associated with reducing uncertainties about the attractiveness of the opportunity, the expected size of the future development effort given the fit with the business strategy and culture, and the decision makers' risk tolerance.

Opportunity analysis may be part of a formal process or may occur iteratively. Business capability and competency are assessed in this element, and sponsorship for further work will be determined. However, despite all of the effort, significant technology and market uncertainty will remain.

EXAMPLES

Opportunity analysis occurred in the nonfat potato chip example when the food company examined the trends in more detail. Did consumers really want a low-fat product, or did they want one that was low-calorie and/or low-cholesterol? How much taste would consumers give up? Was the market mainly a small niche? What were the regulatory issues? In this element the food company also examined the value of such an effort to their portfolio and the competitive threats if they did not develop such products.

Opportunity analysis in the 3M case took place when Silver attempted to find an opportunity for this strange adhesive. Silver visited every division at 3M in his quest to find a business opportunity for this new technology.

Effective Methods, Tools, and Techniques

Many of the same tools used in opportunity identification are used in this element as well. Roadmapping, technology trend analysis, competitive intelligence analysis, customer trend analysis, and scenario planning are all employed in this element. In opportunity identification, these tools were used to determine if an opportunity existed. In this element, considerably more resources are expended, providing more detail on the appropriateness and attractiveness of the selected opportunity. A typical analysis for a large-scale opportunity would include:

- *Strategic framing.* A determination of how this opportunity fits within the company's market and technology strengths, gaps, and threats.
- *Market segment assessment.* A detailed description of the market segment, showing why it represents a great opportunity. Market size analysis, growth rates, and market share of competitors are determined. Economic, cultural, demographic, technological, and regulatory factors that impact the market segment are also evaluated. Often companies will only evaluate opportunities in markets greater than a certain size (such as those with revenue greater than $100 million and growing at 10 percent per year.)
- *Competitor analysis.* Determines who the major competitors are in the identified market segment. Determines the type of new products needed in order to achieve competitive advantage. Evaluates the competitors' strategies and capabilities and the status of recent patents in this area.
- *Customer assessment.* Determines what major customer needs are not being met by current products.

An effective practice in this element for a large-scale opportunity is to assign a specific, multifunctional team whose members work full time to perform the opportunity analysis. The size and makeup of these teams depend on the size, scope, and complexity of the effort and the culture of the organization. Teams typically number three to five people and usually contain a marketing and R&D person. The team effort should begin with a project charter that provides a clear set of expectations, committing resources and outlining the expected outcome. Without such a charter the team will often squander their efforts by evaluating opportunities outside their focus. The content of the project charter is similar to the product innovation charter discussed by Crawford and DiBenedetto (2000) but is focused on identifying new opportunities instead of new products. The team will also benefit from a clear analytical framework for assessing opportunities and the assistance of an experienced analyst. One example of an analytical framework for assessing technical opportunities is the context graph of historical performance, benchmarks, and theoretical and engineering limits that has been used by Alcoa (Turnbull et al. 1992).

   An opportunity analysis for a large-scale opportunity may take approximately sixty to ninety days. Shorter efforts result in assessments from mostly

Most Effective Methods, Tools, and Techniques

◆ Same methods, tools, and techniques used to determine future opportunities, but the effort would be expanded in considerably more detail
◆ Assignment of a full-time specific multifunctional team of three to five people for large projects
   ➢ Creating a charter for the team that points them in the right direction

secondary sources and lack the richness of an in-depth competitor and customer assessment. The level of detail should minimize technical, market and commercial risk and state assumptions used in the opportunity analysis to support the conclusions. This element of NCD is used for identifying the right customer and market segments or for identifying an area of significant technical potential. Further effort in the concept definition element will provide more detail about the opportunity. The desire for great detail in this element must be balanced against the knowledge that the opportunity analysis project will stall if the information collection effort becomes so exhaustive that the project never moves forward.

In many cases the team will loop back to opportunity analysis as new features and constraints are identified in the concept definition stage. Will these new features increase the market, and if so, by how much? If the project cannot deliver on these features, what is the impact? In some cases the team may loop back to opportunity identification to identify entirely new opportunities that were not envisioned at the start of the project. However, the new opportunities should be pursued only if they remain consistent with the team's charter, which should have been defined prior to the start of opportunity analysis.

IDEA GENERATION AND ENRICHMENT

The element of idea generation and enrichment concerns the birth, development, and maturation of a concrete idea. Idea generation is evolutionary. Ideas are built up, torn down, combined, reshaped, modified, and upgraded. An idea may go through many iterations and changes as it is examined, studied, discussed, and developed in conjunction with other elements of the NCD model. Direct contact with customers and users and linkages with other cross-functional teams as well as collaboration with other companies and institutions often enhance this activity.

Idea generation and enrichment may be a formal process, including brainstorming sessions and idea banks so as to provoke the organization into generating new or modified ideas for the identified opportunity. A new idea may also emerge outside the bounds of any formal process—such as an experiment that goes awry, a supplier offering a new material, or a user making an unusual request. Idea generation and enrichment may feed opportunity identification, demonstrating that the NCD elements often proceed in a nonlinear fashion, advancing and nurturing ideas wherever they occur.

EXAMPLES

Idea generation and enrichment occurred in the nonfat potato chip example when several methods of delivering nonfat potato chips were identified. Some ideas involved reducing the total fat content; others were about the development of a fat substitute that could provide the same flavor as fat but would not be absorbed in the body.

Idea generation and enrichment in the 3M example occurred when several product ideas were identified, such as the sticky bulletin board and notepads.

Ideas may be generated by anyone with a passion for a particular idea, problem, need, or situation. Ideas may be generated or enriched by others through the efforts of a key individual or "champion" (Markham 1998; Markham and Griffin 1998). Once the idea is identified, many different creativity techniques can be applied to generate and expand upon it.[9] Those techniques can be used either by individuals or by a team in a brainstorming meeting or other idea-generation session.

Effective Methods, Tools, and Techniques

Understanding the customer and market needs is a consistent theme for successful product development in studies by Bacon and colleagues (1994), Song and Parry (1996), and Cooper (1999). There are many creativity and brainstorming techniques for enriching the idea stream. Other methods for enriching the idea stream utilize TRIZ, the Russian acronym for Theory of Inventive Problem Solving, which is a systematic way for solving problems and creating multiple-alternative right solutions. TRIZ is a methodology that enhances creativity by getting individuals to think beyond their own experience and to reach across disciplines to solve problems using solutions from other areas of science (Altshuller, 1999). Some of the most effective tools and techniques include:

◆ An organizational culture that encourages employees to spend unscheduled time testing and validating their own and others' ideas.

◆ A variety of incentives (e.g., awards, peer recognition, performance appraisal) to stimulate the generation and enrichment of ideas.

◆ A Web-enabled idea bank with easy access to product or service improvements, including linkages to customers and suppliers.

◆ A formal role for someone to coordinate ideas from generation through assessment.

◆ A mechanism to handle ideas outside (or across) the scope of established business units.

◆ A limited number of simple, measurable goals (or metrics) to track idea generation and enrichment. These could include: number of ideas retrieved and enhanced from an idea portfolio, number of ideas generated/enriched over a period of time, percentage of ideas commercialized, value of ideas in a idea portfolio (or idea bank), percentage of ideas that entered the NPD process, percentage of ideas that resulted in patents, and percentage of ideas accepted by a business unit for development.

◆ Frequent job rotation of engineers (Harryson 1997), scientists, and inventors to encourage knowledge sharing and extensive networking.

◆ Mechanisms for communicating core competencies, core capabilities, and shared technologies broadly throughout the corporation.

◆ Inclusion of people with different cognitive styles on the idea enrichment team (Leonard and Straus 1997; Prather 2000).

However, many of these techniques do not lead to breakthrough ideas. Von Hippel (1986) indicated that the actual first user of a product develops over 75 percent of breakthrough inventions. This occurs because the tacit knowledge stays with the user. Von Hippel (1998) refers to this knowledge as "sticky" since it is difficult to transfer from the lead user to others. One method for better understanding the tacit knowledge of the customer utilizes lead user methodology, which involves working with lead and analog[10] users (von Hippel, Thomke, and Sonnack 1999). 3M has utilized the lead user process to develop a way to prevent infections that is less costly and more effective than

---

Most Effective Methods, Tools, and Techniques

◆ Methods for identifying unarticulated customer needs include:
  ➤ Ethnographic approaches
  ➤ Lead user methodology
◆ Early involvement of customer champion
◆ Discovering the archetype of your customer. (Archetype[12] research identifies the unstated "reptilian" or instinctive part of the brain)
◆ Market and business needs and issues continuously interspersing with the technology advances
◆ Identifying new technology solutions
  ➤ Increasing technology flow through internal and external linkages.
  ➤ Partnering
◆ An organizational culture that encourages employees to spend free time testing and validating their own and others' ideas
◆ A variety of incentives to stimulate ideas
◆ A Web-enabled idea bank with easy access to product or service improvements, including linkages to customers and suppliers
◆ A formal role for someone (i.e., process owner) to coordinate ideas from generation through assessment
◆ A mechanism to handle ideas outside (or across) the scope of established business units
◆ A limited number of simple, measurable goals (or metrics) to track idea generation and enrichment
◆ Frequent job rotation to encourage knowledge sharing and extensive networking
◆ Mechanisms for communicating core competencies, core capabilities, and shared technologies broadly throughout the corporation
◆ Inclusion of people with different cognitive styles on the idea enrichment team

EXAMPLES

Idea selection occurred in the nonfat potato chip example when a particular fat substitute molecule was chosen.

Idea selection occurred in the 3M example when the notepad idea was selected for continued development.

traditional surgical drapes. The company has successfully tested the lead user methodology in eight of its fifty-five divisions. Ethnographic approaches involve methods for gaining intimate knowledge of the customer by becoming part of their habitat (Burchill and Brodie 1997).[11] Michaels (2000) notes that Motorola employed anthropological observation to develop two-way pagers for the rural Chinese market.

IDEA SELECTION

In most instances, the problem is not coming up with new ideas. Even when businesses are being downsized, there is no shortage of new ideas. The problem for most businesses is in selecting which ideas to pursue in order to achieve the most business value. Making a good selection is critical to the future health and success of the business. However, there is no single process that will guarantee a good selection. Most idea selection involves an iterative series of activities that are likely to include multiple passes through opportunity identification, opportunity analysis, and idea generation and enrichment, often with new insights from the influencing factors and new directives from the engine.

Selection may be as simple as an individual's choice among many self-generated options, as formalized as a prescribed portfolio management method, or as complex as a multistage business process. Formalized decision processes in the FFE are difficult due to the limited information and understanding that are available early in product development. Financial analyses and estimates of future income for ideas at this early stage are often wild guesses. Idea selection is expected to be less rigorous in FFE than in the NPD portion, since many ideas must be allowed to grow and advance. Additional effort will be invested to define the concept after the idea has been selected.

Effective Methods, Tools, and Techniques

Idea selection often begins as individual judgment, which may occur subconsciously. Often early personal judgments are made at an emotional or "gut" level, with little more than the idea itself to consider. Idea selections within an individual's own mind are almost always the initial part of the selection process.

Although there may not be a single most effective practice for idea selection, our experience has shown that without some formal decision process to commit business resources (time, funding, and people), most new ideas disappear into a kind of black hole. In the authors' experiences, formal processes work as long as there is visible support from management and there is a process owner facilitating the activity.[13] Many people who submit ideas into a suggestion box or other collection process never hear any follow-up. As a result, they are less likely to submit their next new idea, and the stream of new ideas dries up. For this reason, communicating to the originator about what is happening with his or her idea (or simply that it has been shelved) is critical to the process. The process should provide prompt feedback to idea generators on the status of their ideas and periodic reviews of the ideas in the idea bank. The need to have a formal process is consistent with the radical innovation hub suggested by Leifer and colleagues in their book *Radical Innovation* (2000). Their hub would link ideas, opportunity evaluators, and key people in the corporate and operating units.

Most formal processes begin with some person or group looking at a very limited amount of information about an idea (Figure 1-3). They will probably require a number of stages before a final decision to commit significant resources can be reached. In some cases the process may begin with no more than a one-line description of the idea. If the idea is considered attractive, the next step is usually to gather more information. This could be requested of the originator or assigned to someone else. The originator may have a great deal of energy and/or ownership for his or her idea and wish to pursue it further. If further work is assigned to someone else, the originator might feel that his or her idea is being taken away. On the other hand, the originator may not have the time or inclination to do more work on the idea and may view additional work as a burden. Any idea selection process needs to address these possibilities. Once additional information has been gathered and analyzed, the idea usually goes through another decision process. Roles and responsibilities of the

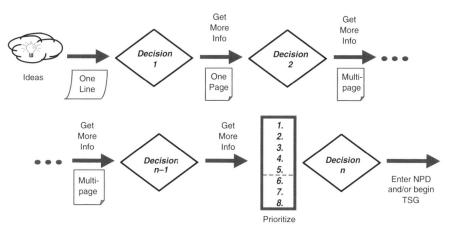

FIGURE 1-3. A typical idea selection process.
In many cases the total number of decisions may be reduced to two.

people involved, and the assumptions and expectations of the process, need to be clearly known and understood by those who own the process and those who rely on it.

In idea selection, decision makers need to adopt a positive attitude rather than to approach the task as a filtering out of less attractive ideas. Decision makers need to ask how an idea can be helped to move forward or how an idea can be modified to make it more attractive, rather than how to determine which ideas to kill. Screening should be done in a way that encourages creativity and should not be so restrictive as to stifle new ideas.

Having decided which ideas are worth further attention, the next step is to prioritize the attractive ideas and select the best ones. Usually a business has many more ideas that it wants to work on than it has resources. It therefore must find a way to determine which ideas are most attractive. Traditional financial measurements, such as sales and profit forecasts and traditional discounted cash flow calculations, are well suited to incremental, short-term product and process development serving well-characterized customer and market needs. However, as the idea becomes novel and the time to commercialization becomes longer and/or more uncertain, metrics such as net present value or internal rate of return break down.

In contrast to the formal people-intensive process discussed above, Nortel has developed an electronic performance support system that allows idea generators the ability to screen their own ideas using an "expert" system—thus eliminating the need for people to screen every submitted idea (Montoya-Weiss and O'Driscoll 2000). Once the idea makes it through each of the three phases (idea qualification, concept development, and concept rating), then a decision maker will electronically receive a standardized form. The overall idea is evaluated based on sixteen dimensions equally divided among marketing, technical, human, and business factors.

In lieu of traditional financial measures, Boer (1999) first suggests considering an idea's "terminal value" (assumed cash flow beyond the finite time horizon of the typical discounted cash flow calculation). He notes that the terminal value may account for 75 percent or more of the value of long-range developments such as new drugs. Second, traditional discounted cash flow metrics burden the project with the total cost of developing and maintaining the business after product launch. Reliance on these metrics contradicts the intention of making small investments of finite duration to encourage rapid screening of ideas and then building the worthwhile ones into business concepts. Third, Boer (2000) indicates that conventional discounted cash flow calculations do not properly treat the dependency of value on risk (beyond that captured by the discount rate or cost of capital). Methods to capture risk are also needed. Cooper, Edgett, and Kleinschmidt (1998) describe numerous examples of techniques used by companies to assess technical and commercial factors that capture "unique risk."

Assessing risk using options theory is yet another approach. Market risk, which options theory represents by the probability distribution of the cash flow stream or its independent revenue and cost components (Angelis 2000), actually

> **Most Effective Methods, Tools, and Techniques**
>
> ◆ Goal deliberation approaches
>   ➢ Time spent on carefully defining the project goals and outcomes
> ◆ Setting criteria for the corporation that describe what an attractive (in terms of financials, market growth, market size, etc.) project looks like
> ◆ Rapid evaluation of high-potential innovations
> ◆ Rigorous use of the TSG for high-risk projects
> ◆ Understanding and determining the performance capability limit of the technology (Foster 1986)
> ◆ Early involvement of the customer in real product tests
>   ➢ Involvement of the customer even before product is completed
>   ➢ Staff up high-potential projects while still in FFE
> ◆ Partner outside of areas of core competence
> ◆ Focus (in contrast to spreading too thin)
> ◆ Pursue alternative scientific approaches
> ◆ Employ product champions if adequate funds are unavailable

Koen 2002). The TSG process may be completely inside, partially outside, or completely outside the NCD. Technology projects that explore fundamental scientific relationships, scout, or evaluate new technology platforms are usually unstructured at the earlier phases and thus are part of the NCD. As the effort escalates, technology risk is reduced to justify further investment. More resources are utilized, and the decisions become more structured, resulting in the later portions of the TSG moving out of the NCD and into the NPD portion. In some cases, the TSG would be completely external to the NCD if the technology activities were mostly structured and with few risks, or if there was a business decision to specifically pursue a particular technology. In contrast, the TSG would remain inside the NCD if these factors were reversed.

CONCLUSIONS

Methodologies, tools, and techniques used in the NPD portion of the innovation process often will not work in the FFE because the FFE is fundamentally different. As a result, the FFE is one of the weakest areas of the innovation process—and so presents one of the biggest opportunities for improvement. There are four significant differences between NPD and FFE. First, FFE work is not structured, but is experimental and often involves individuals instead of multifunctional teams. Second, FFE work is so early that revenue expectations are uncertain, and it is often not possible to predict commercialization dates. Third, funding for FFE work is usually variable. Fourth, FFE work results in strengthening a concept, not achieving a planned milestone.

Our quest started as an attempt to determine the most effective tools and techniques for the FFE. However, this initial effort proved fruitless, since there was no common terminology and vocabulary for the FFE. To this end, our team developed a theoretical construct, the NCD model, which consists of three parts: the uncontrollable influencing factors, the controllable engine that drives the activities in the FFE, and the five activity elements of the NCD. The model highlights the iterative series of activities that may take anywhere from a few seconds in the minds of individuals to many months or years for defining a breakthrough concept.

NCD is not a linear process with specified steps and timing, as is the case of the Stage-Gate™ framework used by many companies for NPD efforts. It is a model that helps us better describe effective methodologies, tools, and techniques for each portion of the NCD. This model, with its common language and terminology, should allow business and technology leaders to better optimize activities in the FFE. That optimization should result in a significantly greater number of highly profitable concepts entering NPD. Further, the common terminology should allow investigators to better focus their research on parts of the FFE while still allowing them an understanding of the whole.

NOTES

1. The FFE is defined by those activities that come before the formal and well-structured NPD process. Even though there is a continuum between the FFE and NPD, the activities in the FFE are often chaotic, unpredictable, and unstructured. In comparison, the NPD process is typically structured, which assumes formalism with a prescribed set of activities and questions to be answered. New product development refers to both product and process development (e.g., a new manufacturing process that provides significant improvement in the product cost).

2. Although the authors prefer to call this portion of the innovation process the "front end of innovation," the conventional PDMA term "fuzzy front end" is used in this chapter. We believe that the latter term implies that the FFE is mysterious, lacks accountability, and cannot be critically evaluated. It is our belief that the term "front end of innovation" more appropriately describes this portion of the innovation process.

3. The authors use the term "effective practices" as opposed to "best practices." The latter term implies that there is a best practice that should be followed. However, certain practices may be "best" only in the particular setting of the company. Thus we use the term "effective"—to imply that these are the effective practices found at the companies studied. Only the company itself can determine what is best for it.

4. The Industrial Research Institute (www.iriinc.org) is a nonprofit organization of over 260 leading industrial companies. The member companies represent such industries as aerospace, automotive, chemical, computer, and electronics; carry out over 80 percent of the industrial research effort in the United States; employ some five hundred thousand scientists and engineers; and account for at least 30 percent of the country's gross national product.

5. Organizational capabilities were placed as an influencing factor since they typically change very slowly and thus are uncontrollable. Alternatively, organizational capabilities could move into the engine to the degree to which they could be modified and controlled by the corporation. In a similar fashion, internal culture was placed

in the engine since it is typically controlled by the corporation, though it could be considered an influencing factor to the extent that it is uncontrollable and changes very slowly.

6. Creative Problem Solving Group (www.cpsb.com), 1325 N. Forest Road, Suite F-340, Williamsville, NY 14421.

7. Center for Creative Leadership (www.ccl.org), One Leadership Place, Greensboro, NC 27438.

8. Society for Competitive Intelligence Professionals (www.scip.org), 1700 Diagonal Road, Suite 600, Alexandria, VA 22314.

9. The reader is referred to *The PDMA Handbook of New Product Development* (Rosenau et al. 1996) and the Industrial Research Institute volume on creativity (*Creativity and Idea Management*, selected papers from *Research Technology Management*, 1987–1996) for a (somewhat dated) compendium of articles on creativity. The reader may also be interested in joining the American Creativity Association (www.amcreativityassoc.org), whose vision is to be the primary association dealing with creativity.

10. Analog users are people who are innovating in areas significantly outside the industry but whose innovations may have direct applicability to providing new insights to the project team looking for breakthroughs. For example, a team looking at new skin creams and issues associated with fissures as the skin cream ages turned to earthquake specialists who were expert at measuring and predicting fissures.

11. Ethnography is a descriptive methodology for studying the customer in relation to his or her environment.

12. An archetype is an inherited idea or mode of thought that is derived from the experience of the race and is present in the unconscious of the individual. For more information about this technique the reader is referred to Archetype Discoveries Worldwide, 14401 South-Military Trail, Suite E203, Delray Beach, Florida 33484 (www.archetypediscoveriesworldwide.com).

13. The process owner is the person who is responsible for maintaining the idea selection process. He or she focuses on the process without becoming overly involved in the content of the meeting or the details of the ideas submitted. The process owner for the idea selection process is typically the same person who is responsible for the Stage-Gate™ process.

14. A disruptive technology is one that does not provide value to the companies' current customers but addresses the need of the company's future customers (Bower and Christensen 1995).

15. The product champion is the person who adopts the project as his or her own and shows a personal commitment to it. They vigorously advocate the project, often at their own political risk, and help the project through its critical times (Markham 1998).

BIBLIOGRAPHY

Ajamian, G. M., and P. A. Koen. 2002. "Technology Stage Gate: A Structured Process for Managing High-Risk New Technology Projects." In P. Belliveau, A. Griffin, and S. Somermeyer, eds., *The PDMA ToolBook for New Product Development*. New York: John Wiley and Sons.

Altshuller, G., 2002. The *Innovation Algorithm: TRIZ, Systematic Innovation and Technical Creativity*. Worchester, MA: Technical Innovation Center.

Amabile, T. M.1998. "How to Kill Creativity." *Harvard Business Review,* September-October, 77–87.

Angelis, D. 2000. "Capturing the Option Value of R&D." *Research Technology Management* 43, 4: 31–35.

Bacon, G., S. Beckman, D. Mowery, and E. Wilson. 1994. "Managing Product Definition in High Technology Industries: A Pilot Study." *California Management Review,* spring, 32–56.

Boer, P. F. 1999. *The Valuation of Technology*. New York: John Wiley and Sons.

Boer, P. F. 2000. "Valuation of Technology Using Real Options." *Research Technology Management* 43, 4: 26–30.

Bower, J. L., and C. Christensen. 1995. "Disruptive Technologies: Catching the Wave." *Harvard Business Review,* January-February, 43–55.

Brown, S. L., and K. M. Eisenhardt. 1995. "Product Development: Past Research, Present Findings, and Future Directions." *Academy of Management Review* 20: 343–78.

Buckler, S. A. 1997. "The Spiritual Nature of Innovation." *Research Technology Management* 40, 2, March-April, 43–47.

Burchill, G., and C. Brodie. 1997. *Voices into Choices: Acting on the Voice of the Customer*. Madison, WI: Oriel.

Capron, B. A. 1997. *Optical Memory Application Assessment*. Boeing Defense and Space Group, Seattle, WA. Sponsor: Army Missile Research Development and Engineering Lab, Redstone Arsenal, AL. Contract F30602–95-C-0270, Project 4594, Task 15, AD-A327 833/0/XAB, RL-TR-96–268, April.

Collins, J. C., and J. I. Porras. 1994. *Built to Last*. New York: HarperCollins.

Cooper, R. G. 1993. *Winning at New Products*. 2nd ed. Reading, MA: Addison-Wesley.

Cooper, R. G. 1999. "The Invisible Success Factors in Product Innovation." *Journal of Product Innovation Management* 16: 115–33.

Cooper, R. G., S. J. Edgett, and E. J. Kleinschmidt. 1998. *Portfolio Management for New Products*. Reading, MA: Perseus Books.

Cooper, R. G., and E. J. Kleinschmidt. 1987. "New Products: What Separates Winners from Losers?" *Journal of Product Innovation Management* 4, 3: 169–84.

Cooper, R. G., and E. J. Kleinschmidt. 1995. "Benchmarking the Firm's Critical Success Factor in New Product Development." *Journal of Product Innovation Management* 12: 374–91.

Crawford, C., and A. DiBenedetto. 2000. *New Products Management*. Boston: Irwin/McGraw-Hill.

Davis, J., A. Fusfield, E. Scriven, and G. Tritle. 2001. "Determining a Project's Probability of Success." *Research Technology Management* 44, 3, May-June, 51–57.

Eldred, E. W., and M. E. McGrath. 1997. "Commercializing New Technology—I." *Research Technology Management,* January-February, 41–47.

Foster, R. 1986. *Innovation: The Attacker's Advantage*. New York: Summit Books.

Fuld, L. M. 1994. *The New Competitor Intelligence: The Complete Resource for Finding, Analyzing, and Using Information About Your Competitors*. New York: John Wiley and Sons.

Griffin, A., and A. L. Page. 1996. "PDMA Success Measurement Project: Recommended Measures for Product Development Success and Failure." *Journal of Product Innovation Management* 13: 478–96.

Grove, A. S. 1999. *Only the Paranoid Survive.* New York: Bantam Doubleday Dell.

Harryson, S. J. 1997. "How Canon and Sony Drive Product Innovation Through Networking and Application-Focused R&D." *Journal of Product Innovation Management* 14: 288–95.

Isaksen, S., B. Dorval, and D. Treffinger. 1994. *Creative Approaches to Problem-Solving.* Dubuque: Kendall/Hunt.

Jackson, N. B. 1997. *Catalyst Technology Roadmap Report.* Sandia National Labs, Albuquerque, NM. Sponsor: Department of Energy, Washington, DC. Contract AC04–94AL85000, DE97009294/XAB, SAND-97–1424, June.

Kahaner, L. 1998. *Competitive Intelligence: How to Gather, Analyze and Use Information to Move Business to the Top.* New York: Touchstone Books.

Khurana, A., and S. R. Rosenthal. 1998. "Towards Holistic 'Front Ends' in New Product Development." *Journal of Product Innovation Management* 15: 57–74.

Koen, P. A. 1997. "Technology Maps: Choosing the Right Path." *Engineering Management Journal* 9, 4: 7–12.

Koen, P. A., G. Ajamian, R. Burkart, A. Clamen, J. Davidson, R. D'Amoe, C. Elkins, K. Herald, M. Incorvia, A. Johnson, R. Karol, R. Seibert, A. Slavejkov, and K. Wagner. 2001. "New Concept Development Model: Providing Clarity and a Common Language to the 'Fuzzy Front End' of Innovation." *Research Technology Management* 44, 2, March-April, 46–55.

Krough, G. V., K. Ichijo, and I. Nonaka. 2000. *Enabling Knowledge Creation.* Oxford: Oxford University Press.

Leifer, R. C., M. McDermott, G. O'Connor, L. Peters, M. Rice, and R. Veryzer. 2000. *Radical Innovation.* Boston: Harvard Business School Press.

Leonard, D., and S. Straus. 1997. "Putting Your Company's Whole Brain to Work." *Harvard Business Review,* July-August, 110–21.

Lynn, G. S., J. G. Morone, and A. S. Paulson. 1996. "Marketing and Discontinuous Innovation: The Probe and Learn Process." *California Management Review* 38, 3: 8–37.

Markham, S. K. 1998. "A Longitudinal Examination of How Champions Influence Others to Support Their Projects." *Journal of Product Innovation Management* 15: 490–504.

Markham, S. K. 2002. "Championing Projects." In P. Belliveau, A. Griffin, and S. Somermeyer, eds., *PDMA Toolbook for New Product Development.* New York: John Wiley and Sons.

Markham, S. K., and L. Aiman-Smith. 2001. "Product Champions: Truths, Myths and Management." *Research Technology Management* 44, 3, May-June, 44–50.

Markham, S. K., and A. Griffin. 1998. "The Breakfast of Champions: Associations Between Champions and Product Development Environments, Practices and Performance." *Journal of Product Innovation Management* 15: 436–54.

McDermott, R. 1999. "Why Information Technology Inspired, but Cannot Deliver Knowledge Management." *California Management Review* 41, 4: 103–117.

McDermott, R. 2000. "Knowing in Community: Ten Critical Factors for Community Success." *IHRIM Journal,* March, 19–26.

McGrath, M. E., and C. L. Akiyama. 1996. "PACE: An Integrated Process for Product and Cycle Time Excellence." In M. E. McGrath, ed., *Setting the PACE in Product Development*. Boston: Butterworth and Heinemann.

McGrath, R. G., and MacMillan, I. C. 2000. "Assessing Technology Projects Using Real Options." *Research Technology Management* 43, 4: 36–49.

Meadows, L. 2002. "Lead User Methodology and Trend Mapping." In P. Belliveau, A. Griffin, and S. Somermeyer, eds., *PDMA Toolbook for New Product Development*. New York: John Wiley and Sons.

Michaels, M. Z. 2000. *Speed: Linking Innovation, Process and Time to Market*. New York: The Conference Board.

Montoya-Weiss, M. M., and T. M. O'Driscoll. 2000. "Applying Performance Support Technology in the Fuzzy Front End." *Journal of Product Innovation Management* 17: 143–61.

Nayak, P. R., and J. M. Ketteringham. 1994. *Breakthroughs*. San Diego, CA: Pfeiffer and Co.

Porter, M. E. 1987. "How Competitive Forces Shape Strategy." *Harvard Business Review,* March-April, 137–45.

Prather, C. W. 2000. "Keeping Innovation Alive After the Consultants Leave." *Research Technology Management* 43, 5: 17–22.

Rosenau, M. D., A. Griffin, G. A. Castellion, and N. F. Anschuetz, eds. 1996. *The PDMA Handbook of New Product Development*. New York: John Wiley and Sons.

Schoemaker, J. H. 1995. "Scenario Planning: A Tool for Strategic Thinking." *Sloan Management Review* 36, 2: 25–40.

Shaver, K. G., and L. R. Scott. 1991. "Person, Process, Choice: The Psychology of New Venture Creation." *Entrepreneurship Theory and Practice,* winter, 23–42.

Slowinski, G., S. A. Stanton, J. C. Tao, W. Miller, and D. P. McConnell. 2000. "Acquiring External Technology." *Research Technology Management* 43, 5: 29–35.

Smith, G. R., W. C. Herbein, and R. C. Morris. 1999. "Front-End Innovation at AlliedSignal and Alcoa." *Research Technology Management* 42, 6: 15–24.

Song, M. X., and M. E. Parry. 1996. "What Separates Japanese New Product Winners from Losers." *Journal of Product Innovation Management* 13: 422–39.

Stevens, G., J. Burley, and R. Divine. 1999. "Creativity + Business Discipline = Higher Profits Faster from New Product Development." *Journal of Product Innovation Management* 16: 455–68.

Stevens, G., J. Burley, and R. Divine. 1998. "Profits and Personalities: Relationships Between Profits from New Product Development and Analyst's Personalities." *Product Development Management Proceedings,* 157–75.

Swink, M. 2000. "Technological Innovativeness as a Moderator of New Product Design Integration and Top Management Support." *Journal of Product Innovation Management* 17: 208–20.

Turnbull, G. K., E. S. Fisher, E. M. Peretic, J. R. H. Black, A. R. Cruz and M. Newborn 1992. "Improving Manufacturing Competitiveness Through Strategic Analysis." In J. A. Heim and W. D. Compton, eds., *Manufacturing Systems: Foundations of World-Class Practice*. Washington, DC: National Academy of Engineering, National Academy Press.

Varnado, S., et al. 1996. *Development of a Technology Development Strategy to Reduce Health Care Costs*. Department of Energy, Albuquerque, NM. Contract MIPR-94MM4592, AD-A318 818/2/XAB, November.

von Hippel, E. 1986. "Lead Users: A Source of Novel Product Concepts." *Management Science* 32: 791–805.

von Hippel, E. 1998. "Economics of Product Development by Users: The Impact of 'Sticky' Local Information." *Management Science* 44: 629–44.

von Hippel, E., M. Sonnack, and S. Thomke. 1999. "Creating Breakthroughs at 3M." *Harvard Business Review,* September-October, 3–9.

Wenger, E., and W. Snyder. 2000. "Communities of Practice: The Organizational Frontier." *Harvard Business Review,* January-February, 139–45.

Wheelwright, S. C., and K. B. Clark. 1992. *Revolutionizing Product Development.* New York: Free Press.

Willyard, C. H., and C. W. McClees. 1987. "Motorola's Technology Roadmap Process." *Research Management* 30, 5: 13–19.

Zien, K. A., and S. A. Buckler. 1997. "From Experience Dreams to Market: Crafting a Culture of Innovation." *Journal of Product Innovation Management* 14: 274–87.

2 Hunting for Hunting Grounds: Forecasting the Fuzzy Front End

Christopher W. Miller

The buffalo are diminishing fast. The antelope that were plenty a few years ago, they are now thin. When they shall die we shall be hungry; we shall want something to eat.

—Tonkahaska (Tall Bull),
chief of the Sioux, 1880

The fuzzy front end (FFE) begins at the point when a need exists that can be identified and a technology exists that could meet that need (Reinertsen and Smith 1991). This fundamental hypothesis suggests that as much as half of the potential lifetime value of a product exists between this hard-to-define "fuzzy" nexus of emerging need and technology and the start point of most formal staged, or phased, product development processes. The gap represented by the FFE is an enormous opportunity. There are few aids to help the product development organization forecast changes that are useful in identifying phase shifts in an industry. Exceptions can be found: Moore's Law has helped the semiconductor industry forecast innovation, and pure demographics can assist in forecasts involving consumer goods and services, but these are rare and offer an incomplete picture. Most industries find themselves caught with little or nothing as a guide and even less experience. The Reinertsen and Smith model might be represented in this way:

|————————Fuzzy Front End————————|—NPD Period—|—Commercial Period—|

Hunting for hunting grounds aggressively fills the fuzzy front end gap. Hunting for hunting grounds has grown from roots in mature consumer goods industries that have aggressively adopted a large variety of voice-of-the-customer data-gathering techniques (Griffin and Hauser 1993). These companies sit on the edge of emerging trends and constantly seek to identify changes that may affect their category. They also have been aggressive in spending on product development and the supporting technical research. In these cases it can be argued that duration of the FFE is much shorter because of their early-stage diligence. The investment in aggressive new product development (NPD) processes has paid off. The mature customer-responsive company may experience less fumbling; NPD processes are well known and understood by most levels of the organization. Many of the senior managers in these firms have NPD experience. The shorter duration of the FFE has created a need in these mature organizations to move to an FFE model that will allow for a forecast of possible needs before they exist in the mass market and the early identification of technologies that have promise to meet those needs well in advance of their effective use. The mature customer-focused company model might be viewed in this way:

|—Fuzzy Front End—|—NPD Period—|——Commercial Period——|

Incremental improvement, the hallmark of a growing category, fails only when the increments are no longer meaningful differentiators to the customer. Prior to this point, incrementalism is preferred over significant leaps in that it allows the value chain to keep pace with innovation. It is the mature category that must seek truly new and unique opportunities. It is also an unfortunate fact of organizational life that it is the mature organization that is least capable of walking away from their extensive capital and human resource investments as well as their investments in a complex and extensive value chain studded with longtime customers (see Figure 2-1). Reducing emphasis on a category, tech-

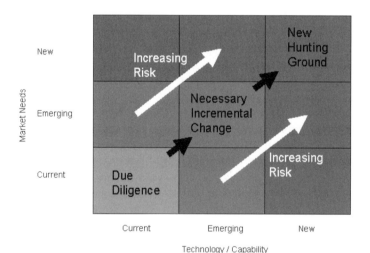

FIGURE 2-1. Need vs. technology matrix.
Using current technology to meet today's need is the business of an aggressive, quality program. Forecasting out from this position toward emerging need and emerging technology is where most NPD activity necessarily resides. Forecasting backward from an identified future opportunity to today's equity and ability world is the province of hunting for hunting grounds. Hunting for hunting grounds provides hypotheses about waves of future growth potential. It also provides the direction necessary to determine which of today's incremental efforts are most appropriate to pursue to support future growth.

EXAMPLES

On the heels of their incredible success with the Snake Light, the Black & Decker Appliances group came to realize that the world did not need another small-appliance manufacturer. While the Snake Light helped to redefine the portable lighting category, it really was the last in a long series of product redefinition efforts in a variety of small-appliance categories. It was clear to Black & Decker that major advances would be farther apart and that when those advances were made, the world's two hundred other manufacturers of similar products would quickly reverse-engineer the Black & Decker innovation. In other words, Black & Decker's very success had attracted many others into their territory. And while new and better would clearly always be part of the Black & Decker charter, they were in danger of going into what the product development team called "the wonder widget phenomenon." This means they were hunting for rabbits and not buffalo. They were hunting for a day, not a season.

---

nology, or customer that has been nurtured by an organization, in some cases for many decades, is difficult and even painful.

Great product introductions that experience a decreasing yield are a symptom of maturation. The strong management team that becomes aware of creeping incrementalism and maturation will use the existing cash flow, organization, and structure to create a renewed flow of revenue. The almost inevitable first steps in fending off the realities of the "S curve"—the flow of a business cycle through emergence, growth, maturation, and decline—are focused on reengineering to ensure that the organization is capable of competing in the existing market. The organizational leadership may see that all reasonable paths still lead downhill. Efficiency and cost reduction cannot in and of themselves build long-term future value. The only absurdity at this

---

EXAMPLES

A *charter* is a contract between the team that has the responsibility to manage the resources of the company and the team being chartered. There are many things that can go into a charter:

◆ A visionary definition of your task (Black & Decker: "We will sell the first dollar of product, on a path to a new $100,000,000 business." Kodak: "Every mother in Latin America will have a picture of her child.")

◆ A background statement that describes why this is important for the organization

◆ Who is on the team and how much time each will need (e.g., five to nine team members, time commitment of 20 to 40 percent of workload)

◆ How much the team has permission to spend without further justification

◆ When the team will report back (three to five months)

◆ Key success criteria—specifically, what constitutes success from several different perspectives:

　➢ A large market with multiple unmet needs

　➢ Protectable position or technology ownership

　➢ Feasible within the planning horizon

---

time is for the management team to continue doing the same thing and expect a different result.

The hunting-for-hunting-grounds process is absolutely dependent on an aggressive and experienced team driven by a focused yet visionary charter.

The charter is a product of a management team that has come to terms with the core business reality of diminishing returns and has decided to change that reality. The charter is given to a cross-functional team of individuals who would be considered indispensable to their regular activities under other circumstances. (For further information on the important issues of charters and teams you are directed to two classics, Crawford and DiBenedetto 2000 and Katzenbach and Smith 1993.)

For product developers the news is good in that in hunting for hunting grounds, many of the stalwart processes of high-quality new product development remain intact. If you are effective at feathering out new product opportunities as an organization, you should be equally good at identifying totally new types of problems for your organization to solve and discovering profitable paths from today's core business to these new revenue areas. What you will have to abandon—customers, capabilities, and long-held strategies—is often the greatest challenge.

Stage	Tools and Activities	Time	Outcome
Preparation	Building a charter and project plan ◆ Team formation ◆ Charter ◆ Charter workshop	Four to six weeks	An experienced team committed to a focused yet visionary charter, supported by their organization and with a process plan to reach the goal
1. Hunting for hunting grounds	◆ Discovery tools (use five or more) ◆ Journaling ◆ Regular team debriefing discussions ◆ Data management	Three to eight weeks	Two hundred to five hundred potential opportunity areas—needs, technologies, emerging trends
2. Model building	◆ Clustering ◆ Cluster testing ◆ Future myth development ◆ Minicharter	Two to three weeks	Three to five problems for your organization to solve
3. Path building	◆ Case development ◆ Portfolio development ◆ Path development	Four to six weeks	Three to five new business opportunities, each supported by twenty to fifty high-probability new revenue-generating ideas

FIGURE 2-2. Stages in the hunting-for-hunting-grounds process.

EXERCISE

Look around the room you are in and make a list of objects that are green. Write these down. After you have finished, close your eyes and name five things that are blue. This perception exercise, taught by the Creative Problem Solving Institute, suggests the problem faced by the new product developer seeking totally new opportunities—that is, hunting for hunting grounds. You have been studying objects that are green; now you are being asked to identify objects that are blue. Your challenge involves your team's perception as much as its process. As Marcel Proust said, "The real voyage of discovery consists not in seeking new landscapes, but in having new eyes."

The purpose of the hunting-for-hunting-grounds process is to discover significant hypotheses for further developmental work and research. There are three major stages in hunting for hunting grounds (see Figure 2-2). First, identify possible hunting grounds, using strategic guidance from within your organization and guidance from the marketplace and technology. Second, build a model that clearly articulates a few potential problems that your organization can solve. And third, define the path between today's core business and the new opportunity in the form of revenue-generating activities and products. This chapter will methodically walk through each stage, providing definition, recommendations, and options. In the end you will be at the beginning.

HUNTING FOR HUNTING GROUNDS

Stage	Tools and Activities	Time	Outcome
Hunting for hunting grounds	◆ Use five or more discovery tools ◆ Journaling ◆ Regular team debriefing discussions ◆ Data management	Three to eight weeks	Two hundred to five hundred potential opportunity areas—needs, technologies, emerging trends (see Figure 2-2)

The hunting-for-hunting-grounds stage has four components: (1) *discovery,* the use of multiple methods to look for and identify opportunity; (2) *journaling,* a personal data management tool; (3) *team debriefing,* regular discussions and thoughtful dialogue among team members; and (4) the *hunting grounds matrix,* an organizational data management tool. These components intermix and should become seamless. For example, a team member is jotting notes in a journal at a team debriefing session over a meal at the end of a day of discovery doing field research. Completing the entire stage within a compressed period is important. This allows the team to draw connections between early experiences and later ones.

Tools That Aid Discovery

The actual act of the hunt creates data that go into your journal; the data are discussed in regular team meetings where understanding is enhanced; the journal notations are enhanced, and from your journal the data go into the matrix. Here are a few recommended approaches to hunting, with a very short definition of each. Note that the tools listed are not completely focused on technology or on need. They cross that boundary, and in many cases a tool will do a bit of data collection on both sides. A minimum of five diverse techniques is recommended over a period of three to eight weeks. If you already use one technique frequently, try an approach that is not so comfortable. There is no intent to be inclusive in this list, nor to limit a team's creativity in building their own unique tools.

As you use each of these approaches keep in mind that there are four general sources of innovation to look for:

- Discontinuities in system patterns, something that is breaking up a traditional pattern—for example, a move to on-line purchasing and bidding by a traditional customer
- Disequilibrium or lack of balance in a system, when one partner appears to be getting more than other members of the value chain—for example, retailer domination of traditional brand names
- Disintermediation opportunities, usually emerging because of the above—for example, customers doing substantial research without approaching the dealer or retailer
- Compensatory behavior, signaling that a system member is experimenting while waiting for a market innovation—for example, a fan duct-taped to an operator's station, or for that matter, any use of duct tape

1. *Secondary research*. Use information that exists. Be cautious about past research that is based on common assumptions. It is more than likely correct and more than likely rather ordinary in the conclusions it will direct you toward. Now you know what everybody else already knows. Secondary data should be part of the preparation for every ideation or other market research effort (Malhotra 1999). Secondary data include all data not specifically generated for this particular task. This would include data internal to the company, such as company records and past research, as well as data external to the company, such as trade publications, census data, and so on. Secondary data will serve several critical purposes. They will:

- Spark questions as well as provide answers
- Be the input material for further analysis
- Provide a good picture of the status quo
- Clarify gaps where primary data may be needed
- Lead to new ideas and other sources
- Help define the problem more clearly
- Serve as a reference base

2. *Audit prior work*. Study reports and project archives for ideas old and new. Seek out the authors of this prior work and ask additional questions. Share the nature of your current effort and ask for their insight. Whether or not you discover anything, you will win huge organizational points (Aaker, Kumar, and Day 2000; Kinnear and Taylor 1996; Malhotra 1999).

3. *Technology identification brainstorming or audit*. Inform the technical community in and around your organization of your effort. Ask them to empty their file drawers and come loaded with possibilities and "almosts"—projects that didn't quite make it in the past. That was then. This is now. Assume that each of these projects has an observation or approach that could be valuable to your team. It is your job to find it (Millett and Honton 1991).

4. *Operational capabilities brainstorming or audit*. What are you good at that you can sell? Safety-Kleen emerged from the ability of Chicago Leather to manage tanning fluid. Do you have expense items or waste products that can be harnessed or approached in a new way? Look at all of your functions as potential core competencies in your new hunting ground (Prahalad and Hamel 1990; Leonard-Barton 1992; Porter 1980, 1985).

5. *Technology identification consultancies*. If you have not had a technical effort in your organization, you probably cannot build it overnight. You may wish to form a subproject around identifying emerging technical opportunities that might impact your organization and provide ideas for hunting grounds.

6. *Market trend brainstorming or audit*. Invite your entire marketing organization to speculate on opportunities and issues. This is a good chance to include vendors, ad agencies, and interested others.

7. *Futures/trend tracking*. Collect as many cutting-edge periodicals as you can, or subscribe to *The Futurist* (the magazine of the World Future Society) or *Technology Review* (a monthly publication of MIT, previewing developments in science and technology). Like many professions, futurism seems simple until you start to peel away the layers of the onion. Do not be deceived by apparent simplicity (Slaughter 1999).

8. *Demographics research*. Demographics knowledge should cut across all other thinking. Good sources are the publications of *American Demographics* and the U.S. Government Printing Office. Be a bit cautious about depending on industry data and company affiliation groups. They tend to be based on consensus definitions of terms and markets and are the "common wisdom." You are not looking for something common.

9. *Delphi interviews—the traditional approach*. Identify a dozen or more thought leaders. Secure their agreement to participate. Send them a set of open-ended questions, such as: "What are the most exciting and disruptive trends in manufacturing? What are leading-edge manufacturers doing differently from the rest of us? If you were a plant manager, what would be your biggest concern today? In five years?" Ask them to respond in writing. Consolidate their responses. Send out the consolidation with a second round of questions. Usually the payoff for the thought leader is to receive the report (Kerstin and Knut 2001; Matt and Penny 2001; Rowe and Wright 1999).

10. *Delphi interviews—an alternative approach*. Conduct telephone and in-person interviews with thought leaders and interesting people. Schedule

sixty- to ninety-minute phone conferences; longer than ninety minutes and a break will be needed. It may be better to schedule a follow-up call. Have a lead interviewer. Allow the entire team to listen in and ask follow-up questions. Audiotape the interview and your team's debriefing following the interview (permission for the audiotaping should be obtained when scheduling the interview). Start by saying who you are, describe your charter, and say, "We aren't even smart enough yet to ask good questions. Help us." And they will. Finish by asking them who else you should interview.

11. *Slingshot groups.* Involve noncompetitive new product development, marketing, and technical professional peers in a facilitated discussion. This process mimics what one team saw going on at Product Development and Management Association conferences (see technique 15 below in this section, and Chapter 10 in this volume, "Lead User Research and Trend Mapping"):

1. Recruit three to five noncompetitive professional peers who may also be consumers of your product (prosumers).
2. Recruit four creative consumers.
3. Use a focus group facility to add ease to the taping process.
4. Moderate a discussion of deep needs in the morning.
5. Break for lunch. Everyone shares together over the meal—consumers, customers, prosumers, and you. Then thank and dismiss the consumers.
6. Take the gloves off and listen as the prosumers "slingshot" from their consumer experience to their professional side.
7. Debrief the whole experience.

Krueger and Casey (2000) provide a detailed, systematic roadmap of how to design, prepare for, implement, analyze, and report on focus groups. They give detailed discussions of issues to consider, tips, and warnings to help prepare for the expected and unexpected. If you use focus groups of your current customers, be cautious in designing them. Review the lesson from the green/blue exercise in the introduction to this chapter.

12. *People not like us.* Find communities where the problems might need to be solved differently. Black & Decker spent time in the Amish community of Lancaster County, Pennsylvania. The Amish avoid the use of electricity and have devised other ways of performing the tasks that Black & Decker's electric appliances and tools perform. Other companies have spent time with people with special needs that force them to create special "abilities." What do the physically or visually challenged know about your product category that you don't? Learn about hands-free operation from people who don't have the use of their hands. (See also Chapter 12, on universal design, for more on this topic.)

13. *Ethnography.* Conduct on-site interviews of core customers in the context of the product being discussed. You should conduct eighteen to twenty interviews lasting about two hours each (with more time if appropriate). The interviews frequently include a guided tour of product usage and the environment. In the context of hunting for hunting grounds, you are looking at activ-

ities and behaviors related to your core product, competencies, and equity that may open up new avenues of opportunity (Woodland 2001).

14. *Immersion.* Perhaps the most promising method for gaining a deep understanding of the user's world is for the developer to immerse himself or herself in the user's environment, much like an anthropologist. Spend thirty-six hours or longer in a single intriguing environment (Epstein 2000; Stevens 1999). This is more properly called participant observer methodology. If you want to improve computer usage in the classroom, spend a week assisting a teacher in a single class (eat the lunch, do playground duty, prepare a lesson plan, and deliver the lesson). If you want to understand your tractor and the average age of a farmer is fifty-eight, assign a fifty-eight-year-old team member to do the actual work for eighteen hours a day during peak season. These techniques imply that you, the researcher, can only truly understand by "going native." The result is to become the customer and therefore produce "empathetic design" (Leonard-Barton and Rayport 1997).

15. *Lead user.* (See Chapter 10 for a more complete method.) Compensatory behavior can often be identified by lead user analysis. According to Eric von Hippel, lead users are customers who (1) experience needs months or years before the majority of the marketplace encounter them and (2) stand to benefit substantially from innovations (von Hippel 1988). From the manufacturer's perspective, involving lead users allows for time to develop the technology, evaluate prototypes, and have a first-generation product available when the majority of the market begins to realize the need and seek a solution.

Lead users have some additional characteristics that make them complementary partners. Lead users frequently can articulate the emerging need. They may have already worked on a solution and at the very least have identified implementations that do not work. Furthermore, they have an application that can be used to prove and demonstrate the benefits of the new technology (von Hippel 1988; see also von Hippel 1982; von Hippel et al. 1999).

Tool Use and Design

Your team should be writing in their journals at every possible moment and having team debriefing discussions on a regular basis. Keep the total elapsed time as short as possible; three to eight weeks is a reasonable minimum. To get the most benefit from the project, use a set of at least five different approaches. Don't worry about redundancy at this early stage; the same information from

EXAMPLES

The Black & Decker team clearly understood the issue of time from many different sources. However, hearing the themes of time compression repeated by futurists (in the Delphi format), by children (in an ethnographic format), and in the Amish community (in the immersion format) gave them insight into the nature, complexity, and pervasiveness of the issue.

EXAMPLES

The Castrol team was doing an excellent job of gaining market share in a declining market. (Who changes their own oil anymore, after all?) In the hunting-for-hunting grounds process they chose (1) internal interviews, (2) a technology brainstorming session, (3) a market trends brainstorming session, (4) slingshot groups, (5) immersion efforts, including an in-market show and an out-of-market show, and (6) opinion leader Delphi interviews.

Each team member accepted the responsibility to coordinate one aspect of the process. This made a daunting administrative task a bit more equitable. One of the keys to success was good documentation and open communication with team members who focused their team time on just a few multiday team meetings. During these meetings they identified over 450 hunting grounds.

---

different sources can have different implications. In fact, the same information seen by different eyes on your team may have different implications.

Internal methods (technology and market trend identification—see Chapter 10) can be done either as a brainstorming group effort or as an audit. Experience suggests that audits engender a certain amount of organizational concern and defensiveness. This may work against your ability to get at the hidden opportunities. Brainstorming, when done well, allows you to involve people whose efforts you would otherwise be auditing, and to do so in a less formal, more participatory way. The bias of this author is to involve more people more openly whenever possible.

External methods can raise concerns about confidentiality and idea ownership. Clearly this is the province of your legal team. Advice runs from the extreme of frightening, multipage legal documents to simple statements that acknowledge that all conversations are public domain, describe the importance of the work to the organization, and express the wish that the discussion be kept in confidence for a period. Openness is the friend of innovation.

Processing the Data

Perhaps the most important tool you can have is your journal. A large unlined sketchbook can be purchased for just a few dollars at most bookstores and works well. The purpose of the journal is to help you to:

◆ Observe your world with the "new eyes" Proust suggests
◆ See the details (success is often in the details)
◆ Compare observations with team members and across time

The more a team uses journaling, the more value it will get from the overall experience. Jot notes, add articles, insert photographs, and doodle as you go (see Figure 2-3). Your journal is a mental compost pile—the more you toss in and the more you stir it, the better it will be at doing its job. Remember there are four primary sources of innovation opportunity: discontinuities in system

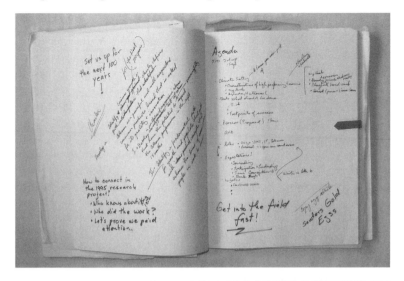

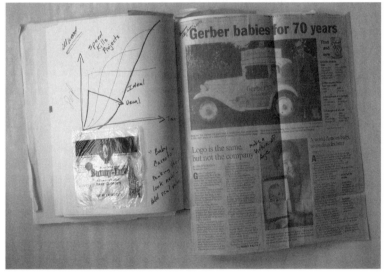

FIGURE 2-3. Journaling is a personal data management tool.
The personal journal can be either structured or unstructured. While it is important that all team members use a journal, allow personal flexibility in style. A journal can be more than notes. It can also be pictures and artifacts.

patterns, disequilibrium or lack of balance in a system, disintermediation opportunities, and compensatory behavior.

Draw conclusions. The journal is not just a simple recording device. Use your brain in conjunction with your eyes and ears. Build beginning hypotheses—theories about events, behaviors, and technologies, and how they may be expected to work together in the future. These observations may even take the form of product ideas and designs.

Team Debriefing Sessions

Review where you have been. Share your observations with your journaling teammates. Frequent team meetings to exchange information discovered by different team members will assist in the process. Where possible these discussions should be connected to the data-gathering experience. A few minutes of review done while sitting in the car after a site visit may have more value than hours spent the following week. The emphasis should be placed on sharing knowledge gained, discussing the possible implications, finding synergy between data points, and of course adding to the hunting grounds matrix.

Building and Managing Your Data with the Hunting Grounds Matrix

The key to your success will be your ability to organize and eventually reduce a large amount of disparate data to a single, clearly articulated "problem worth solving." Start by building a hunting grounds data management matrix. Each entry, record, or card should reflect several things:

- ◆ Name of the hunting ground you have discovered
- ◆ The category into which it appears to fit (e.g., trend, market, need, technology)
- ◆ The hunting ground description (the nugget of information)
- ◆ The source (type of source and how to track back)
- ◆ Specific opportunities that the hunting ground suggests (product, service or venture)

Here is one way it might look. (Please note that teams have been successful with both low-tech index cards and high-tech Web-based systems.)

Hunting Ground Name	Category	Description	Source	Opportunities
1. Back to Nature	Attitudes and ideals	People want to feel connected to nature	Delphi interview (Susan/team)	◆ Products that are made from all-natural materials ◆ Products and services that help to bring nature into the home
2. Who's Cooking?	Kids/youth	Over 75 percent of kids occasionally or regularly fix meals for themselves	Newspaper article (Bob/team)	◆ On-line cookbooks for kids ◆ Meal ideas and easy-to-understand directions on product packaging

3. Mega Materials	Materials technology	Anodized aluminum that looks like aged copper	Vendor interview (Tim/ purchasing)	◆ High-tech, lightweight, low-cost traditional look

CHOOSING YOUR HUNTING GROUNDS

Stage:	Activity:	Time:	Outcome:
Model building	◆ Clustering ◆ Cluster testing ◆ Future myth/ scenario development ◆ Minicharter	Two to three weeks	Three to five problems for your organization to solve (see Figure 2-2)

Finding Problems Worth Solving

Nikola Tesla said, "There are so many things one could do with electricity. I chose to do something worthwhile." The objective of this stage is to organize and reduce the large amount of data so that the team can prioritize and find a problem worth solving—something that is worth your effort. There are four steps: (1) *clustering*, finding significant identifiable data sets within the larger data set; (2) *cluster testing*, discovering initial ways your organization can extract value for the identified problem; (3) *future myth development*, building a story of a future where the problem is solved; and (4) *minicharter building*, creating a subteam and supporting charter to investigate the opportunity further.

Castrol identified 450 hunting grounds, Black & Decker over 300, and a pharmaceutical company over 400. However, the definitions of each hunting ground are imprecise and redundant. Think of the hunting grounds matrix as a tree canopy. As you look up, the leaves and branches overlap and intertwine. Together they create complete coverage. Your objective at this stage is to find a few of the sturdiest and most intriguing branches upon which to build your future.

The *qualitative cluster analysis* is a process for pulling clusters of hunting grounds out of the matrix and reducing them to a usable structure. Quantitative cluster analysis (Hair et al. 1998; Malhotra 1999) finds "naturally occurring" groupings. A danger is that the process will always create clusters regardless of whether they actually exist. In hunting for hunting grounds, we are using a qualitative approach as a data reduction technique. Interdependent relationships within the whole set of variables are examined. The primary objective of cluster analysis is to classify objects into relatively homogeneous groups based on the set of variables, the hunting grounds being considered. Objects are similar within groups/clusters but different between clusters.

The Model-Building Session

Drawing the team together for several intense days of discussion is essential to the hunting-for-hunting-grounds process. Prior to the session each team member should review the entire matrix of hunting grounds and select those hunting grounds that are most intriguing based on his or her understanding of the charter.

The agenda for a model-building session might include these activities on the first day:

◆ Set the tone of the meeting. Reconnect with the charter. Discuss your team's purpose and what you have learned up to this point.

◆ Vote individually for those hunting grounds each team member feels have the highest probability of providing sustainable waves of opportunity. A simple vote quickly exposes areas of agreement and disagreement.

◆ Discuss the results of the vote and nominate leading hunting grounds. These leading hunting grounds will become the central points of each theme. Suggest other hunting grounds that fit with or have implications for the lead theme. The leading hunting ground plus the connected ones creates an initial cluster.

◆ Break into small groups to discuss emerging hunting ground clusters. Identify the higher-level themes that hold the cluster together.

◆ Go back through the matrix and identify second- and third-tier hunting grounds that fit with the cluster. You are building the hunting ground cluster branching out from a central theme, "a problem worth solving." Consider doing the cluster-building exercise on a four-by-eight-foot sheet of foam-core board using cards or collage (see Figure 2-4).

◆ Present the results of your exercise to one another and repeat the process with other potential core areas.

◆ Allow substantial time for conversation. It is important to follow a converging conversation with a period of divergence where additional possibilities can be explored. Interconnections between clusters are significant. These represent potential paths for future growth.

◆ Select three to five clusters or themes to move forward. It is better to leave an area behind than to overcluster. All of your work should be carefully documented so that you can return to this point in your process and select additional clusters at a later time. There must be a realistic match between the number of clusters the team wishes to explore further and the available resources. Each cluster represents a possible new business unit. This implies that the resources required to establish and sustain it will be substantial.

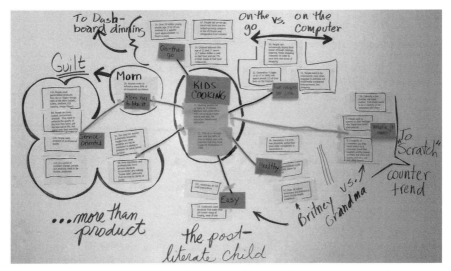

FIGURE 2-4.   The cluster is a collection of organized and interconnected hunting grounds that suggest a theme or "problem worth solving."
In the example above, "Kids cooking" is the central issue. "Kids cooking" is surrounded by trends and countertrends, and it is also connected to another theme/problem cluster on another board called "Dashboard dining."

Building Second-Tier Charters

As the clusters start to stabilize and clarify, team members should emerge who prefer one cluster or another and are willing to champion additional exploration. If a champion does not emerge from the team, drop the cluster. The emerging champions should build minicharters for their clusters. A very small, very short-lived team will be moving to explore this territory. Building the minicharter is the focus of your second day of model building:

♦ Test each cluster by brainstorming how your organization could solve the identified problem.
  ➢ State the nature of the problem.
  ➢ List one hundred to two hundred beginning product and service ideas. Look back through your journal and the hunting grounds matrix. There should be many useful ideas.

Characteristics of a Good Hunting Ground

A good hunting ground describes an emerging intersection of a variety of trends and needs. For Coleman, the "Coleman Back Home" line of business emerged from a set of hunting grounds titled "Bringing the Indoors Outside." Subordinate hunting grounds included issues such as: people want to spend more time outside than they do; no time to camp; bugs are bad; people erased the line between the kitchen, dining, and family rooms and now want to erase the line that represents the back of the house; reduced tolerance of temperature variations as we age; security issues (fear) keep people indoors.

Sample Future Myths

(Excerpt from a future myth developed by a technology company team)

Pat and Terry awake with anticipation of an exciting day. They will be going to Europe. Pat jumps up and collects one of their PCDs (personal communication devices) and checks for diamond prices on foreign markets as well as any news related to the diamond market. Pat notices that several news articles have been automatically flagged. The electronic catalog used by Pat's customers is simultaneously updated.

Terry, in the meantime, wants to know what the weather is in Venice, their first stop in Europe. Terry is concerned that the latest heat wave in Venice will cause unpleasant odors. Terry's new PCD with artificial sensory perception lets Terry see, feel, and smell the air in Venice. Good, the air is clear and pleasant.

(Modified from the Coleman team's hunting ground)

Jim pulled into his garage at close to 7:00 P.M. He was tired and felt he still had a lot of work to do. As the garage door went down the whole property security screen went back up. Jim remembered when he had had to carry a house key and punch in codes. He walked through the kitchen, grabbing the newspaper on the way, and out onto his deck. It was only a little warmer outside than it had been in his house or car. He knew that the temperature was still nearly 90 and the humidity almost as high, but his amazing outdoor climate control gave him the feeling of a cool June evening, not the reality of August in Chicago.

He pulled a beer out of the deck cooler and sat down in his favorite chair. Immediately he spotted a red-winged blackbird sitting on the reeds growing in his little swamp. The swallows swooped and dived overhead. In the back of his mind he knew it was the huge mosquito population that drew these graceful creatures to his yard, but it had been years since he'd been bitten, thanks to his bug-control system.

The phone rang. It was his wife, Jane. She had just picked up the kids on her way home from the office and asked if he'd mind cooking that night. Jim said, "Sure," and picked up the remote. He hit his preprogrammed "superdad" meal. The grill started to heat and the burgers went into a fast thaw cycle as he read the paper.

> Select a dozen or two of the most interesting and develop simplified product concepts: a product name, what the idea is, how it would work, core benefits.

◆ Build the minicharter using the cluster and the beginning product concepts. Use a format similar to that presented earlier in this chapter, but in this case the team is sponsoring you for a short-run effort. *Mini-* refers to the very short duration of this effort, not to the length of the document. The criteria should be the same as those for the project charter.

◆ Build a future myth that fits the new charter. The myth is a story, a creative description of what the world would be like when the problems you have chosen to solve have been solved using the product and service concepts you developed. How is life different, better, scarier, safer? Make it personal and descriptive. Fun and fantasy are important at all stages of this work, but more so now, as you come close to expanding your circle of team members.

BUILDING A PATH TO YOUR HUNTING GROUND

Stage:	Activity:	Time:	Outcome:
Path building	◆ Case development ◆ Portfolio development ◆ Path development	Four to six weeks	Three to five new business opportunities, each supported by twenty to fifty high-probability new revenue-generating ideas (see Figure 2-2)

It is unreasonable to expect an organization to leap from an existing core business area that has significant harvest value to a new opportunity without substantial plans, proof, and support. That support may take one of many forms, which can be categorized on three levels:

1. A plan with substantial revenue potential
2. Research that suggests the potential is real
3. Early-stage experiments—product introductions, ventures, and acquisitions that generate revenue

Failure to provide a path that allows the organization to learn and fund second-generation activity will destroy the program. The path to your new hunting ground must be filled with enough early-stage opportunity to excite the organization and enough later-stage opportunity to prove that what you have found is a path to new growth. The most tangible way of exciting an organization is with a customer order. The customer order is the only unimpeachable evidence of value.

In the hunting-for-hunting-grounds phase, breadth was the essential element. In phase two, the qualitative cluster analysis, depth was key. Now you are looking for the stones to place along the path between today's equity and the place to which you hope to move your company. The critical measure is volume—breadth times depth. There are three steps: (1) case development, when you build a case for the opportunity using available knowledge; (2) portfolio development, in which you create a recommended set of product opportunities; and (3) path development, where you align the product opportunities. Each step is accomplished for each hunting ground separately.

Case Development

Each of the three to five hunting ground teams should have an informal case, minicharter, scenario or future myth, and supporting qualitative data from the first two phases of the work. This becomes a briefing document for introducing the hunting ground to others. Added research now can enhance the program as long as it is done in parallel and does not add to the elapsed time. Usually secondary research is sufficient to establish enough knowledge to move forward. Perhaps the most critical information is who knows more than you do about the proposed hunting ground. Who knows the most about this area and who can be temporarily invited on to the team to aid in the process? A native guide dramatically enhances the chance of success. Key components of an informal case:

◆ Short summary of the opportunity

◆ Value chain description and segments within each link if understood

◆ Targeted customer and consumer needs for each segment (real or hypothesized)

◆ Value proposition (how your organization can add value)

◆ Competitive set (how this problem is being solved now)

◆ Strategic advantage (why customers will win because you enter this market)

Time and resources permitting, *scenario analysis* (reverse trend extrapolation) is an optional tool that can be used to enhance the case. The team members become forecasters. Typically an expected scenario, a worst-case scenario, and a best-case scenario are developed. The scenarios have three purposes (Thomas 1993): (1) they display the interaction between several trends and events in order to provide a holistic picture of the future, (2) they help check the internal consistency of the set of forecasts on which the scenario is based, and (3) they predict the future situation in a way that is readily understandable by the nonspecialist in the subject area. To develop a scenario, the team forecasts the individual underlying trends to the specified point in time in the future. The team then develops the list of particular events or conclusions that must be true if the individual trends are true. The team plots specific events with dates back to the present time. Then a scenario or narrative describing the events is developed. These are typically narrower than the "future myth" but provide an important opportunity to look at the negative as well as positive implications of a hunting ground.

Portfolio Development

Move quickly from the case to creating the products that will make the case tangible—the portfolio. Take three steps: (1) *ideation,* creating a large set of possible ideas, (2) *concept building,* developing leading ideas further, and (3) *concept screening,* selecting those concepts that have the strongest potential value.

Ideation

This is serious product concept building, not the simple test done during model building (Miller 1997). Numbers are your friends (Stevens and Burley 1997). Between 500 and 1,500 ideas and 100 to 150 product concepts are reasonable

The Ideation Agenda

Preset a room with a semicircle of chairs with five or six flip charts at the front of the room. Make sure the room is large, bright, and conducive to fun, creative thought. Consider adding toys, music, and lots of food and beverages. Before guests arrive they should have had a chance to review your case and ask questions. They should also have appropriate legal agreements in place. You must have assigned facilitators who manage the process and a task owner who speaks for the team.

One-day session agenda

7:30	Welcoming continental breakfast and name tags.
8:00	Climate-setting introductions. Each individual should give his or her name, expertise, and why the task is important personally.
8:15	Discuss the rules of brainstorming and the process.
8:25	Statement of the task.
8:30	Brainstorming. Target two to three ideas per minute. All are written on flip charts and captured live by a stenographer. Insert exercises to enhance ideation, such as a fishbowl discussion with the consumers.
10:00	Vote and break. Use stick-on dots to highlight about 10 percent of the total number of ideas.
10:15	First-round concept building. Work in pairs, with each pair aiming to produce two concepts.
10:45	First-round report back. Place an annotation of all concepts on the wall
11:00	Break into smaller groups, each targeting an aspect of the task. Each group is facilitated. Concepts can be written over lunch.
1:00	Report back
1:30	Second-round breakout groups and report back.
3:30	Build concepts alone and report back.
4:00	Vote on all concepts. For one hundred concepts give each person ten to fifteen votes.
4:15	Guests should write a letter of advice.
5:00	Closing conversation with thanks to participants.

numbers per hunting ground. Your team should expand to be more inclusive. Each team should have four types of people involved:

◆ *Decision makers*. You need decision-making ability in the room. A decision-maker is one who has the ability to significantly impact funding sources.

◆ *Implementers*. These are the people in your organization likely to be needed to act on making these ideas into a reality.

◆ *Relevant expertise*. These are your "native guides." If your hunting ground is outside your current field of expertise, you may wish to involve those whose expertise is more closely related.

◆ *Diverse value chain representation.* This group includes suppliers, customers, and consumers. Customers—people whose need you seek to fulfill—are particularly valuable; these people understand the problem and, if properly managed, are great idea generators.

Although most groups, including outside resources, tend to include from twelve to twenty-five individuals, the right number of people has more to do with your need for different knowledge sets, the politics of getting the right people involved, and your process preference than with anything inherent in ideation. (The Black & Decker team designed a process that enabled them to involve more than 160 different internal people in a series of half-day brainstorming sessions. A large financial services firm seeking to serve women with low incomes not only involved those they sought to serve in the ideation but brought them back to present the concepts created to management.) Be creative with your approach to ideation.

Concept Building

Within the ideation process, ideas must develop to the concept stage. The product concept does not need to be fully developed; however, there should be a clear statement of what the idea is, how it might be expected to work, the value of the idea, and the unresolved issues. More detail may hinder your ideation process (Miller 1999). Ideally you will be working with a hundred or more concepts per hunting ground.

Concept Screening

Any screening should have several distinct selection stages and not be dependent on a single review or pass. A multistage process is preferred:

1. A simple advisory vote from the ideation team. In a vote about one-third of your concepts will rise to the top. About one-third will drop away. Often newer thinking falls into the middle territory and may need special attention.
2. After an initial review, an evaluation by criteria is appropriate. Go back to the charter. Your original charter should have a set of criteria stated or implied. Screening by criteria should help you find key ideas. Here are some examples of rough-cut screening criteria:
   ◆ Meets a real need for a growing demographic; value chain excitement
   ◆ Fits with corporate identity, strategy, internal excitement
   ◆ Is feasible within the planning horizon ("we think we can do this")
   ◆ Is platformable, with many ways to use the investment (it is not a "wonder widget")
   ◆ Provides competitive insulation—ownable, patentable, protectable
3. Initial portfolio development—the nomination of the leading concept candidates that will be used in building your path. Focus more on solving the real problem than on feasibility or current need.

EXAMPLES

In the mid-1980s Hershey Foods Corporation identified a potential "killer application" in merging the needs for portion control, complex mouth feel, more flavor complexity, and familiarity—all needs associated with a mature, aging consumer. The concept, later called Hugs, was a Hershey's Kiss with white confection stripes. When it was first envisioned it was beyond the technical capability of the organization and the consumers then present clearly did not want it. Measured by the standards of that time, it would have ranked very low. Measured as a place to get to over a period of years, it rated very high (Ward and Miller 2000). From the outside the changes made to the Kiss appear incremental. However, a close look shows that the path from Kiss to Hugs and now beyond not only added $300 million in incremental value to the brand but also created the new category of "candy as decorator item" (see Figure 2-5).

Given three to five hunting grounds, three hundred to five hundred concepts gives substance to the territories you are seeking to populate. Selection followed by clustering helps to identify the key points for your planning process. The screening process must deliver concepts that offer immediate value; others should provide value tomorrow and some not for years to come. Don't give in to numerical tyranny. These numbers emerged out of your team's "informed intuition" (Miller 1993)—in other words, you guessed. Ideas are infinitely flexible. A key to your success will be to get back to your most exciting concepts and rebuild them.

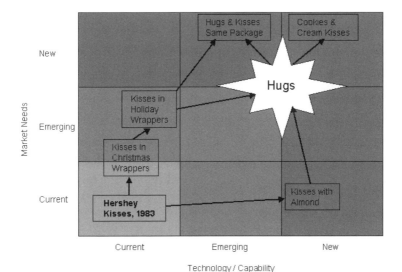

FIGURE 2-5. Define the path from the killer application back to your current equity and capability.
In the case of Hershey, Hugs was both a product concept and a hunting ground. Hugs was neither feasible nor desired in the market of the time. Hugs was dependent on a new capability that required invention and a market that did not yet exist.

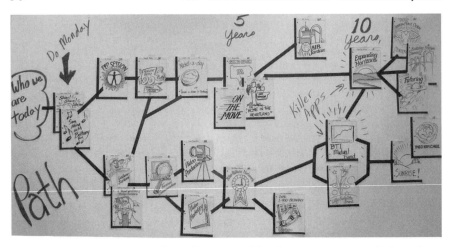

FIGURE 2-6. Plot a path back from the most exciting future concept to today's equity and ability.
This rough draft sample path was based on work done by Bethany Theological Seminary in 1991. Each image represents a product or service concept. The path started with killer applications that were perceived to be ten years out. Backward planning identified critical services and projects along the path. This path included a significant change made on the following Monday. Eventually intensity of focus allowed them to cut the ten-year plan to three years. (Note that the concepts are taped on a four-by-eight-foot sheet of foam-core board. The connecting lines are done with electrical tape.)

Path Development

Finding and getting to the "killer application(s)" in your hunting ground is your goal. Interestingly, the ability to deliver on the application is less important than the direction it provides. If your killer idea is immediately possible, one has to wonder why it has not already been done or what kind of competitive insulation there might be in the idea. If a hunting ground does not have at least one killer idea, it may have to be dropped or become part of another hunting ground. It may be a great opportunity but not for you at this time.

Use your most exciting idea or ideas, no matter how fantastic, as a stake in the ground of the future. In path development your objective is to find out how to get from that future point back to where you are now. This approach can be described as backward planning. Instead of working out from where you are, work back from where you wish to be.

Your leading concepts or concept clusters are your killer applications. Usually killer applications appear to be fairly long-range opportunities. Look at your secondary and tertiary concepts and plot a path back to today's equity and capability (see Figures 2-6 and 2-7).

Use a wall or get a large sheet of foam-core board. Place your killer application at the extreme right. At the extreme left put up a card that says "Today's equity and ability." Now use your concept set to mark a path between where you are today and where you will be. To get to your killer application, you

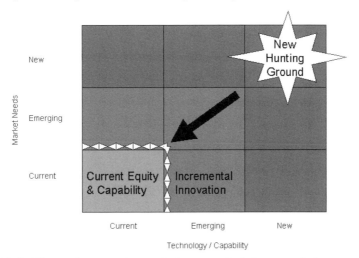

FIGURE 2-7. The hunting ground can only be approached along a path that supports the organization financially and culturally as it migrates away from its traditional territory. *This path is best defined by working backward from the hunting ground you eventually want to reach.*

need to build core technologies and platforms. Identify what earlier-stage concepts are going to allow you to experiment with key platforms and fund their development one step at a time. Determine which concepts are going to introduce you to customers in the new value chain in a way that makes good sense to you and to them.

A path is as long and as complex as it needs to be. Some appear to be only a few years long, and in other cases they can stretch over decades. Some paths may have only a dozen concepts; others may have scores. Some paths are primarily products; others are goods, services, marketing programs, alliances, and acquisitions. The best paths integrate these approaches to growth. Every concept on the path must be logically linked to those before and after it. This logical link must imply customer need and perception, technical ability, and corporate will.

Paths to different hunting grounds can share concepts, platforms, and equity. This is good news. It means that you may be able to travel in parallel for a period of time and delay the ultimate decision to select a specific path and make the necessary commitment.

In the planning stage perhaps three to five hunting grounds can be investigated at a time. An organization should not expect them all to survive careful scrutiny. The ideal scenario is to split up the core team now and create exploratory business units. Two members from the original core plus two to five new members with appropriate expertise will get the ball rolling. You should involve potential candidates for these positions in the ideation effort. A sense of excitement and idea ownership significantly enhances the team.

For each selected hunting ground, the champion, team, and of course a sponsor now must create a new charter and go their own way. Your hunting ground is your guide, a path studded with a score or more of possible revenue-generating ideas to provide the raw material for your work plan. You are at the beginning.

CONCLUSION

The hunting-for-hunting-grounds process has the potential to identify totally new business opportunities for your organization. It can also effectively shorten the front end of the NPD process by inserting an aggressive time-bound structure into the fuzzy front end. In either case, the goal is to provide sustainable, competitively advantaged waves of profitable innovation. The hunting for hunting grounds process ends with a set of actionable information-based hypotheses about what opportunities the future will hold. The objective is to anticipate needs and technologies rather than to react after they have become certain. Hunting for hunting grounds surfaces and organizes opportunities; it does not replace or eliminate any aspect of an integrated planning or product development process. Nor does hunting for hunting grounds eliminate incremental product development. It should provide a strong, positive voice that directs planning and investment (see Figure 2-8).

The most effective hunting for hunting grounds projects will reduce or eliminate funding and effort. Areas that are perceived not to have a future path become targets for elimination. Resources can be redirected to speed movement along an agreed-upon path. Quickly moving into an area where there is room

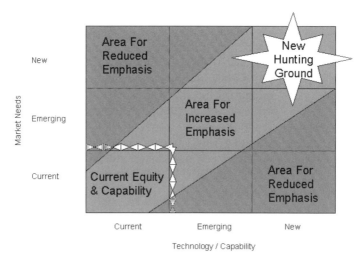

FIGURE 2-8. There is a significant opportunity to redirect resources away from projects that are no longer on the critical path.
Focusing these resources on the path may allow the organization to reach its objectives more quickly. A key issue for moving toward a new hunting ground is creating the organizational will to abandon territory previously considered strategic.

to grow, more intense need, and less competition will allow for stronger margins necessary for the long-term investment needed by the healthy organization.

Organizations seeking aggressive incrementalism can also benefit from a process that provides direction. Too often we think in terms of the "flavor of the month." In fact, there is nothing wrong with the flavor-of-the-month mentality when it is connected to a strategy to own strawberry. Incrementalists must still choose which of their many options are most exciting and appropriate. Without such a process the organization is driven by the tyranny and whims of responding only to today's customer needs. Critical as these are, today's requirements are subordinate to the powerful changes that will hit your customers in the future. Your job is to provide both what your customers need today and the security of knowing that the trusted equity that your organization represents is working to meet their needs in the future.

BIBLIOGRAPHY

Aaker, David A., V. Kumar, and George S. Day. 2000. *Marketing Research,* 7th ed. New York: John Wiley and Sons.

Crawford, C., and C. Anthony DiBenedetto. 2000. *New Products Management.* Homewood, IL: Irwin/McGraw-Hill.

Epstein, Eve. 2000. "The Human Touch." *InfoWorld* 22: 26–27.

Griffin, Abbie J., and John R. Hauser. 1993. "The Voice of the Customer." *Marketing Science* 12, 1: 1–27.

Hair, Joseph F., Jr., Rolph E. Anderson, Ronald L. Tatam, and William C. Black. 1998. *Multivariate Data Analysis,* 5th ed. Upper Saddle River, NJ: Prentice Hall.

Katzenbach, J. R., and K. S. Smith. 1993. *The Wisdom of Teams.* Boston: Harvard Business School Press.

Kerstin, Cuhls, and Blind Knut. 2001. "Foresight in Germany: The Example of Delphi '98 or: How Can the Future Be Shaped." *International Journal of Technology Management* 21: 767–780.

Kinnear, Thomas C., and James R. Taylor. 1996. *Marketing Research: An Applied Approach,* 5th ed. New York: McGraw-Hill.

Krueger, Richard A., and Mary Anne Casey. 2000. *Focus Groups: A Practical Guide for Applied Research,* 3rd ed. Thousand Oaks, CA: Sage Publications.

Leonard-Barton, Dorothy. 1992. "Core Capabilities and Core Rigidities: A Paradox in Managing New Product Development." *Strategic Management Journal* 13 (special issue): 111–25.

Leonard-Barton, Dorothy, and Jeffrey F. Rayport. 1997. "Spark Innovation Through Empathic Design." *Harvard Business Review* 75, 6: 102–118.

Malhotra, Naresh K. 1999. *Marketing Research: An Applied Orientation,* 3rd ed. Upper Saddle River, NJ: Prentice Hall.

Matt, D. Dawson, and S. Brucker Penny. 2001. "The Utility of the Delphi Method in MFT Research." *The American Journal of Family Therapy* 29: 125–140.

Miller, Christopher W. 1993. "Informed Intuition: Take One Giant Step In." *Visions* 17, 3: 10–15.

Miller, Christopher W. 1997. *The Focused Innovation Technique: A Creative Problem Solving Process.* Ed. Linda S. Crill. Lancaster, PA: Innovation Focus.

Miller, Christopher W. 1999. *A Workbook for Innovation: Developing New Product Concepts.* Harrisburg, PA: Pennsylvania Chamber of Business and Industry Educational Forum.

Millett, Stephen M., and Edward J. Honton. 1991. *A Manager's Guide to Technology Forecasting and Strategy Analysis Methods.* Columbus, OH: Battelle Press.

Porter, Michael E. 1985. *Competitive Advantage: Creating and Sustaining Superior Performance.* New York: Free Press; London: Collier Macmillan.

Porter, Michael E. 1980. *Competitive Strategy: Techniques for Analyzing Industries and Competitors.* New York: Free Press.

Prahalad, C. K., and G. Hamel. 1990. "The Core Competence of the Corporation." *Harvard Business Review,* 68, 3: 79–92.

Reinertsen, Donald G., and Preston G. Smith. 1991. "The Strategist's Role in Shortening Product Development." *The Journal of Business Strategy* 12, 4: 18–23.

Rowe, Gene, and George Wright. 1999. "The Delphi Technique as a Forecasting Tool: Issues and Analysis." *International Journal of Forecasting* 15: 353–375.

Slaughter, Richard A. 1999. "A New Framework for Environmental Scanning." *Foresight: The Journal of Futures Studies, Strategic Thinking and Policy* 1, 5: 441–51.

Stevens, Greg A., and James Burley. 1997. "3,000 Raw Ideas = 1 Commercial Success!" *Research Technology Management* 40, 3: 16–27.

Stevens, Tim. 1999. "Lights, Camera, Innovation!" *Industry Week* 248: 32–38.

Thomas, Robert J. 1993. *New Product Development: Managing and Forecasting for Strategic Success.* New York: John Wiley and Sons.

von Hippel, Eric. 1982. "Get New Products from Customers." *Harvard Business Review* 60, 2: 117–123.

von Hippel, Eric, Stefan Thomke, and Mary Sonnack. 1999. "Creating Breakthroughs at 3M." *Harvard Business Review* 77, 5: 47–57.

von Hippel, Eric, Stefan Thomke, and Mary Sonnack. 1988. *The Sources of Innovation.* New York: Oxford University Press.

Ward, Kirk, and Christopher Miller. 2000. "Innovating in New Product Development with In-Store Data." Paper presented at Consumer Sciences Conference, Institute for International Research, Boston.

Woodland, Cara L. 2001. "Researching Your Customers: Tools from Anthropology and Ethnography Applied in New Product Development." Paper presented at the conference Ethnographic/Observational Market Research, Institute for International Research, New York.

3

Telephoning Your Way to Compelling Value Propositions

George Castellion

◆ Why should customers buy your new product?
◆ How do you know prospective customers will accept the product as fast as champions of the product idea have forecast?
◆ Will you be able to market the product at a profit?

A compelling value proposition can provide a strong foundation for answering these and other marketing questions in the fuzzy front end of a business-to-business product's development. A compelling value proposition is a short, clear, and simple statement of how and on what dimensions the product concept will deliver value to prospective customers. The proposition serves two important groups of people—the prospective customers and the product developers.

DEFINITION

Fuzzy front end: *"The first three stages (strategic planning, concept generation, and especially, pretechnical evaluation) constitute what is popularly called the fuzzy front end of the new product process. This doesn't mean our minds are fuzzy; it means the product concept is." (Crawford 1997)*

In this chapter we describe a tool that can help build compelling value propositions for new product ideas in the fuzzy front end, quickly and inexpensively. This tool—in-depth, qualitative telephone interviewing—gathers information, knowledge, and insights from key individuals in the product concept's potential market. The collected results of the interviewing can then be recast into a value proposition.

We've used the telephone interviewing tool to help build value propositions for more than thirty years. Our first use of the tool occurred when we served in the role of inventor and manager of R&D groups. Here we needed to understand what certain product concepts contained, beyond some obvious technical features, that aroused such enthusiasm from prospective customers. After we made a career move into marketing, we used the tool to collect information on why customers might buy a product based on an innovative product idea. We also extended it to gather information on the pace of marketplace acceptance and profitability. Finally, for the past sixteen years as a service provider to North American and European companies, we've employed the tool to resolve marketing questions such as the three listed above.

We've also trained individuals and teams in the use of the telephone interviewing tool. During these workshops we've seen some of the common traps inexperienced interviewers encounter. These traps prevent realization of the full potential of telephone interviewing. The traps and tips for avoiding them will be described later.

PURPOSE, PROCESS, AND PAYOFF

The purpose of this chapter is to describe the in-depth, qualitative telephone interviewing tool so anyone can begin to use the tool effectively for building compelling value propositions when dealing with fuzzy product concepts. The process for doing this involves discussing four topics: (1) characteristics of compelling value propositions, (2) in-depth, qualitative interviewing, (3) using the telephone to do this interviewing, and (4) recasting the collected information, knowledge, and insights into a compelling value proposition. The payoff is proficiency in a powerful tool that can help discover the way a new product concept's critical attributes are seen by prospective customers.

Characteristics of Compelling Value Propositions

At the heart of turning a new product idea into a profitable product is discovering what position the concept occupies in the minds of prospective customers (Urban and Hauser 1993). The developer determines early in the development work who these customers are for the product and what product features, uses, and benefits are critical to them. Targeted customers should need and want the product—and should be able to buy it. A company's strategy must enable it to deliver a *value proposition,* or a set of benefits, different from those of competitors. It is not an effort to be all things to all customers. It defines a way of competing that delivers unique value in a particular set of uses or for a particular set of customers (Porter 2001).

No developer can position a product idea in the minds of prospective customers. Only the customers can take what they know about the concept's attributes and position that product idea in their minds. What the developer can do is to discover what the critical features, uses, and benefits are for prospective customers. Then the developer can group the customers into business segments—defensible competitive arenas within which market leadership is valuable (Koch 1994). Next the developer targets a segment where the product attributes of the concept are needed, wanted, and valued by prospective customers. Finally, through a compelling value proposition, the developer focuses the development work on delivering the critical attributes of the product to the target customers.

DEFINITION

Product attributes: *"The characteristics by which products are identified and differentiated. Usually comprises features, functions (uses) and benefits."* *(Crawford 1997)*

A compelling value proposition needs to be short, clear, and simple. Because the concept is new, its features, uses, and benefits must be quickly understood by the prospective customers. In the first few seconds, the proposition should give the customer enough information to begin to view the new product in his or her mind and compare it to competing products. Prospective customers are busy people, and they want to understand quickly how they can gain by buying a new product or can lose by not buying it. This means the product attributes included in the proposition must be narrowed down to no more than four attributes that are critical—critical to the customer, not the developer.

Urban and Hauser (1993) give the following examples of value propositions: American Express Travelers Cheques (accepted everywhere, prompt replacement, complete protection if lost, and prestige) and Hewlett-Packard LaserJet (quietly prints documents with excellent print quality on several media and in several fonts).

Entrepreneurs seeking venture capital to develop new products have learned that short, clear, and simple value propositions pass the "elevator test" (Moore 1991; Reinertsen 1997). In this theoretical test, the entrepreneur enters an elevator on the twentieth floor of a building. Unexpectedly, a venture capitalist also enters and pushes the down button. The entrepreneur has until the elevator reaches the ground floor to convince the venture capitalist to fund development of the product concept. What does the entrepreneur say to the venture capitalist? The entrepreneur articulates the concept's compelling value proposition.

At Microsoft, salespeople are expected to pass this elevator test—to be able to explain the benefits of their product to a prospective customer in thirty seconds. When they do, they are considered to have proven they can "boil the product down to its most compelling essence" (Strauss 2001). Geoff Allen, president of AnyStream, recently passed a two-floor elevator test in answering a question posed to him by Gary Arlan, a columnist from Washtech.com (Arlen 2001). The instant the elevator doors closed, Arlan said to Allen, "Describe what you do before the doors open at the next floor." Allen's answer to Arlen was: "AnyStream can handle any transmission speed for streaming media delivery."

Note that Geoff Allen's value proposition for AnyStream contains only ten words. (It helps that the name of Allen's company describes the service it provides.) Our experience is that compelling value propositions for successful product concepts contain twenty-five words or less. Boiling the value proposition down to its most compelling essence—twenty-five words or less—is important. The fewer words a developer needs to make a prospective customer in a target business understand a product concept's critical attributes, the easier it is for customers to visualize what position the idea occupies in their minds. The customers can then explain to the firm's decision makers why the firm should consider evaluating the finished product. As the concept moves out of the fuzzy front end, fewer critical attributes need to be balanced in a product developer's trade-off decisions, and the concept moves faster through the later steps in development work.

Boiling down the value proposition to twenty-five words or less is hard work. The first four common denominators of successful new product projects cited by Cooper (1999) speak to the work needed to remove the fuzz from a product concept before leaving the front end.

1. Up-front homework pays off.
2. Build in the voice of the customer.
3. Seek differentiated, superior products.
4. Demand sharp, stable, and early product definition in the fuzzy front end.

Another productive guide to building a compelling value proposition is to ask what the proposition excludes. If no prospective customer, no possible application, and no feature has been excluded, then the value proposition has no information content. In one development project Reinertsen (1997) was associated with, "a perceptive VP of engineering commented that it might be easier to define who we were *not* trying to target." A developer must make choices to concentrate on no more than four product attributes that are critical to the target customer's purchase decision.

In-Depth, Qualitative Interviewing

A rabbit stew recipe's beginning sentence, "First, catch the rabbit," contains a worthwhile caution for product developers assembling a compelling value proposition. Often the demand for action is so great—to move beyond the product concept and into technical development—that the information used to decide on the critical attributes is hastily put together and assumed to be acceptable. Often, however, such information is outdated or mistaken—the developer

---

DEFINITION

Qualitative marketing research: *"Marketing research techniques that use small samples of respondents to gain an impression of their beliefs, motivations, perceptions, and opinions. Qualitative marketing research is used to show why people buy a particular product, whereas quantitative marketing research reveals how many people buy it."* (Dictionary of Business 1996)

is taking on the equivalent of making rabbit stew without first catching the-rabbit. Research shows that solid homework drives new product success (Cooper 1999). Getting timely and accurate information about a concept's critical attributes is the keystone of up-front homework. In-depth, qualitative interviews with prospective users and customers provide such current information.

To interview is to question in an effort to discover the opinions or experiences of another. An interview needs at least two people, each in a different role. The interviewer asks most of the questions. The respondent answers them. Most of us have been in the role of a respondent. The interviewer seeks to learn something useful to them by interviewing us. Perhaps the interviewer, a doctor, is trying to narrow the number of diagnostic possibilities and make the best choice of medicine to treat the symptoms you display. Perhaps a reporter interviews you, tapping into your expert knowledge and insight to give credibility to an article being prepared for broadcast or publication. Perhaps the interviewer is a detective seeking information to solve a crime you saw or of which you have some knowledge.

Doctors, reporters, and detectives are experienced, and often trained, in the use of in-depth, qualitative interviewing to gain information, knowledge, and insight from respondents. A key skill in qualitative interviewing is inquiry, "holding conversations where we openly share views and develop knowledge about each other's assumptions" (Senge et al. 1994). Columbo, a fictitious detective in a TV series, often asks inquiry questions: "Sir, can you help me understand your thinking here?" Another key skill is balancing inquiry questions with advocacy statements. Columbo also balances inquiry questions with advocacy statements: "Here's what I think, sir, and here's how I got there."

Interviewers experienced in qualitative interviewing elicit information from respondents using inquiry questions and advocacy statements in a flexible, semistructured way. In the following section on telephone interviewing, an effective elicitation technique, "bringing out the professor," is described (Brown 1986). One of the essentials for gaining information while bringing out the professor is for the interviewer's manner always to remain in the role of a nonthreatening "intelligent pupil."

Prospective customers have a position in their minds—a mental model—into which a product concept fits. This mental model fixes the critical attributes of the concept and determines whether the customers are likely to buy the product developed. Surfacing these critical attributes can be done through in-depth, qualitative interviews with members of a small, purposeful (not random) sample of at least twenty-five respondents in the product concept's distinctive user and customer chain in the target business segment.

DEFINITION

The user and customer chain: *"The primary purpose of generating a user and customer chain is to outline who each of the key parties is in the target segment, to identify the needs of each party, and to highlight how users and customers make their purchase decision." (Wilson 1996)*

Experienced in-depth, qualitative interviewers have learned to do at least twenty-five interviews before drawing conclusions. "The combination of learning curve issues . . . linguistics, degree of comfort with the interview script, and the need to network through respondents to obtain the best sources . . . requires a minimum of twenty-five interviews. Every time we have short cut the process due either to time or budget limitations, either we or our client have been disappointed with the result" (Lax 1992). Lax terms this the "rule of 25 interviews." A research study at MIT by Griffin and Hauser (1993) provides some background as to why doing twenty or more interviews is important. Griffin and Hauser demonstrated that qualitative interviews with twenty to thirty potential customers would identify 90 to 95 percent of all the attributes in a new product concept. After only five interviews less than 50 percent of these attributes were identified.

DEFINITION

Rule of twenty-five interviews: *Experienced in-depth, qualitative interviewers have learned to do at least twenty-five interviews before drawing conclusions. (Lax 1992)*

Using the Telephone for In-Depth, Qualitative Interviewing

Benefits

QUICK AND COST-EFFECTIVE   Compared to face-to-face interviewing, telephone interviews can be done quickly and more cost-effectively. Face-to-face interviews require the interviewer to make an appointment to meet the respondent. Most product ideas are directed at a target business spread over a wide geographical area, often worldwide. Setting up the interviews and coordinating the interviewer's travel to keep the appointments are time-consuming. Direct travel costs are high, and the time lost while the interviewer travels from appointment to appointment can be costly.

ACCESS TO TARGET RESPONDENT   Another of the benefits of telephone interviewing compared to face-to-face interviews is that the interviewer interviews the target respondent, not someone drafted to fill in after the original face-to-face respondent's appointment schedule changed at the last minute. When a telephone interviewer's opening remarks catch a potential respondent's attention, the respondent enters the professor role for an interview that sometimes can last more than an hour. If it is not convenient for the respondent to talk then, most often they will give the interviewer a time to call them back when the interview can continue at length.

REFERRAL TO POTENTIAL RESPONDENTS   At the close of the interview, the respondent often mentions new potential respondents for the interviewer to telephone. Often this occurs if a critical insight has been unearthed during the interview. The telephone interviewer can change the priority for the next three to five calls and interview the new potential respondents to verify or negate the critical insight.

MORE EFFECTIVE IN ELICITING UNIQUE INFORMATION   Telephone interviews can be more effective in eliciting critical information, knowledge, and insights than face-to-face interviews. On the telephone, respondents cannot see the interviewer and are often more willing to talk openly and reveal information and insights. As the respondent becomes conversational and relaxed, the interviewer can probe even deeper (Beckwith 1997; Brown 1986). McLuhan's definition of the telephone as a "cool" medium explains why telephone interviews can be so productive (McLuhan 1994). A cool medium transfers less information than needed for the mind to receive a complete message. The respondent's mind is uncomfortable with an incomplete message, so, in a professor role, it fills in the missing information to complete the message to the interviewer. Much of this information has not been recorded previously. It is known only to knowledgeable respondents who are either creating the information or watching it being created. Most are target customers working in the product concept's user and customer chain.

Drawbacks

NOT BEING THERE   The telephone interviewer does not feel as much in touch with the respondent as in a face-to-face interview, and this can lead to shorter interviews (Weiss 1994). Weiss found that qualitative interviews conducted by telephone ran about forty-five minutes or less, while face-to-face interviews ran an hour and a half or more. Weiss's comment: "A team that has done a great deal of telephone interviewing describes it as 'the next best thing to being there.' "

MAKING COLD CALLS   Often individuals with excellent potential as interviewers but little training in in-depth, qualitative telephone interviewing shy away from using this tool because almost all productive interviews must start as cold calls to strangers. These individuals' reluctance to use the tool may be due to their experiences of being cold-called by strangers, such as brokers touting the latest hot stock. They know what it's like to have strangers demanding the time of busy people. Individuals who shy away from making cold calls are also concerned about being rejected by the person on the other end of the phone. They may be more familiar with granting the favor of an informational interview, rather than asking for the favor of getting one.

DEFINITION

Cold call: *An interviewer's telephone call to an (until now) unknown person during which the interviewer tries to enlist that stranger as a respondent.*

Practice is the best way for a novice interviewer to overcome reluctance to make cold calls. Since the interviewer doesn't know the stranger being called, the interviewer can't possibly know what they are going to say. This keeps the interviewer mentally alert for the mention of new and different product attributes made by the respondent about the product concept. Once novice interviewers get into the rhythm of making cold calls, it can become enjoyable as the number of interviews increases toward twenty-five or more.

Seven Steps to Minimize Rejection When Cold-Calling Potential Respondents

The seven steps are listed here and then discussed below in detail.

1. Research secondary resources to define provisional business segments—defensible competitive arenas within which the product idea could provide market leadership.
2. Identify—as potential respondents in the business segments—innovators, early adopters, and early majority users.
3. Build a first telephone calling list of fifty or more prospects.
4. Compose an opening script that all prospects will hear, to act as a frame for the inquiry questions and advocacy statements to follow.
5. Put the respondent in the role of the professor. The interviewer always remains in the role of the intelligent pupil.
6. Place calls consistently and steadily.
7. Begin to uncover eligible profitable target businesses.

1. Research Secondary Resources

Until a few years ago, a researcher needed to spend a day or two in a good business library to define provisional business segments and potential respondents. In the library the researcher did this by skimming through hard copies of published secondary resources. Now a researcher with access to the Internet can find and view many of these published documents online. As well as these documents, the Internet contains a vast number of documents created solely as Internet resources. Many of these documents contain up-to-date information for provisionally defining the target business segment and its user and customer chain. Among such documents are Web pages of companies in the target business segment, pages of associations—trade, technical, and marketing—serving the segment, and pages announcing conferences, workshops, and seminars on important issues within a business segment.

Of equal importance to the researcher using the Internet is the emergence of effective search engines to examine sites and return only those pages of relevance to the researcher. One such search engine, Google (www.google.com), runs on a combination of advanced hardware and software and rapidly searches more than one billion Web sites. It then displays for the researcher, in less than a minute, the most relevant Web pages—based on the researcher's query and Google's PageRank algorithm. Another valuable search engine, Northern Light (www.northernlight.com), searches both the Web and Northern Light's special collection of over seven thousand full-text publications that are not available to other Web search engines. Our experience has been that searching the Web using both the Google and Northern Light search engines is an effective way of compiling relevant information from secondary resources.

2. Tentatively Identify Potential Respondents in the Target Business

Rogers (1995) studied, for more than forty years, how innovations are adopted. He classified adoption of an innovative product based on the characteristics of five different adopter categories. Rogers found that the cumulative number of adopters of an innovative product plotted against time passed since the product was launched forms an S-shaped curve (Figure 3-1a). He also found that plotting the mean number of adopters per year against the time since product launch forms a bell-shaped curve. This bell-shaped curve can be partitioned into five adopter categories. The curve and partitions are formed by plotting the number of units sold per year against the time since product launch when an individual in a category begins to use the product (Figure 3-1b).

Individuals in each of the adopter categories have dominant characteristics. Rogers termed those first to adopt an product *innovators*. They account for about 2 percent of the cumulative adoptions of the product. Innovators play a gatekeeping role in the launch of an innovation and are able to cope with a high degree of uncertainty about the product as it emerges from development. In general they are on the fringes of the network inside the business they are part of. *Early adopters* (13 percent) are more integrated into the business's network than innovators. They decide soon after the product is developed that its use will improve the business they manage. The *early majority users* (34 percent) advance with a deliberate willingness in adopting innovative products. They are important for the success of the new product since they make up a third of the total cumulative adoptions of an innovative product and adopt fairly early in the life cycle of the product.

The final two categories, *late majority users* (34 percent) and *laggards* (16 percent), are not useful potential respondents for interviews in the fuzzy front end. Late majority individuals are skeptical about innovations. All the misgiving about using a new product must be removed before late majority adopters feel it's safe to adopt. Laggards are always the last in a business to adopt. Their decision to adopt is lengthy and lags far behind early awareness of the product's critical attributes.

Moore (1991), working with clients in developing technology-based products having short product life cycles, pointed out a chasm between early adopters and early majority users in their view of critical product attributes. Early adopters expect the attributes to give them a jump on their competitors, whether from lower costs, faster time to market, better service features, or some other advantage in the business. Early adopters are comfortable with handling the organizational problems that always occur when a new product is intro-

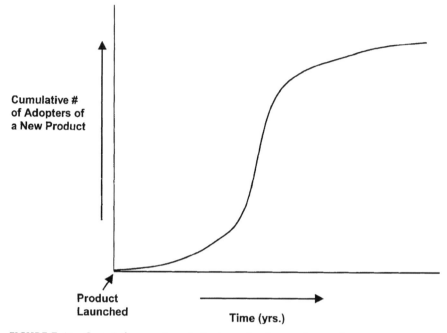

FIGURE 3-1a. Cumulative number of adopters of an innovative product vs. time since launch.

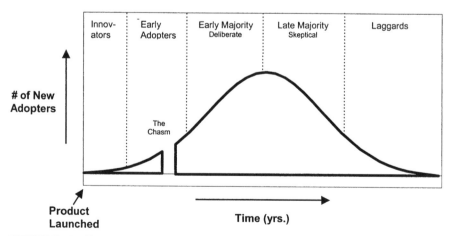

FIGURE 3-1b. Number of adopters each year vs. time since launch.
Adapted from Microsoft Visio clip art.

duced into a business. Early majority users, however, in contrast to early adopters, maintain a strong position in their mind for critical product attributes that will increase value of their business but not cause start-up problems. If attributes at launch do not address the early majority users' concern about integrating the product into their business without disruptions, then cumulative adoptions will fall rapidly. There can be a virtual chasm between adoption by early adopters and early majority users. Early adopters and innovators are necessary for the launch and first sales of a product based on a product concept emerging out of the fuzzy front end. However, if the value proposition is built only on critical attributes elicited in interviews with innovators and early adopters, the odds are against adoption by early majority users. (A rule of thumb we have found useful is to get 50 to 60 percent of the total number of interviews from early majority users, 30 percent from early adopters, and the rest from innovators.)

3. Build a First Calling List of Fifty or More Potential Respondents

Don't start calling for interviews until research with secondary sources results in a list of at least fifty names, with telephone numbers, of potential respondents. Once calling begins, keep calling to be most effective. It is common in the beginning stage of interviewing to make as many as twenty to thirty telephone calls without gaining an interview. The interviewer will, however, get referrals, leave messages on voice mail or with administrative assistants, learn when the potential respondents will be in the office, and gather direct-dial numbers for respondents to reach them outside usual business hours.

One method we've used for productive calling is to put into a database record the potential respondent's name, phone number, and organization name. Then we sort the potential respondents' records in either ascending or descending order on one of the fields and print them out. (For the layout we use in Filemaker software, each printed page includes the contact information for eight potential respondents.)

We start at the top of the first page and call all the potential respondents

---

TRAP: Fewer than fifty potential respondents on the first calling list. With fewer than fifty on a first calling list, the interviewer may be tempted to concentrate on reaching those respondents the interviewer somehow feels will have the most valuable information. The interviewer becomes discouraged and disappointed when constant calling does not result in an interview, and gaining twenty-five interviews seems to be an impossible task. (No matter how innovative the product idea is, there are more than fifty individuals reachable for interviews by telephone who have valuable information on its critical product features, uses, and benefits.)

TIP: Interviewers can't possibly know before an interview what useful knowledge a potential respondent holds. Even with a first list of more than fifty prospects, the interviewer needs to be alert to possible bias on his or her own part about which prospect to call next.

on that page, in order (assuming they are in suitable time zones), before moving on to the next page. When we've finished calling through the complete first list, we code the records of those respondents interviewed so they subsequently don't sort or print. We then add potential respondents referred by phone contacts to the list and others found in new searches of the Web. Then we sort and print out the expanded list and start calling again. (We find that by the time twenty-five interviews are completed, the list contains more than 150 records.)

4. Build an Opening Script

We recommend that interviewers write out the opening script, which all prospects will hear, and have it in front of them always. Do not try to do it from memory. When a potential respondent calls unexpectedly in answer to a message left on voice mail, the interviewer needs to be able to frame inquiry questions and advocacy statements in the same manner used in other interviews.

The same manner includes the same pattern in tone and emphasis in the interviewer's voice. Otherwise the interviewer may give unconscious bias to the answer the interviewer expects from the respondent. This doesn't mean the interviewer delivers the opening script like a robot. The whole interview should finish with the respondent feeling that he or she has had a satisfying conversation with someone who was alert, interested, and appreciative.

Following is a generic opening script that can be customized for a specific product concept. In italics are comments about essential phrases and why our experience signals they should be included. Customized versions of this generic script have worked well with respondents in businesses as varied as chicken farming, high-speed fiber-optic switches, and toothpaste.

Generic Opening Script

"Good morning [or afternoon, or evening—depending on the prospect's time zone]. My name is [first name, last name] from [name of interviewer's organization]."

(The "My name is . . . from . . ." phrase comes from a project twenty years ago when a group of three of us were doing an intensive set of in-depth, qualitative, telephone interviews with potential customers and competitors. We had the same opening script, but one of us slightly adapted the second sentence to the "My name is . . . from . . ." phrase. He gained more interviews than the other two of us. When we adapted our opening sentences to track with his, our rate of gaining interviews jumped to his level.)

If the administrative assistant answers the call:

"Is it convenient for me to talk with [prospect's name] now?" If it isn't: "Can you suggest a better time?"

If the prospect answers the call:

"Is this a convenient time to call?" or "Is there a better time?"

(Interviewers should always ask if it is a convenient time to call. Prospects hear this courtesy so rarely that many busy prospects give the interviewer a more convenient time in the next breath, often with a direct-dial telephone or cell phone number that will reach them after office hours.)

TRAP: Taking notes on a computer during the interview.

TIP: The interviewer should develop shorthand that will allow him or her to capture the essence of the respondent's remarks without prejudicing the conversational nature of the interview. The noise of a computer keyboard clicking away in the background drops many respondents out of the professor role. In addition, few interviewers can interview and handle a computer without dropping out of the intelligent pupil role.

TRAP: Taping an interview either with or without the respondent's permission.

TIP: It's illegal in many states to do so without the respondent's permission. It's always unethical to do so without permission. Even if the interviewer receives permission, taping acts to keep respondents out of the deepest and richest arenas of information when they are in the professor role.

Next the interviewer truthfully describes what he or she does and the reason for the call.

(*For my calls the following is suitable. Interviewers should adapt their scripts appropriately, always being accurate in their description*).

"I'm a consultant working with clients in the fuzzy front end of new product development. I'm on assignment now from a client that is working on an innovative [five-word description of the product idea]. On a small scale they have been able to demonstrate that with this new [concept in one or two words] that [describe what the small-scale work has demonstrated]. I'd like to ask you some questions based on the understanding that I'm beginning to build about the use of [the product idea in the target market].

"My client is a [U.S., European, Far East]-based company. They have a good basic knowledge about the [technology, service features] going into the product concept but little knowledge about its application in [the target market].

"What I'm having a tough time understanding. . . ."

(*At this point, if the interviewer has followed the opening script, he or she will find that potential respondents will go into the role of the professor. What follows then is a series of inquiry questions and advocacy statements on possible product attributes. Interviewers need to write these out and have them where they can see them always. The interviewer should use essentially the same wording of each question or statement with each respondent. The order in which they are asked depends on where the professor wants to go. Throughout the interview the interviewer stays in the role of the intelligent pupil.*

When the phone conversation has exceeded twelve minutes and the respondent has remained in the role of the professor, we count it as one of the twenty-five or more interviews. For shorter conversations we capture in shorthand the information given by the phone contact but do not count this conversation as an interview.)

5. Place the Respondent in the Role of the Professor

"The professor" is the name for that part of all of us that loves to discuss a subject we're expert in with someone whom we can trust and who is seriously

TRAP: Anticipating what the professor is going to say and moving on to another question on the interviewer's agenda.

TIP: Don't anticipate, participate. Interviewers must remain in the intelligent pupil role and not anticipate what the respondent is going to say next. This can be hard for an interviewer after he or she has completed twenty interviews, collected a hundred or so product attributes, and believes he or she understands which two or three are critical. First, when an interviewer anticipates, he or she falls out of the intelligent pupil role and the respondent falls out of the professor role. Second, and more importantly, this is usually the point in the interview where the professor, filling in missing information, mentions a wanted attribute that previous respondents hadn't crystallized as clearly. If the pupil asks, "I don't understand. Why is that so important?" the answer often uncovers the key critical attribute for a successful and profitable product idea.

interested (Brown 1986). To encourage the respondent to continue in this role, the interviewer should always remain in the role of the intelligent pupil.

Knowledgeable individuals in the professor role limit the discussion with an intelligent pupil to critical product attribute information normally thought to be hard or impossible to get. This is valuable to the pupil not because it is secret information (it's not) but rather because it's current but unwritten information on the needs of respondents in the target business.

Intelligent pupils listen intently to every word. They may be the best audience the expert has ever had. In turn the pupil gives some value in exchange for what the pupil has received—for example, trends the pupil has noted emerging out of previous interviews.

In-depth, qualitative telephone interviewing is ideal for this exchange. There are no clues, as there are in face-to-face interviews, of the interviewer's appearance. Often such clues can trigger an unconscious bias in the professor's remarks.

6. Call Consistently and Steadily

One simple practice that works for many interviewers is to keep a log in which the interviewer continuously tracks the following items:

- ◆ The number of calls made in a given time period
- ◆ The number that result in interviews
- ◆ The number of messages put on potential respondents' voice mail
- ◆ The number of calls intercepted by potential respondents' administrative assistants
- ◆ The number of calls reaching busy signals
- ◆ The number of potential respondents' names given to the interviewer as referrals by the person who answered the phone

Tracking the results over a long period smooths out the roller-coaster swings that characterize the short term. For example, our results over the past sixteen

The professor limits the interview to information and knowledge that are:

• Current

• Usually thought to be hard to obtain (little of it is available in secondary resources)

FIGURE 3-2. Bringing out the professor.
Adapted from Microsoft Word clip art.

years are consistent no matter what the industry or the job role of the potential respondent. For every four to six calls we make, we will gain one interview with a knowledgeable respondent. The log helps keep us on track, especially at the start of most interviewing projects, when we may make several calls without gaining an interview.

(We make only two calls to a potential respondent. On the second call we leave a message to return our call with the prospect's administrative assistant or on the prospect's voice mail. More than 85 percent of these respondents do call back, and the log shows that most callbacks result in an interview. Increasingly these voice mails are returned when the respondent is in an airport. Often they've used their elite status with an airline to board their flight well before its takeoff time. They activate their cell phone, pick up their voice mails and assistant's messages, and return calls in the half hour or so before the plane's door is closed. An interviewer may have more uninterrupted time with this respondent than many of the respondent's subordinates have had in days. And the interviewer is not bringing problems to the respondent. The interviewer is inquiring about something the respondent is expert in, and few other people have shown such an interest. Reaching a potential respondent's administrative assistant can be as productive as voice mail in gaining callbacks. By speaking to the assistant, the interviewer may also find the prospect is not a suitable one to interview about the subject, and the assistant may give referrals to one or two more suitable prospects.)

7. Begin to Uncover Eligible Profitable Segments Within the Target Business

Every target business can be divided in many ways. There is not one "correct" segmentation. As interviewers accumulate interviews, they will find that the division becomes much richer than when they began. It will become richer because the interviews trigger insights about unexploited segments, pieces of the business where prospective customers will pay for the attributes a product concept can deliver.

To uncover eligible profitable pieces, interviews need to be spread across knowledgeable individuals in relevant links of the user and customer chain. One way to decide which are the relevant links is to reserve judgment until the interviewer has made a few calls into each link. Often there are unobvious but useful links, for example, equipment or service suppliers to the business's operating facilities. Respondents in such links often have a good knowledge of what early majority users such as operating managers need in the way of critical product features, uses, and benefits and how they have failed to get them in competitive products.

Putting It All Together

The interviewer, especially a service provider, is not the proper person to build a compelling value proposition for the product idea. Choices involved in the building task belong to the development team who will take a focused, and no longer fuzzy, product concept out of the front end and transform it into a new product. The interviewer's task is to meld the raw material of the interviews into a clear, concise written report so the developers can grasp the essential points easily and quickly and use them to build the value proposition. This raw material includes all the shorthand notes on the information, knowledge, and insights acquired from respondent interviews. It also includes all information the interviewer learned from secondary sources, and all the interviewer's insights after reflecting on the patterns of product features, uses, and benefits emerging from the interviews.

Recasting the raw material into a finished report is step-by-step and often tedious work. The interviewer goes back over the material, time and again, looking for patterns and reflecting on what they might mean. Often more interviews must be conducted to check the interviewer's assumptions and theories. Upon reflection, a profitable opportunity in an until-now-unrecognized business may crystallize, and more interviews must be made to flesh out the wanted attributes. When respondents, even after several checking and rechecking interviews, cannot come to agreement on the critical attributes, the writer must judge which respondents to trust or ignore. Even if the interviewer is the only person who will build the value proposition, transforming the raw materials into a clear, concise written report helps in choosing the right words.

THREE EXAMPLES OF THE TOOL'S USE

Marketing and Processing Improvements

A European firm had been trying for years to increase its share of the North American market for a formulated specialty product (code-named TUFF) used in paper mills. A United States–based competitor with a similar product held the major share. This competitor was beginning to have some success with the product in the firm's home market. The firm had a concept for a new formulation, TUFF+, with improved technical features. It wished to build a compelling value proposition for TUFF+ that would meet prospective North American customers' needs.

In collaboration with the firm, the generic opening script was customized to provide the frame for open-ended questions on TUFF+'s attributes. Spending a day in the local library resulted in a first calling list of more than fifty potential respondents. Among the library's sources was a publication listing the location and phone number of every paper mill in North America and each mill's papermaking capacity, equipment used, and key personnel. This source contributed most of the names on the list. Others came from the personnel changes columns in several trade publications specializing in the paper industry. (The secondary research work for this example was done before the Web became so useful for finding secondary resources.)

After about fifteen interviews two patterns emerged. Respondents were mildly interested in the improved technical features of the TUFF+ product concept. However, they also were not clear why they were using the current product, TUFF, no matter whether their supply was from the firm or from the competitor. TUFF was more expensive than another class of formulated product (code-named RUFF), which, according to some respondents, performed the same role as TUFF. Early on the interviewer found that a few mills used TUFF all the time, some mills used TUFF most of the time and RUFF part of the time, and many mills used RUFF all the time.

From this point on, the interviewer concentrated on identifying and interviewing potential respondents who were responsible for running the multi-million-dollar, high-speed papermaking machines—early majority users. A shift superintendent in a mill in Maine remarked, as an aside twenty minutes into the interview, that when TUFF was used there was less unscheduled maintenance of the machine. His experience was that TUFF prevented the machine from going into uncontrollable vibration when producing certain grades of paper. He did not know the reason, but his experiences over twenty years showed the TUFF was much better than RUFF at preventing the onset of vibration. (When high-speed papermaking machines are shut down for unscheduled maintenance, paper production lost during shutdown represents thousands of unrealized revenue dollars.)

The shift superintendent referred the interviewer to a staff scientist at a papermaking machine manufacturer who he thought might be in interested in the superintendent's observations. The interviewer contacted the scientist within the hour. During the interview the scientist disclosed that unexpected vibration was an expensive problem at mills worldwide. He suggested reasons why the superintendent's observations were valid. He also gave the names of possible respondents in mills across North America who might add to the interviewer's knowledge about this problem.

Later interviews with superintendents at these mills suggested that those using RUFF because of its lower price did experience more unscheduled maintenance shutdowns than those using TUFF. Respondents in these interviews who were using TUFF did not mention TUFF's benefits in decreasing the rate of unexpected vibration-related shutdowns, only its higher costs. Then, while interviewing a mill superintendent, the interviewer learned not only that this mill was using TUFF but also that they were a test site for the competitor's new version of TUFF, TUFF++. The superintendent said the new version's price would be substantially below both that of current TUFF and that of RUFF. This was the first signal to the interviewer—and the firm—that such a product even existed.

With a few interviews, the interviewer found the site of the competitor's plant producing TUFF++. Then in an interview with a respondent in the local government's environmental impact office near the plant site, it was discovered there were public records of the environmental impact statement for the new TUFF++ plant. Included in the records were a description of the production process and a diagram of the plant. When the interviewer received a copy of the plans from the environmental impact office, it was obvious that TUFF++ was physically and chemically identical to TUFF. However, it was produced by dropping two expensive steps in the TUFF production process. Through telephone interviews with more respondents these conclusions were confirmed.

In a short, concise written report and a daylong spoken presentation at the firm's office, the interviewer highlighted the findings on TUFF++ with choices for actionable decisions for the development team on building a compelling value proposition. Within two months after presentation of the report, the firm was able to drop processing steps and produce TUFF at a much lower cost. The firm built a short value proposition highlighting TUFF's new low price and its ability to decrease unscheduled maintenance shutdowns. Using the new value proposition, the firm blunted the competitor's success in the firm's home market and increased the firm's share of sales of TUFF in North America.

A Marketing Improvement and a Changed Product

A U.S.-based firm supplied a wide range of specialty products to carry out a critical task in a wide range of industries. This was achieved through three distinct classes of products. Classes A and B were high-volume, high-margin products. Class C products were, for the firm, low-volume and low-margin. The firm's salespersons were responsible for selling all three classes to the firm's customers but spent most of their time on Class A and Class B and little on Class C products. Over the years the salespersons reported back to the firm's R&D group on technical features that customers said they needed to induce them to try improved Class C products. However, when new Class C products with these features were developed by the firm and launched, cumulative sales were low and mainly to innovator customers. Prospective early majority users viewed them as costing much more but without enough incremental benefits to warrant adopting the new product.

The firm occupied a good position in the minds of prospective customers as a technical leader and supplier of Class A and Class B products. It wanted to occupy a good position in the minds of customers for Class C products. Once again the firm's R&D group came up with a new Class C product idea containing some new technical features. As with all ideas at the firm, for this concept to move out of the fuzzy front end and into development, the new product had to exceed a minimum gross margin hurdle.

The interviewer's objective in a project for the firm was to find answers to the following questions: What did customers and users of Class C products want in an improved product idea? What would they buy? What was the firm's position in the minds of prospective customers for Class C products? The opening script described only the new technical features of the product concept. A first calling list of more than fifty people was assembled mainly by seining Web sites for potential respondents in trade associations for Class C products and in the businesses that were prospective customers and users of Class C products.

After twenty interviews, three important patterns emerged; two expected by the firm, one unexpected. The first expected pattern was that prospective customers felt the new technical features would improve operations using Class C products. The second expected pattern was that they would pay no more than 5 percent extra for the improved Class C product. (It wasn't that prospective early majority users didn't understand the value of the improved product. Rather, the operations in which Class C products were used made prospective customers deliberate in adopting any new Class C product with a higher price.) The third and unexpected pattern was that when prospective customers ranked the top seven competitors in Class C products in favorable position in the customers' minds, the firm was number two. Before the interviewing project the firm believed and acted as if it ranked no higher than sixth. It appeared that the excellent reputation of the firm for Class A and B products "leaked over" in the minds of prospective customers for Class C products. Twenty more interviews confirmed these emerging patterns. The firm was well regarded by prospective customers and users for Class C products. Many respondents expressed their bewilderment the firm was not a major participant in the business. But the firm, because of its gross margin hurdle, just didn't have a new product that would meet this business's price needs.

After putting it all together in a written report, an oral report was made at an all-day presentation. The interviewer made three key points: (1) The firm had an excellent position in the minds of prospective users and customers in the target business. (2) Key decision makers in the business were early majority users. They communicated the success or failure of new products in their operations at trade association meetings where they were on the board or on standards committees—an old-boy network. The firm had an "old boy" in its ranks, one who was associated with their Class C products—a female who was well regarded by early majority user respondents because of her experience in the business before she joined the firm. A choice for an actionable decision in the final report was for the firm to have this employee represent it on boards and standards committees. (3) The internal gross margin hurdle coupled with the prospective customers' unwillingness to pay a significant premium for an improved product had to be resolved before a compelling value proposition could be built. At this point in the final oral presentation, the firm's general manager interrupted. One of his plants produced a product never before used in Class C products but used in Class B products. He now suspected if this product was added in small amounts to the ingredients composing the product idea, it could add a wanted technical feature. This potential additive had a high gross margin. Adding this product to the ingredients of the product concept would meet the firm's gross margin hurdle while meeting prospective customers' cost requirements.

Within ten months a product concept incorporating the new additive was developed into a product and launched in the North American market with a sales force dedicated to selling Class C products. (It was more effective to train a dedicated sales force than to retrain the salespersons selling Class A and Class B product to sell the new Class C products.) The old-boy employee was appointed to represent the firm on the boards and standards committees of important trade associations. The value proposition for the improved product incorporated the high respect early majority users had for the firm's technology skills. It also included the favorable effects the improved product could produce in these users' operations, and testimonials from the early majority users. Two years after the successful North American launch, the firm launched the product globally, well on the way to aligning its market share with the position it occupied in the minds of customers for Class C products.

Difficult Decisions

This is an example of a product idea for which interviewing uncovered no prospective customers who saw value in the concept's attributes. In this project a firm in the fiber-optic communications industry requested that the interviewer help the firm evaluate marketplace acceptance, in a noncommunications business, of some fiber-optic technology for which the firm had been approached as a possible investor. The opportunity to invest was offered to the firm by a venture capitalist group with which the firm had good relations.

The technology had been incorporated into a product concept developed and reduced to a crude prototype by a principal in the venture capital group. The concept was to use fiber-optic sensors to detect gasoline in contaminated soils. Fiber-optic sensors can detect gasoline with a greater sensitivity than the sensors currently used. After completing fifty interviews with prospective users and customers—federal, state, and local government environmental law enforcement personnel; technical officers at environmental activist organizations; and service station equipment manufacturers—a final written report put it all together. An oral report was presented at a full-day session at the firm's offices. When the report was finished, the firm chose the main option for an actionable decision—not to invest in the technology.

The reason the firm chose not to invest to develop the technology was that the interviewer found no prospective customers for the product were it to be developed. No one, not even environmental regulation enforcement personnel, wanted a more sensitive sensor of the type represented by the concept, for two reasons. First, emphasis in the business was being placed on preventing spills or leaks from contaminating the surrounding soil. (Even a few drops from the nozzle of a gasoline pump hose reaching the surface of some soils spread rapidly and can be detected at the perimeter of a service station for several years after the spill. Thus, half of the service station sites opened since the 1920s have been abandoned and not used for any other purpose.) Second, the business was making large investments in modern storage tanks and sensing equipment placed inside the tanks to detect leaks. Sensing equipment in modern double-hulled tanks detects changes of volume in the inner tank or leakage into the space between hulls. Detection of volume changes gave another, and valuable, benefit for service station operators. By telemetering information on volume changes from a service station to a distribution storage center, deliveries of gasoline by tank truck to a service station can be efficiently scheduled.

Two months after receiving the final report, the firm requested that the interviewer make the same presentation to the key principal at the venture capital firm. The interviewer traveled to the venture capitalists' city with some misgivings. It was clear the principal was the champion of the concept. Champions often are essential for shepherding product concepts through the white water of the fuzzy front end. However, some champions are not pleased to be told that their idea, in the view of prospective customers, does not have a compelling value proposition. The interviewer need not have worried. After meeting the firm's representative in the lobby of the venture capitalist's building, the interviewer was ushered into a conference room where the champion and another senior partner were waiting. They then heard the same presentation and read the same written report as the firm had two months earlier. At the end the champion said, "Thank you for such a good report. It strengthens what I've been feeling for some time. We must do much more work with positioning it in the market so they recognize the innovative features of our sensor and how to apply it." Eight years later, this product idea was "still under active development" at the venture capital firm, with prospective customers still to be found.

A PRACTICE EXERCISE FOR APPLYING THE TOOL

Assume you are responsible for evaluating new product opportunities for your firm, which produces toothpaste. One of your firm's executives read the September 2000 news accounts of the success that researchers at the University of Augsburg in Germany had in making commercially useful wires and cables for high-temperature superconductors. The executive recognizes that the calcium-based material being used at Augsburg to change the superconductor to achieve breakthrough current density is close in composition to the patented calcium-based material in your firm's leading brand of toothpaste. Your firm holds the patent, which has fourteen more years to run.

You and key development decision-makers at your firm have worked out an opening script. Your objective is to use the telephone interviewing tool to build a compelling value proposition to guide your firm's decisions to maximize the value of the patent in the superconductor business. Through your interviews you need to find out the marketplace acceptance of the Augsburg discovery. You also need to identify laboratories that could quickly and objectively decide the value of your patent in the superconductor business. Finally, you need to find and evaluate firms who might license your technology for superconductor applications.

In less than half a day, you produce a first list of more than fifty potential respondents through a search of Web resources. Using Google, a search using the words "high-temperature superconductors" results in more than thirty thousand links to sites containing these three words. In the first hundred of these sites, listed in order of relevance, there are several links to companies manufacturing high-temperature superconductors and to programs and proceedings of conferences on commercializing high-temperature superconductors. Clicking on these links, you seine out, as potential respondents, forty names from the executives at the companies and the several hundred presenters of papers at these conferences or on conferences' organizing committees. A search on Northern Light using the words "high-temperature superconductors Augsburg" gives you more than three hundred links, many to full-text pages of the work of rivals and collaborators with the Augsburg superconductor group. From these you select twenty names to add to your first list of potential respondents.

In your database program, you sort the records of potential respondents by time zone and then by organization name and print out the list. You begin calling, early in the morning, potential respondents at time zones ahead of you and keep moving down the list page by page until by evening you are calling potential respondents in time zones behind yours.

After five or more days of telephone calling, you've completed forty to fifty in-depth, qualitative telephone interviews Looking over your shorthand notes on the interviews, you write a rough draft of the final written report. You make confirming interviews to new respondents, to make sure your key choices for actionable decisions are not biased. After adapting the draft, you let it sit for a day or so before putting it in final form. Finally you stand up in front of your colleagues, each of whom has a hard copy of your presentation for annotation, and present the report, page by page. At the end you and the group are ready to move forward to build a compelling value proposition for use of your firm's proprietary calcium-based material in high-temperature superconductors.

SUMMING UP

In the three examples given above, using the tool gathers accurate and timely information, knowledge, and insights for a compelling value proposition. In the first example, "Marketing and Processing Improvements," only two product attributes were critical—decreased unscheduled maintenance shutdowns and lowest cost. The second example's proposition had three critical attributes—a superior additive, an affordable price, and the fact that it was from a trusted company. In the final example, "Difficult Decisions," the product concept had no compelling value proposition in the eyes of prospective customers. This conclusion may be hard for a product concept's originators to accept. Much harder, in our experience, is moving into the later stages of development without knowing if target customers need and want the product—and will buy it.

BIBLIOGRAPHY

Arlen, G. 2001. "Elevator Speech in a Two-Story Building" (www.washtech.com, accessed May 21, 2001).

Beckwith, H. 1997. *Selling the Invisible: A Field Guide to Modern Marketing*. New York: Warner Books.

Brown, H. 1986. "Tricks of the Information Trade." *Planning Review*, March 22–27, 40–41.

Cooper, R. G. 1999. "From Experience: The Invisible Success Factors in Product Innovation." *Journal of Product Innovation Management* 16: 115–33.

Crawford, C. M. 1997. *New Products Management*, 5th ed. Chicago: Irwin.

Dictionary of Business, 2nd ed. 1996. Oxford: Oxford University Press.

Griffin, A., and J. Hauser. 1993. "Voice of the Customer." *Marketing Science* 12: 1–27.

Koch, R. 1994. *The Financial Times Guide to Management and Finance: An A-Z of Tools, Terms and Techniques*. London: Pitman Publishing.

Lax, J. R. 1992. "Correcting for Disaster: Preventing Competitive Intelligence Mistakes." *Journal of AGSI (Association of Global Strategic Intelligence)*, November, 127–33.

Moore, G. A. 1991. *Crossing the Chasm: Marketing and Selling Technology Products to Mainstream Customers*. New York: HarperBusiness.

McLuhan, M. 1994. *Understanding Media*. Cambridge, MA: MIT Press (reprint).

Porter, M. E. 2001. "Strategy and the Internet." *Harvard Business Review*, March, 71.

Reinertsen, D. G. 1997. *Managing the Design Factory: A Product Developer's Toolkit*. New York: The Free Press.

Rogers, E. M. 1995. *Diffusion of Innovations*, 4th ed. New York: The Free Press.

Ross, R., and C. Roberts. 1994. "Balancing Inquiry and Advocacy." In P. M. Senge et al., eds., *The Fifth Discipline Fieldbook*, 253–59. New York: Doubleday.

Senge, P. M., C. Roberts, R. B. Ross, B. J. Smith, and A. Kleiner, eds. 1994. *The Fifth Discipline Fieldbook*. New York: Doubleday.

Strauss, S. 2001. "Small Business Center: Ask an Expert," *USA Today*, January 19.

Urban, G. L., and J. R. Hauser. 1993. *Design and Marketing of New Products*, 2nd ed. Englewood Cliffs, NJ: Prentice Hall.

Weiss, R. S. 1994. *Learning From Strangers: The Art and Method of Qualitative Interview Studies.* New York: The Free Press.

Wilson, E. 1996. "Market Analysis and Segmentation Issues for New Business-to-Business Products." In M. Rosenau, A. Griffin, G. Castellion, and N. Anschuetz, eds., *The PDMA Handbook of New Product Development.* New York: John Wiley and Sons.

4

Focusing NPD Research on Customer-Perceived Value

Charles Miller and David C. Swaddling

EXAMPLES

Loctite Corporation launched a product, designated RC-601, that was a pastelike substance used for repairing worn or broken metal machine parts. As a paste, the compound could be molded to the required shape and then cured to a steel-like hardness. This was a new concept for machinery repair, so to encourage trial usage, the product was priced at a modest $10 per tube. The company knew that production engineers might be skeptical, so they prepared marketing materials that provided detailed technical specifications. Despite these steps, sales of the new product were miserable (Best 1997).

Less than a year after its introduction, RC-601 was withdrawn from sale. Loctite was obliged to reimburse distributors for their unsold inventory. While figures on the magnitude of those reimbursement costs are not available, some conservative estimates can demonstrate the potential losses from this failed product launch. If only 500,000 units were returned, at a $5 manufactured cost per tube, about $2.5 million would have been refunded.

Following this recall, appropriate customer research was finally conducted to determine what went wrong. It was discovered that the production engineers weren't the right target market. They resisted new, unproven solutions, and the low price per tube of RC-601 probably added to their fears of the product's quality. With this in mind, Loctite decided to target sales of RC-601 to maintenance workers and renamed the product Quick Metal. Loctite simplified the instructions for its use and raised the price to nearly $20 per tube. This price was still within the budget authority of maintenance workers, but better conveyed the value of the product. The newly introduced product was, and is, a huge success.

Most NPD professionals would agree that customer input into the NPD process is important, but the current state of NPD practice tells a different story. Robert Cooper observes, "Recent studies reveal that the art of product development has not improved all that much—that the voice of the customer is still missing, that solid up-front homework is not done, that many products enter the development phase lacking clear definition, and so on" (Cooper 1999). Each of these shortcomings can be directly or indirectly tied with customer research (or lack thereof) done in conjunction with new product development.

From a practical standpoint, it's clear why customer research does not play a more prominent role in NPD. First, customer research is time-consuming, and virtually all industries are being challenged to significantly decrease their new product development cycle time. There's not much incentive to add steps that appear to extend rather than shorten NPD cycle time. Second, customer research requires additional resource investments. NPD professionals are often challenged to effectively allocate existing resources and have a hard time justifying additional efforts and expenditures on something they are not already doing. Third, the very nature of NPD does not lend itself to customer input. Typically, technical specialists are responsible for developing products. It is then left to the marketers to try to sell those products. There's usually little coordination between these two groups (Butscher and Laker 2000).

Given these practical considerations, how can customer research be justified? The answer is simple. Customer research reduces the inherent risk associated with NPD to an extent that exponentially exceeds its direct or indirect costs. The opening vignette for this chapter provides just one such example. Additional examples presented later all clearly demonstrate that the effective use of customer research in the NPD process just makes good sense.

This chapter presents an approach for systematically improving the effectiveness of NPD customer research by focusing that research on the bottom line of NPD success—delivering customer-perceived value (CPV). Understanding what customers will—and won't—pay for is at the heart of new product acceptance. The approach described in this chapter allows product developers to make decisions that increase the probability of product success by understanding how customers are likely to view the relative value of new offerings. This approach can effectively be used in both consumer and business-to-business markets.

Focusing NPD research on CPV forces product developers to truly understand what drives customers to embrace or reject offerings. This is accomplished by:

1. Identifying customer wants and needs early in the NPD process
2. Identifying factors, or attributes, that influence customer judgments of a product's value
3. Determining the relative importance of these attributes
4. Determining how offerings are viewed on each of these attributes relative to the customer's alternatives

As Figure 4-1 illustrates, components of this research take place throughout the NPD process—from the fuzzy front end through release of the product—and beyond. This chapter describes how research focusing on customer-perceived value can be integrated throughout the NPD process and how results can be applied to guide the development of products that offer superior value to customers.

FOCUSING ON CUSTOMER PERCEPTIONS OF VALUE

Most companies are finding themselves in the midst of a fundamental shift in the way they must do business. The Information Age has brought with it the

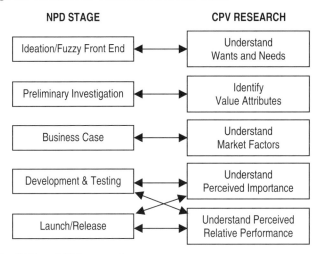

FIGURE 4-1. NPD and CPV research.

capability for customers to know more than ever about potential suppliers and what they have to offer. This capability has resulted in a fundamental shift of power in the buyer/seller relationship—from the sellers holding the upper hand to buyers now having most power (Swaddling and Miller 2001).

What is the result of customers seizing power from suppliers? It means that successful competitors must do some things differently. What serves as the basis for competition when companies are unable to differentiate themselves through product or service quality? "Customer focus" is a good term, but it is not explicit, complete, or robust enough to describe the new playing field. Swaddling and Miller suggest the basis of competition today is customer-perceived value, which is defined as follows:

DEFINITION

Customer-perceived value *is the result of the customer's evaluation of all the benefits and all the costs of an offering as compared to that customer's perceived alternatives. It's the basis on which customers decide to buy things.*

It is convenient to envision an old-fashioned balance scale as a depiction of customer-perceived value (see Figure 4-2). On one side of the scale, the customer stacks up all the perceived benefits of a particular offering. Those benefits will be tangible and intangible, product-related and service-related, externally (product) and internally (process) oriented, and so forth. Actually, there are no rules for what goes on the balance scale—it's whatever the customer wants.

On the other side, the customer stacks up all the costs associated with the supplier's offering. Certainly price is one of those costs, but only one. Required training, changes in processes, or the uncertainty of a new supplier could very well go on the minus side of the balance scale along with the price. Again, there

BENEFITS
Increased Revenue
Reduced Costs
Time Savings
Increased Safety
Prestige

COSTS
Purchase Price
Training
Process Changes
Taxes
Insurance

FIGURE 4-2. CPV balance scale.

are no set rules—the customer judges what should go on this side of the scale as well.

The customer establishes a balance scale like this for each offering or alternative under consideration. Whichever one tips most toward the positive side is the customer's choice. That's who gets the business. That's why it is important for new product developers to use this model and to manage the development of products based on customer-perceived value.

All too often, products are developed with no regard to how the customer will view the benefits and costs and how they will tip the value scale. It is important, therefore, that product developers know what benefits and costs are being considered by the prospective customer and how the prospect will weigh them on the balance scale of customer-perceived value. Since that information exists only in the mind of the customer, obtaining it requires good research. There is no way to gain this understanding without talking to, or at least observing, customers. Once information about customer perceptions is acquired, the smartest product developers design their offerings to emphasize the most important benefits, as perceived by the customer, and mitigate the perceived costs. This throws the buying decision to that supplier's advantage, which is, after all, the ultimate goal of new product development.

A clear perspective of customer-perceived value requires consideration of the four characteristics of CPV:

1. Customer value is market-perceived.
2. Customer value is complicated.
3. Customer value is relative.
4. Customer value is dynamic.

The first and most important rule to remember about CPV is that the *customer's* perception of value is the only thing that matters. That is why a fundamental characteristic of customer-perceived value is that it is *market-perceived*. Too many companies have learned the hard way that it doesn't matter if they have the best new whiz-bang product since the personal computer if the customer doesn't see it that way. Or conversely, for example, it would be a problem if the customer thinks warranty service is very important but a company feels it's not worth worrying about something that takes place so long after the sale.

The reference to evaluation of *all* the benefits and costs in the definition of CPV alludes to the complex nature of customer-perceived value and the fact that *customer value is complicated*. Using the balance scale analogy, it's apparent that a lot goes into the customer's evaluation of the value companies deliver—probably much more than immediately meets the eye. Dealing with this complexity means more than simply understanding the customer's balance scale. Companies actually have to influence what's on the scale. This is accomplished by designing offerings based on what the customer indicates would shift his or her balance scale in the supplier's favor. This is why customer research is such an important part of any legitimate product development process. Before taking a great advance from the R&D group to market, companies should be sure they know how potential customers are going to evaluate the offering using their balance scale. Perhaps that no-cost frill would mean a lot in the marketplace. Or, more importantly, is the supplier sure that expensive feature will be properly valued by customers? Remember, it doesn't matter how you feel about your offering; it only matters how the customer will judge it. And how the customer judges it is complicated.

The fact that *customer value is relative* is reflected in the definition of CPV by the phrase, "as compared to alternatives." Every time a customer gets out the balance scale to make a purchase decision, he or she is evaluating the offering as it compares to alternatives. Those alternatives obviously include your competitors' offerings, but may also include options available to the customer to address his or her immediate problem without buying anything from anyone (e.g., through reallocation of resources that are already available, or even deciding to do nothing and just live with the problem). All of those alternatives are relevant and need to be considered in an attempt to understand how potential customers perceive value.

Finally, to make matters worse, customers keep changing their minds because things that go onto their balance scale keep changing. This is why *customer value is dynamic*. Perhaps the customer's own circumstances have changed, so what used to be a valuable benefit isn't so important anymore. Competitors' offerings change as well, and that changes how other offerings stack up in a comparison. The point to remember is that customer perceptions of value constantly evolve, so it's not safe for companies to assume that they have their customer's value equation all figured out for very long. World-class competitors are learning to consistently and continually listen to their marketplace and proactively track the changes that impact their customer-perceived value.

So the bottom line is that CPV is "where it's at" for informed product development and management. Managers interested in growing revenue, increasing market share, or improving profitability will focus on delivering increased CPV. Since CPV is the basis upon which customers make their buying decisions, understanding CPV is the most direct route to improving a company's hit rate when buying decisions are made.

TOOLS TO UNDERSTAND CPV

As stated earlier, a commitment to delivering customer-perceived value must be built upon an investment in quality customer research. While companies may have some internal information sources that can be used in helping to monitor and manage the delivery of CPV, (e.g., transaction records, sales information), there's no getting around the fact that effective measures must involve the true judges of value delivered—customers. The three types of tools for measuring customer perceptions are exploratory research, secondary research, and confirmatory research.

Exploratory Research

Those starting from scratch to address specific CPV issues require the use of exploratory research techniques. The purpose of exploratory research is to become more familiar with an issue, to gain new insights, and to help formulate more specific research questions or hypotheses. This type of research is typically qualitative in nature, focusing on gathering responses to open-ended questions. There are few constraints placed on the nature of the data gathered. Customers are allowed to express and explain their perceptions as they choose. Exploratory research is most important in the early stages of the NPD process—from the "fuzzy front end" through the preliminary investigation stage.

Exploratory research, typically being qualitative in nature, provides the tools necessary to effectively understand the feelings and emotions that drive customer behavior. Understanding why customers buy, or don't buy, is at the crux of providing superior CPV and successful NPD. Qualitative research also serves as a necessary prerequisite for good quantitative research as the refinement of a new product or service progresses.

Qualitative research does have some notable weaknesses. It is rarely safe to assume that qualitative results provide a highly accurate representation of the customer group of interest, so there is inherent risk in making decisions based on qualitative research results alone. Many managers view nonstatistical data as being too soft. The link between what the manager can do in developing a product and how it will impact what the customer thinks is often unclear. A third problem with qualitative research is that it looks easy to do—just ask some questions. This fact makes it easy for almost anyone to claim he or she can do qualitative research. In actuality, being able to ask questions no more makes one a qualitative researcher than being able to operate a microwave makes one a chef.

Secondary Research

Secondary research relies on existing sources of information. Instead of gathering data from scratch, the research questions are answered with data or information that has already been captured by another party. This information is available through a variety of general business information resources, company-and industry-specific information resources, news sources, governmental statistics, and previous research.

There are a number of advantages to using secondary research in the NPD process (McDaniel and Gates 1999). Secondary research provides a means for clarifying or redefining the definition of a concept as part of the exploratory research process. It also has the capability to answer some research questions without the investment and effort required for confirmatory research. In instances where confirmatory research is not practical (e.g., short time frame, limited budget), secondary research may offer an acceptable approach to secure needed information. Secondary research can provide necessary background information on markets to enhance the interpretation of findings from other sources, and it can alert researchers to potential problems that might be encountered in moving forward with development of a new product.

NPD professionals should find secondary research to be particularly useful in the business case development stage of new product development. This type of research allows product developers to investigate factors that set the context in which buying decisions are made. These factors can obviously influence customer perceptions of value.

Confirmatory Research

As mentioned earlier, exploratory techniques and secondary research lay the groundwork for confirmatory research that provides the accurate and generalizable results necessary to advance product developments with greater confidence of success. Confirmatory research answers—with a known degree of precision—the who, what, when, where, and how questions. Confirmatory research is also used to describe correlations, or relationships, among factors. Ultimately, confirmatory research can explain or predict how customers perceive value. It is quantitative in nature and typically involves either descriptive survey research or experimental research. Confirmatory research is most valuable in the later stages of the NPD process—from development and testing through launch.

Once qualitative research identifies how customers perceive value, determining the importance of value attributes and the performance of potential product offerings in providing value can be done with a reasonable degree of accuracy. This is where primary quantitative research is necessary. Primary quantitative research is best described as statistically based research that gathers data directly from customers and generates numerical (quantitative) results with a known degree of accuracy.

The fact that quantitative results have a known degree of accuracy allows managers to reduce the risk they take in making decisions based on these results. The same does not hold true when considering qualitative results alone. Another key advantage of quantitative research is that it can accurately describe a situation or identify causal relationships. From a practical standpoint, this is invaluable information for managers, since linkages between actions that NPD professionals take and the impact of these actions on customer perceptions become much clearer.

While quantitative information has very desirable characteristics, its weaknesses should not be ignored. Quantitative results are usually assumed to be representative, valid, and reliable. Unfortunately, the rules that must be followed to ensure these qualities are often ignored. For example, it is common practice among many researchers to report that a sample provides results that are accurate to within plus or minus three points (or five points, or whatever) based on the size of the sample. However, they ignore the fact that the sample may not be representative because the respondents were not randomly sampled or nonrespondents were ignored. If the sample is not representative, any indicators of accuracy are meaningless. Why should we assume that an 8 percent response rate on a survey is any more representative of a population's perceptions than the performance on a single monthly P&L statement is representative of financial performance for a full year?

Acceptance of quantitative research results must be based on evidence that the researcher has made every effort to ensure that the data are valid, reliable, and representative. Evidence that appropriate analysis and interpretation of the data are being done should be obvious as well.

Having considered the principal characteristics of CPV and the three research tool kits available to NPD professionals, we can now focus on specific CPV research tools and their application in the NPD process. Utilizing an appropriate combination of exploratory, secondary, and confirmatory research will allow NPD professionals to answer virtually any question about CPV they might have.

UNDERSTANDING THE CUSTOMER'S NEEDS

The study of buyer behavior is a science all its own. Many have studied and written about the psychological motivations that drive buyer behavior (Hawkins et al. 1995; Howard and Sheth 1969; Assael 1987; Engel, Blackwell, and Miniard 1994). Let's now consider the practical aspects of understanding buyer behavior gleaned from these works and relate these aspects to the CPV model. To understand customer needs from a CPV perspective, a continuum of issues must be addressed. These issues, ranging from the most abstract to the most tangible, include the following:

1. Personal needs
2. Buying needs
3. Benefits
4. Features

Personal Needs

Guiding all human behavior, including taking the necessary actions to buy a particular product or service offering, is motivation. Why do people do the things they do? The answer, fundamentally, is because of what's in it for them. They satisfy some kind of need by taking that action.

Abraham Maslow's hierarchy of needs explains that categories of personal needs build upon one another in a set sequence. First, basic needs such as food, clothing, and shelter must be met. When that's been done, people move on to the higher-level needs, such as socialization, ego, and self-actualization. Potential customers may be attempting to satisfy needs at virtually any of these levels. Precisely identifying the level of psychological need prospects are seeking to satisfy is a very difficult undertaking. Fortunately, it's rarely necessary to do so.

Buying Needs

Among other things, personal needs drive buying needs. The concept of buying needs moves us closer to what NPD professionals must understand in order to know how to deliver CPV through an offering. For example, a prospect may have a child preparing to attend an expensive college and doesn't want to lose his or her current job. This personal need to keep the job places the prospect under a great deal of pressure to meet his or her operating unit's budgeted profit goals. Consequently, the prospect's buying need to increase profitability is, in fact, driven in large part by the personal need to keep his or her job.

Buying needs, like personal needs, exist at different levels. In fact, they form a continuum from general to quite specific. If the prospect's fundamental buying need is to increase profitability, he or she may have a more specific need to reduce costs. Beyond that, the prospect may have an even more specific need to cut labor costs. This type of information is a lot closer to something that the NPD professional can use and act upon.

Benefits

Generally speaking, prospects go shopping for benefits in order to fulfill their needs. Benefits are the results that customers realize from the use of a product or service. To rephrase McKenna's famous reference, people don't buy drill bits, they buy the ability to make holes. Making holes is the "benefit" that drill bit manufacturers are selling.

Note that benefits are related to a specific offering. Continuing with the example presented above, the prospect may perceive that an improved computer system would make a manufacturing fabrication process more efficient and, therefore, help to meet the need of cutting labor costs. In that case, the benefit perceived by the prospect would be saving workers' time. Outsourcing part or all of the fabrication process, however, might offer the same prospect a method of reducing labor costs by reducing the cost of hiring and administration of employees. In this case, the benefit of the offering is reduced overhead costs. Both solutions address the prospect's need to reduce labor costs but present very different perceived benefits designed to meet that need.

Features

Finally, then, are the features of the product or service. Features are those aspects of an offering that create the benefits; they are typically a focal point of NPD. The drill bit's feature is that it has very sharp cutting edges. Those edges enable the benefit of making holes better, faster, and cheaper.

Continuing the example presented above, features of the improved computer system for the manufacturing fabrication process may include an intuitive user interface and extensive use of automated process monitoring and equipment calibration. Both of those features would contribute to the benefit of saving workers' time. Other features of the computer program might be the ability to create an output log and feed that information into a larger enterprise resource planning program. Those features might be attractive but do not directly contribute to the benefit of reducing workers' time.

The difference between the features of an offering and the benefits perceived by prospects is an important one. Understanding this difference is the point of transition from a product-focused organization to a customer-focused one. Companies produce product and service features, but prospects buy product and service benefits. This is why understanding customer-perceived benefits is a prerequisite to ensure the delivery of CPV through the NPD process. In other words, focusing on customer needs and benefits should take place early in the NPD process. Once this has been accomplished, then it makes sense to focus on features in later stages of NPD.

Researching Needs and Benefits

As Figure 4-1 illustrated early in this chapter, understanding customer needs and perceived benefits is ideally a component of the ideation and "fuzzy front end" stage of NPD. Many notable inventors and entrepreneurs have been successful at intuitively understanding needs and benefits. This, however, appears to be the exception rather than the rule. Figure 4-3 lists specific research tools that offer an alternative to a total reliance on intuitive customer insight. These research tools permit the systematic investigation of customer needs and potential benefits.

As Figure 4-3 indicates, a number of commonly used qualitative techniques can be integrated into the earliest stages of NPD. The most commonly used are customer interviewing approaches, including one-on-one interviews and focus groups. Qualitative research is certainly not limited to conducting interviews,

Research Technique	Application in Understanding Need Relative to CPV
One-on-one interviews— in person, by phone, via the Web (depth interviews)	Used to identify and discuss needs of small or important targeted customer groups.
Focus group interviews, small group interviews	Used to discuss and identify needs of larger populations (e.g., market segments). Can be traditional or Web-based.
Nominal group sessions	Small groups of customers independently generate ideas about a subject and then discuss these ideas as a larger group. Used to generate ideas based on individual needs.
Lead user interviews	Dialogue with selected users who have exhibited advanced product usage abilities and inclinations. Reasons for adoption reveal needs.
Customer advisory panels	Structured customer interactions to periodically gather feedback directly from cohort group of customers. Can be used to discuss underlying needs.
Web-based interactive	Provides opportunity to share information and identify major issues and concerns.
Contextual inquiry	Systematic observations of representatives of target market in ordinary settings, allowing the researcher to identify needs that may or may not be realized by potential customers.
Customer role playing, customer shadowing (mystery shopping)	Systematic evaluation of the customer's experience involving a product and/or service. Provides the researcher with an opportunity to identify needs that may or may not be realized by potential customers.
Customer events	Informal customer gatherings focusing on ownership and use of specific products. Allows researcher to understand underlying needs associated with selection and use of the product.
Zaltman metaphor elicitation (ZMET)	A projective technique. Customers create collages representative of their experiences and feelings about a product. They then meet with researchers to explain the connection between the images selected and their experiences with the company and its products. Allows researcher to investigate underlying perceptions, including needs.
Sentence and story completion	Technique to study customer reactions to situations. Allows researcher to investigate motivators, or needs, that drive specific reactions.

FIGURE 4-3. Research tools to investigate customer needs.

however. Some techniques involve very nontraditional approaches, including observation of customers in their "natural habitat" (anthropological approaches) and gathering perceptions completely outside of the context of the product and/or service being offered (projective techniques).

For example, S. C. Johnson Wax, a leading manufacturer of a wide range of household products, almost always begins the idea generation process by sending researchers out across the country to visit homemakers in their own homes. These researchers spend entire days with these consumers, simply following them around the house and watching as they perform common chores. In one such study, researchers recognized the problem consumers were having with ladders, pails, and hoses when trying to wash the outside of their second-floor windows. As a result, the idea of a no-rinse, streak-free, spray-on window cleaner was born, and another customer need was eventually met.

Health-care giant Kaiser Permanente offers an example of a model Web-based interactive approach for studying customer needs. At www.kponline.org, members are offered the unique opportunity to take part in moderated discussions on a variety of specific medical topics. Kaiser doctors and psychologists read every posting and offer advice or correct misinformation as warranted. Not only do these forums provide members with easy access to information, they also provide Kaiser with a wealth of information about its members, including their concerns and any potential medical challenges that might warrant new services (Stewart 1999).

Other examples of the successful, proactive investigation of customer needs abound. Those presented above illustrate the kinds of steps that NPD professionals can take to gather invaluable customer-based information early in the NPD process.

IDENTIFYING ATTRIBUTES CUSTOMERS USE TO JUDGE VALUE

Once customer needs are identified to support the advancement of a new product idea to the preliminary investigation stage, the next piece of the CPV puzzle to tackle is a clear understanding of what things customers will put on their balance scale when evaluating the offering. These criteria are called benefit and cost attributes (Gale 1994).

Attributes

The basic building block of a CPV model is the attribute. An offering attribute is defined as follows:

DEFINITION

An **attribute** *is a benefit or cost of a specific product or service offering as perceived by the customer.*

As described earlier, customer-perceived benefits go on the left side of the CPV balance scale and customer-perceived costs go on the right side.

Benefits

CPV benefit attributes are the same benefits discussed earlier in the context of understanding customer needs. Benefits are usually the best point in the customer needs continuum on which to focus conversations with customers because they represent customer-oriented perceptions but are still close enough to supplier-oriented features to permit that linkage to be made by the product developer.

AirNet Systems operates a one-hundred-plane fleet of small business jets and twin-engine propeller driven aircraft primarily serving the banking industry by moving documents overnight from one institution to another. The company's service features include late-night departure schedules with early-morning delivery, six-hour access to all the major money markets in the United States, and a ground delivery network coordinated with the airline. The banks' buying needs are to clear the cash represented by millions of canceled checks into their own accounts as quickly as possible. The link between features and needs is customer-perceived benefits. In AirNet's business, those benefits include matters such as how well the airline's schedule matches the banks' check processing procedures, the speed of final delivery, and the reliability of the service. These are the benefits that would appear on the banks' CPV balance scale.

It is possible to simply ask customers what benefits they perceive a potential offering provides and get a fairly lengthy, and accurate, list of attributes. Usually, in fact, this is the very approach that is used to begin to construct an image of the CPV scale. Some care is necessary, however, to be sure that a complete list is obtained, as some benefit attributes may not be foremost in the prospect's mind at the time of the interview. Sometimes some "homework" is necessary on the part of the researcher (prior observation of or preliminary research with the prospect) in order to enable skillful yet unbiased extraction of a complete list of benefits that may be considered by the prospect.

Costs

There are also attributes, of course, on the right side of the CPV balance scale—all of the costs associated with the offering as perceived by the prospect. The most obvious of those costs is the purchase price for the offering, but there are many more.

When the natural gas industry was deregulated, the existing distribution companies were required to announce to their customers that they now had a choice of from whom to buy their gas. New entrants for that business sprang up everywhere and mounted major promotional campaigns to attract this business. And the price advantage to consumers was clear—the new companies charged less for the same commodity. Yet after several years most consumers still purchased their natural gas from the distribution company that served them prior to deregulation. Why? Because there was a hassle involved in switching to a new supplier. There were forms to be filled out and the uncertainty of not knowing which of the new companies to select. Those consumers had placed another item, typically referred to as "switching costs," on the right side of their CPV balance scale. In many cases, it offset the weight of the incumbent seller's higher price and any other benefits the new entrants could suggest.

Some experts exclude price from the attributes of an offering, leaving all the benefits and other costs to define the quality of the offering, and then compare that quality to the price. Although it is almost always a very important attribute, others view price as being only one of several cost attributes that customers might place on their CPV balance scale. Following this logic allows managers to make sure that the purchase price is managed as one of several product attributes and is not mistakenly assumed to be the primary focus of the prospect.

Identifying Customer-Perceived Value Attributes

Figure 4-4 lists tools for identifying CPV attributes. Most of these research techniques are the same as those listed earlier for understanding customer

Research Technique	Application in Determining CPV Attributes
One-on-one interviews— in person, by phone, via the Web (depth interviews)	Used to identify CPV attributes for small or important targeted customer groups.
Focus group interviews, small group interviews	Used to identify CPV attributes for larger populations (e.g., market segments). Can be traditional or Web-based.
Nominal group sessions	Small groups of customers independently generate ideas about a subject and then discuss these ideas as a larger group. Pros and cons identified in larger group discussion reveal attributes.
Lead user interviews	Dialogue with selected users who have exhibited advanced product usage abilities and inclinations. Reasons for adoption reveal value attributes.
Customer advisory panels	Structured customer interactions to periodically gather feedback directly from cohort group of customers. Can be used to discuss attributes used to judge value of offerings.
Zaltman metaphor elicitation (ZMET)	A projective technique. Customers create collages representative of experiences and feelings about a product. They then meet with researchers to explain the connection between the images selected and their experiences with the company and its products. Allows researcher to investigate underlying perceptions, including value attributes.
Sentence and story completion	Technique to study customer reactions to situations. Allows researcher to investigate value attributes driving specific reactions.

FIGURE 4-4. Research tools to identify CPV attributes.

needs. The difference is the focus of the research. As product development moves into the preliminary investigation stage, a product concept now exists that will allow potential customers to react to a theoretical offering. This context allows targeted customers to consider the pros and cons of the proposed offering relative to their needs. These pros and cons translate into benefit and cost attributes.

UNDERSTANDING POTENTIAL MARKETS

With an understanding of the criteria that customers will use to judge a potential offering, next it is important to consider these criteria in the broader context that will influence the customer's judgment. Buying decisions are made in the real world, not in a vacuum. It is incumbent on those involved in NPD to have a grasp of how factors beyond their control will influence the perceived worth of an offering.

Understanding the market environment is an inherent characteristic of work done in the business development stage of NPD. Focusing on CPV, however, requires that the new product business case reflect a thorough understanding of the market in which CPV will be judged. To understand markets from a CPV perspective, it is necessary to answer three questions:

1. How will the CPV attributes be judged in the marketplace?
2. What alternatives to the potential offering exist?
3. How might competitors offering alternatives attempt to influence the customer's balance scale?

How Customers Judge CPV Attributes

It was noted earlier that it is important to secure a complete list of attributes in the preliminary investigation stage of NPD. Caution should be taken in reducing this multitude of attributes down to a few that seem to be most important. Specifically, it should not be assumed that any individual CPV attribute is of little or no importance to the customer. This is because attributes progress through a life cycle that impacts how they are judged by customers in the marketplace (see Figure 4-5).

A *potential attribute* is one that no supplier is currently able or willing to provide. These are the attributes in the category of "I wish someone would offer me this benefit." (For example, owners of full-size SUVs would like to have the fuel economy of subcompact cars without sacrificing vehicle size, utility, or power.) Such attributes might appear on a prospect's CPV balance scale but do not actually come into play in the decision between suppliers because the attribute is unavailable.

Theoretically, potential attributes could even be benefits that are unstated and unrecognized by the prospect and, therefore, aren't yet on the prospect's

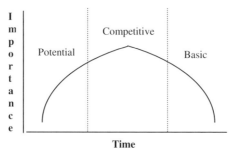

FIGURE 4-5. CPV attribute life cycle.

CPV balance scale. They would be there, however, if an aggressive new product developer found a way to deliver them. Potential attributes, then, are the basis of innovation.

Attributes introduced through a product or service offering, and accepted as such in the perception of the marketplace, become *competitive attributes*. These are the most obvious attributes to the prospect and the ones upon which buying decisions are currently made.

Over time, competitive attributes evolve into *basic attributes* as competitors continually improve their offerings and deliver better and better (as well as more similar) attributes. These are the benefits and costs of a product or service that are the same across all of the prospect's alternatives.

For example, a basic attribute in the commercial airline industry would be the safety with which each airline carries its passengers. Because these matters are regulated by the Federal Aviation Administration, the maintenance and safety practices of most airlines are the same and the perception of flyers is usually that one commercial airline is just as safe as another.

Basic attributes may appear on the CPV balance scale because they remain important to the prospect and customer. They are typically not a basis on which the buying decision hinges, however, because everybody offers the prospect the same thing in connection with this attribute. Unfortunately, since the customer has come to expect and accept the attribute as something deserved, sometimes it won't even be considered as being very important. That perception changes dramatically when a basic attribute is not provided, however.

An unfortunate example of what occurs when customers perceive that basic attributes are not being provided is the tragic story of ValuJet Airlines. Safety was a basic attribute in the airline industry until the disastrous day in May 1996 when ValuJet's Flight 592 crashed into the Everglades. Suddenly, safety became an important *performance attribute*. From the day of that crash, at least for a period of time, safety was a prime consideration of buyers of airline service. Few chose to patronize ValuJet because of a perceived shortfall in what had become the critical performance attribute of safety. Within a short period of time, ValuJet was forced to reorganize and change its name rather than face complete bankruptcy.

Alternatives to Potential Offerings

A second critical question that must be answered when investigating potential markets is "What alternatives exist to this offering that we're developing?" It is common to consider similar offerings by competitors, but CPV judgments are not always limited to obvious alternatives. Customers may choose among offerings that are obviously in direct competition, find a completely different solution to meet their needs, or choose to buy nothing at all.

Continuing with an air travel example, we recognize that the basic need of airline passengers is to get from point A to point B. Obvious alternatives are provided by any carriers that have flights out of point A and into point B. However, other options may include travel by car, bus, or train. In the case of a businessperson, no travel at all may be a very feasible option if teleconferencing or videoconferencing can be arranged. We see, therefore, that both apparent and unapparent alternatives to a potential offering should be investigated in the course of developing a business case that accurately considers customer's perceptions.

Competitor's Actions

A third question that should be answered when investigating potential markets is "How will competitors react to the introduction of this offering in the marketplace?" In other words, what will competitors do to attempt to influence customers' CPV balance scale?

Typically, market investigations will include activities such as competitor identification and profiling followed by SWOT (strengths, weaknesses, opportunities, and threats) analyses. These provide valuable information but often fall short of providing the depth of knowledge necessary to understand how competitors might influence CPV. For example, it's one thing to know who the largest players in a market are and the strengths and weaknesses of these players, but it's another to learn what tactics they are likely to pursue in an effort to provide superior CPV. Is the leader's branding so strong that the customer associates the brand itself with superior value? Will a competitor use a reduced pricing strategy in an effort to provide value that is superior to your new offering? Will they attempt to tip the customer's CPV balance scale by offering an extended warranty, or reduced cost service, or free service for the product? These are the types of questions that should be considered and can be answered with some degree of certainty based on past behavior by competitors.

Researching Potential Markets

Most of the questions that guide the investigation of the ways markets will influence CPV can be practically addressed through secondary research. Much of the data required to answer these questions is available through observations already made and recorded, or data already gathered, by others. The secret is identifying these sources of information.

Figure 4-6 lists secondary research information sources available to NPD professionals. This list includes traditional resources such as print media and industry-specific information sharing, as well as Web-based resources coming into more common use (Kassler 2000). These sources provide a broad per-

Type	Notable Sources
Print media	◆ Business periodicals (*Wall Street Journal, Harvard Business Review, Fortune*, others in *Business Periodicals Index*) ◆ *Encyclopedia of Business Information Resources* (Gale) ◆ *Business Information Sources* (Daniells)
Commercial online services	◆ Lexis-Nexis ◆ Dow Jones ◆ Dialog
Internet	◆ Search engines (e.g., Northern Light, Google, Look Smart) ◆ www.ceoexpress.com
Company-specific resources	◆ *Million Dollar Directory* ◆ *Directory of Corporate Affiliations* ◆ *Standard & Poor's Register of Corporations, Directors & Executives* ◆ *Ward's Business Directory* ◆ EDGAR (www.sec.gov) ◆ Public Register's Annual Report Service (www.prars.com) ◆ *Thomas Register of American Manufacturers* ◆ Hoover's (www.hoovers.com)
Industry-specific resources	◆ *Business Information Sources* (Daniells) ◆ *U.S. Industry & Trade Outlook* ◆ Hoover's (www.hoovers.com) ◆ *Directory of Business Periodicals Special Issues* (Wycoff, 1995)
News resources	◆ Newsgroups ◆ Alerting services (e.g., www.scoop.com, www.onesource.com)
International resources	◆ *International Business* (Pagel and Haperin, 1998) ◆ Kompass (www.kompass.com) ◆ *Principal International Business* (Dun and Bradstreet)
Governmental resources	◆ Federal Web Locator (www.infoctr.edu/fwl) ◆ STAT-USA (www.stat-usa.gov) ◆ U.S. Congress official Web site (thomas.loc.gov)
Patent resources	◆ U.S. Patent and Trademark Office (ww.uspto.gov) ◆ Intellectual Property Network (www.patents.ibm.com) ◆ World Patents Index (www.cas.org) ◆ European Patent Office (www.european-patent-office.org)

FIGURE 4-6. Secondary research information sources.

spective of markets, factors affecting those markets, and company-specific information on market players.

As with any research effort, the collection of data and analysis of secondary research results are best guided by clear research questions. "Rooting about" using the various tools indicated above is inefficient and unlikely to generate very meaningful results. The key, therefore is not only to know what type of information each resource can provide, but also to begin with a clear definition of what type of information is needed to answer specific questions.

UNDERSTANDING THE IMPORTANCE OF CPV ATTRIBUTES

Up to this point, this chapter has discussed how NPD professionals can clearly discern the attributes that customers will place on a CPV balance scale when evaluating a new offering. Now the focus will shift to learning how the new offering can be designed to tip the balance scale in its favor. In order to accomplish this, it is necessary to first understand the relative importance that customers place on CPV attributes. This can be accomplished during the development phase of the NPD process.

A thorough investigation into all of the possible benefits and costs that a prospect might perceive in an offering will usually result in a lengthy list of attributes—thirty, forty, or even more, depending on the offering. Even relatively small acquisitions will involve numerous attributes that a buyer might consider.

Recent observation of a company buying a small telephone system for a remote office location illustrates this point. The price of all systems under consideration was significantly less than $10,000. Yet the buyer had identified thirty-one different attributes to be compared, some of which had numerous, complex characteristics. The attributes considered included items such as purchase price, subsequent upgrading prices, service prices, expandability, availability of training, and payment terms—in addition to product features such as programmable keys, LED displays, number of speed dials, separate volume controls, voice mail ports, message storage capacity, and so on. This was not a complex purchase decision—the phone systems in the price range were considered to be all pretty much alike. Still, the number of attributes the buyer could place on the CPV balance scale was large.

It is well established that a large number of variables is usually beyond the reasonable capability of most people to process. Most prospects, therefore, will probably place only the attributes that are most important, *in their own judgment*, onto the CPV balance scale. Identifying those attributes deemed most important by the customer, and determining how the customer views the relative importance of these attributes, is both more descriptive of how the prospect is thinking about the buying decision and more practical for the product developer to consider in her own decision-making process.

Measuring CPV Attribute Importance

Figure 4-7 presents research tools for evaluating the customer-perceived importance of attributes. From a practical standpoint, selection of a research approach for this purpose is usually dependent upon NPD investment resources. The more detail and the higher the level of confidence in the results desired, the greater the resource investment required. This would explain why survey research is more popular than choice modeling or market experiments for new product developers with limited resources.

The most common type of confirmatory research used in the later stages of the NPD process is survey research. The flexibility of survey research, and the fact that it's usually less expensive to use than some of the more advanced research approaches, makes it a preferred research tool. The appropriate type, or types, for use in the NPD process are largely dependent on the types of customers and customer contacts being targeted. For example, while mall intercept surveys may make perfect sense for a clothing designer, it's hard to imagine a case where an industrial equipment manufacturer would use this technique.

While survey research can often play an important role in most NPD efforts, Figure 4-7 illustrates that it is not the only primary confirmatory

Research Technique	Application in Determining CPV Attribute Importance
Surveys ◆ *Mail* ◆ *Telephone* ◆ *Mall intercept* ◆ *Direct computer (used in mall environment)* ◆ *Point-of-service touch screen* ◆ *Self-administered (comment cards, captive audience surveys)* ◆ *Interactive Voice Response (IVR)* ◆ *Fax* ◆ *E-mail* ◆ *Intranet/Internet-based*	One-way collection of data (primarily quantitative) from individual representatives of a target market. Provides a relatively inexpensive means for gathering importance perceptions from a sample of prospects.
Choice modeling	Presentation of designed choices to representatives of target market. Based on preferences selected by prospects when choosing between options, statistical analyses are used to determine the relative importance of attributes.
Market experiments	Presentation of offering to prospects with changes in levels of the CPV attributes. All other extraneous variables are controlled. Results can determine attributes that do or do not significantly impact perceptions.

FIGURE 4-7. Research tools to determine the importance of CPV attributes.

research approach available. Experimental research is an option that goes beyond the basic description of customer perceptions provided by survey research. Experimental research offers a means of clearly measuring the relative importance and worth of value attributes to customers. This makes it a particularly appropriate research methodology to use when designing new products and/or service offerings when CPV attributes are already defined and validated. This, of course, would be the case if the approach described in this chapter has been followed.

In the choice modeling methodology involving conjoint analysis, customers indicate relative preferences among combinations of value attributes at a variety of levels. This approach is an example of a powerful research tool for understanding CPV. The reason this approach is not more commonly used is the practical challenges and costs associated with data collection. As the number of levels of value attributes to be considered increases, there is an exponential increase in the time required of customers to evaluate and respond to the available options. This makes experimental research more expensive to conduct than survey research. Note, however, that additional investment results in additional accuracy in the research results.

Whatever research method is used, the objective is to place customer-perceived attributes on the CPV balance scale according to customer-perceived weights of importance. This provides the product development professional with a customer-oriented statement of the offering's value proposition. It says, "Here are the benefits and costs associated with this offering, each weighted according to its respective importance to the customer."

UNDERSTANDING PERCEIVED RELATIVE VALUE

Even a completed CPV balance scale, with benefits and cost attributes well defined, weighted for importance, and stacked up on both sides, has limited utility. What it portrays is how the prospect is likely to behave if he or she has no alternatives other than to buy nothing at all. In that case, if the balance scale tips toward the benefits side (the left side), then the purchase is likely to be pursued. It may also be useful in that product developers can use it to understand an offering's value proposition for the first time.

In most cases, however, the prospect is selecting from among several different offerings. When that is the case, the CPV balance scale for the new offering is meaningful to the NPD professional only when it is placed next to a comparable CPV balance scale for its competitive offerings. It really doesn't matter how attractive the offering under development is to a prospect. To predict buying behavior, it matters only that the offering is *more* attractive than the prospect's alternatives.

Understanding the customer-perceived value of an offering relative to alternatives is best done in the later stages of the NPD process. However, this evaluation of value offered can be continued through product launch, and even throughout the course of the offering's life cycle. It is important to remember that the context value is judged in does not remain static. Customers face a myriad of factors that can influence both the importance of value attributes and how the relative value of offerings compared to alternatives are viewed. Each time this determination is made for a product, it is only a snapshot in time. So continual monitoring of relative value provided by an offering is desirable.

Measuring Perceived Relative Performance

Figure 4-8 presents the research techniques that can be used to gather data on the relative performance of an offering. Many are the same used for determining importance, with a shift in the focus, application, and/or analysis of the technique to performance measurement. All of these are primary research techniques that generate quantitative results. At this stage in the NPD process, these are the types of research results necessary to make the final critical go/no-go decisions and reduce the risk of possible failed launches.

Note that an absolute necessity in any CPV performance measurement is the customer's evaluation of alternatives—not just the proposed offering itself. Without an understanding of the ways prospects view alternatives compared to the new offering, the only measurement being made is probable satisfaction with the new offering—not its CPV. So doing research to find out how potential customers like attributes X, Y, and Z of a proposed offering without finding out how they feel about alternatives is going to provide little information of real value to the product developer.

Analyzing Customer-Perceived Value

There are a number of ways in which an NPD professional can take the information gathered through CPV-focused confirmatory research and apply it to specific product development issues. At a minimum, the information should be analyzed in a couple of different ways.

A starting point for evaluating relative performance is a stacked bar chart, representing the ratings on each CPV attribute for each of the competitive offerings. A sample chart of this type is shown in Figure 4-9.

The bar chart on the right side of the diagram reflects the actual ratings (this time on a scale of 0 to 10) on each of the CPV attributes of the potential offering, as well as that of the prospect's "next best alternative." We show these absolute ratings because it is important, for some purposes, that the NPD professional know how well his or her offering, and those of competitors, is meeting the expectations of the customer. This customer satisfaction data is a part of the picture, just not the entire picture. That's why we show it on the right side of the diagram and depict the relative performance ratings on the left side.

It is not difficult to convert the absolute ratings on the right side into relative ones on the left side. All it takes is some simple arithmetic to convert the

Research Technique	Application in Determining CPV Relative Performance
Surveys ♦ *Mail* ♦ *Telephone* ♦ *Mall intercept* ♦ *Direct computer (used in mall environment)* ♦ *Point-of-service touch screen* ♦ *Self-administered (comment cards, captive audience surveys)* ♦ *Interactive voice response (IVR)* ♦ *Fax* ♦ *E-mail* ♦ *Intranet/Internet-based*	One-way collection of data (primarily quantitative) from individual representatives of a target market. Provides a relatively inexpensive means for gathering perceptions of importance from a sample of prospects.
Choice modeling	Presentation of specific choices to representatives of target market. Based on preferences selected by prospects when choosing between actual choices (new offering vs. actual alternatives), statistical analyses are used to determine relative performance on specific attributes.
Market experiments	Presentation of the offering and alternatives to prospects. All other extraneous variables are controlled. Results can determine where significant differences in perceived performance do or do not exist.
Use testing	Observation of representatives of the target market using the proposed offering in a controlled setting. Follow-up to this observation provides an opportunity to gather data on the perceived performance of the offering relative to alternatives.
Beta testing	Use of the proposed offering by representatives of the target market in their own environment. Periodic evaluations during the beta test process provide an opportunity to gather data on the perceived performance of the offering relative to alternatives.
Test marketing	Pilot, or controlled, introduction of the offering into the marketplace. Prospects' reaction to the introduction can serve as a secondary indicator of relative performance.

FIGURE 4-8. Research tools to assess relative performance on CPV attributes.

actual ratings into a ratio. Doing this creates an easy-to-read chart. Every bar extending to the right side of the vertical axis means that the manager's product is outperforming the competitor's on that attribute, in the perception of the customer. Bars extending to the left side of the axis mean that the manager's product is perceived as not performing as well on that attribute as compared to the competitor's. Where there's no bar indicated, it means the customer per-

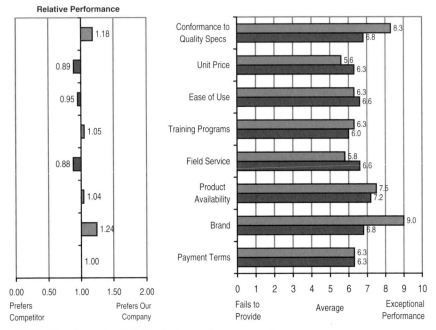

FIGURE 4-9. Example of CPV relative performance ratings.

ceives no difference in the performance on that attribute of the alternatives from which he chooses.

A CPV matrix is a diagram that offers a second, more powerful approach for studying CPV results. An example of this type of diagram is shown in Figure 4-10. The CPV matrix incorporates the importance weightings of each attribute with the performance ratings to help the product developer make priority decisions about how to improve the overall CPV of the offering.

In this chart, the vertical axis represents the customer's relative importance weighting. The more importance attributes will be found, therefore, toward the top of the matrix. The horizontal axis reflects the relative performance ratings, directly from the relative performance rating chart just described. The attributes on which the offering is perceived as performing better than the competition's will show up toward the right side of the chart, and those attributes where the offering is at a disadvantage to competitors appear toward the left side.

McDaniel and Gates (1999) suggested labels for the four quadrants of this diagram. Attributes that fall into the upper left quadrant are termed "threats." These are the attributes that are more important to the customer, but the offering is perceived as performing less well than that of the competitors. These are attributes that need to be improved in order to increase competitive advantage.

The lower left quadrant also contains attributes that are perceived as performing poorly in comparison to the competition, but they are less important attributes to the customer and, therefore, probably of somewhat less concern to the product developer. This quadrant is referred to as "weaknesses."

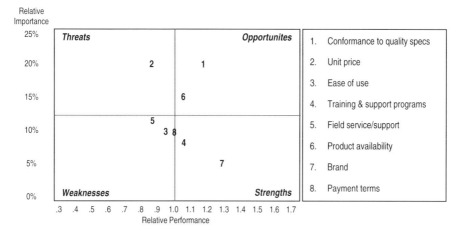

FIGURE 4-10. Example of a CPV matrix.

The right side of the diagram contains the attributes that the customer perceives to be performing at a higher level than the competitors'. The lower right quadrant contains the attributes that are less important to the customer, so this quadrant is called "strengths." These attributes probably should be maintained, but they will have less impact on CPV than the attributes appearing above them.

Finally, the upper right quadrant contains those attributes that are perceived as the strengths of the offering and are also more important to the prospect. This quadrant is called "opportunities," because these are the attributes that provide the competitive advantage. If leading with your strengths is the strategy, then these are the attributes upon which the strategy may be based.

The two diagrams provided here are only the beginning of how CPV can be depicted. The important point is that CPV, which is a very complex, intangible, and transitory concept, can be described in very simple, tangible, and useful ways in order to help managers make difficult product development decisions. That process of simplifying and making tangible something that is so very complex is not to be taken lightly. It requires care, expertise, and judgment. Properly done, however, it provides management with a tool of immeasurable worth.

THE PAYOFF FOR FOCUSING ON CPV

Much of what has been described in this chapter probably makes intuitive sense. However, the intuitive appeal of a management approach does not always guarantee its success. To illustrate that focusing on CPV pays off in handsome, quantifiable returns, consider an example from the annals of 3M, among the most well known of all product innovators.

3M's programs to encourage creativity and innovation among all of its people are legend. Yet by the mid-1990s, top management was concerned that few breakthrough ideas were being produced and that most of the new product revenue was coming from less rewarding changes to existing products. Among other actions it took, top management set an organizational goal that fully 30 percent of sales should come from products t created no more than four years earlier. That challenge spurred a number of 3M employee groups to look for new ways to come up with new, breakthrough product ideas (von Hippel et al. 1999).

One group was charged with finding a breakthrough in the area of anti-septic surgical drapes. Surgical drapes are the materials used to cover the area surrounding the site of surgery with the purpose of preventing the spread of infection. 3M was the dominant supplier in this high-technology business, but its position was beginning to stagnate. Changes were continually occurring in the surgery field, such as in the use of catheters or tubes, but little had changed about the surgical drapes. If some breakthrough wasn't found to revitalize sales growth, 3M would sell off the product line. A task force was formed to address the problem.

In this case, the team developed an aggressive program of meeting with lead users to explore product innovation ideas. The cross-functional group consisted of six members who dedicated half their working hours to this effort. Over a six-month period, they networked their way into discussions with the world's leading authorities on surgical procedures and bacterial and viral infections. As a result, the team recommended four new product thrusts for 3M, all of which were accepted and implemented by top management. These were innovations that were directly the result of moving outside of this very creative company to talk to customers. The results were spectacular.

What did this cost? Let's estimate that the salary costs of the project team on this work amounted to about $150,000. Add that same amount for travel expenses and support, and the total investment in this customer research must have been around $300,000. Now compare that investment with the total cost of developing and introducing four new products. Let's say that at $1 million each (a nice round number), the product development costs totaled $4 million. That's what the company is putting at risk in introducing four new products. If the customer research increased by only 10 percent the likelihood that these four products will be successful, that increased the expected value of the product introduction investment by $400,000. Already 3M is ahead!

Is a 10 percent improvement in new product success probability a reasonable figure? It's probably too conservative. Surveys indicate that nearly 80 percent of all new products will fail to meet commercialization criteria within one year. And experts point out that the leading cause of new product failure is lack of input into the development process by the intended market (Cooper 1999). So good customer research shouldn't have much trouble improving success rates by only 10 percent.

The calculations above do not even consider the lifetime of profits from new product innovations that wouldn't have been conceived at all except by good customer research. This offers further evidence that the return on investment in product innovation customer research is embarrassingly high.

CLOSING THOUGHTS

This chapter has presented approaches to integrate research into the NPD process through focusing on customer-perceived value (CPV). Most of the research techniques described are familiar to NPD professionals. It is the focus of the research, and its execution at appropriate points in the NPD process, that optimizes the worth of the research done in conjunction with NPD.

Beginning with exploratory and secondary research in the early stages of the NPD process and advancing to the use of confirmatory approaches in the later stages of the offering's development, this chapter has suggested the use of specific research tools at specific points in the NPD process. Obviously, there are no clear lines of demarcation to say when one type of research should stop and another type should start. For example, it is likely that secondary research will be conducted in the preliminary investigation stage of the NPD process in addition to its use in the business case development, as described in this chapter. The point is that the research components in the NPD process should complement one another, with each effort building upon the findings of earlier research. Using a haphazard approach to integrate research components into the NPD process virtually guarantees that the worth of any research done will not be maximized.

The most useful NPD research will focus on the delivery of value through the offering based on balancing benefits and costs as perceived by the potential customer. Following this approach reduces the inherent risks associated with the NPD process and generates tremendous, quantifiable returns for the research investment. The approach described here shifts the focus to the ultimate arbiter of NPD success—the customer for which the offering is targeted.

BIBLIOGRAPHY

Assael, Henry. 1987. *Consumer Behavior and Marketing Action*. Boston: Kent.

Best, Roger J. 1997. *Market-Based Management: Strategies for Growing Customer Value and Profitability*. Upper Saddle River, NJ: Prentice-Hall.

Butscher, Stephan A., and Michael Laker. 2000. "Market-Driven Product Development." *Marketing Management*, summer, 49.

Cooper, Robert G. 1999. "The Invisible Success Factors in Product Innovation." *Journal of Product Innovation Management*, March, 115–33.

Engel, James E., Roger D. Blackwell, and Paul W. Miniard. 1994. *Consumer Behavior*. Fort Worth: Dryden.

Gale, Bradley T. 1994. *Managing Customer Value: Creating Quality and Service That Customers Can See*. New York: The Free Press.

Hawkins, Delbert, Roger Best, and Kenneth Coney. 1995. *Consumer Behavior: Implications for Marketing Strategy,* 6th ed. New York: Irwin, 1995.

Howard, John A., and Jagdish N. Sheth. 1994. *The Theory of Buyer Behavior.* New York: John Wiley and Sons.

Kassler, Helene. 2000. "Information Resources for Intelligence." In Jerry P. Miller, ed., *Millennium Intelligence,* 97–120. Medford, NJ: CyberAge Books.

McDaniel, Carl, and Roger Gates. 1999. *Contemporary Marketing Research,* 4th ed. Cincinnati: South-Western College Publishing.

Stewart, Thomas. 1999. "Customer Learning Is a Two-Way Street: E-Commerce at Dow, GE, Kaiser, and Xerox." *Fortune,* May 10, 158.

Swaddling, David C., and Charles Miller. 2001. *Customer Power: How to Grow Sales and Profits in a Customer-Driven Marketplace.* Dublin, OH: Wellington Press.

von Hippel, Eric, Stefan Thomke, and Mary Sonnack. 1999. "Creating Breakthroughs at 3M." *Harvard Business Review*, September-October, 47–57.

Part 2

Project Leader Tools to Use Anytime

Project leaders will find the tools in Part 2 useful across the entire life of a project. They can apply any of these at any time, although projects likely would benefit most from starting to use these tools earlier rather than later in the project. At their base, all four chapters deal with predominantly people issues. Each presents a different way to enable the people working on the project to be more effective and increase the probability of project success.

As Chapter 5, "Product Champions: Crossing the Valley of Death" shows, one people-based solution to improve the probability of delivering a successful product to the marketplace is to use and support a product champion for the project. Champions are particularly useful in helping projects move from technology invention to formal physical new product development. Firms have one set of structures, resources, and people in place for creating technology inventions. They have another (different) set of structures, resources, and people in place for new product development and commercialization processes. What is missing in many firms is an effective transition from one set of structures and processes to the other. Champions are one way to move through this "valley of death" between the two situations. The chapter defines and provides insights into nine activities champions must complete to move projects successfully through the valley of death and into commercial development. Using champions thus is likely to benefit project teams for both newer-to-the-world and newer-to-the-firm projects more than it will benefit those teams making straightforward improvements, or iterations, to current products or extending a product line in a clearly linear manner.

Chapter 6, "Managing Product Development Teams Effectively," continues in the theme of people-based solutions to increase the probability of project success. This chapter introduces two principles of NPD team management that lead to effective team performance: cooperation and integration. It then provides a series of analytical steps a team leader can take to diagnose how well their team is performing. It then provides a number of organizational design, support structure, and management-style actions that leaders can take to improve team performance in the near term. The chapter closes with a series of longer-term actions that an organization can take to create a context that will help future teams consistently function better. This chapter will be especially important for those managing larger, more complex multifunctional teams.

Good product development decisions are timely, complete, appropriate, high quality, actionable, and expeditious. Yet, because of the strong emotions usually present around NPD projects, decisions are frequently troublesome, emotionally charged, and, much too often, flawed. Chapter 7, "Decision Making: The Overlooked Competency in Product Development," defines each of these decision characteristics and provides techniques to improve the gate decision-making processes used at the end of stages of new product development. The chapter provides a self-assessment tool for quantifying your organization's decision-making effectiveness, proposes an agenda to use in structuring successful gate meetings, and details a number of specific actions organizations can implement to improve gate decision making.

For many of the same reasons that project decision making is difficult, facing risk for a specific project is also fraught with difficulties. Chapter 8, "How to Assess and Manage Risk in NPD Programs: A Team-Based Risk Approach," presents a ten-step process for managing risk in new product development projects. This chapter describes in detail how to complete each step in the process, from preparing to begin a risk analysis for a project through executing the action plans that reduce risk, and learning from the risk management activities. While the team leader usually prepares the initial plan for the risk program, the rest of the team is involved in the other steps, sometimes in group tasks and sometimes in individual actions. Using the team to develop the risk plan increases the collective ownership and thus the likelihood that the team members take the actions necessary to reduce risks during development. This process can be applied by the team at any point in the project, but provides the most benefit if applied as part of the business case development stage, and then reviewed at each subsequent stage to ensure that the issues identified as risks are being managed appropriately.

5 Product Champions: Crossing the Valley of Death

Stephen K. Markham

No one wakes up in the morning and says, "Hmm, today I think I'll be a product champion." Universities do not award degrees to champions. Firms generally do not hire a person as a champion, nor do firms establish official positions known as "champion." Though the role of champion is well known in new product development (NPD), champions' real contributions are often outside the formal product development process. Some authors even argue that a project dies without a champion (Schon 1963).

Champions are informal leaders who emerge in a somewhat erratic fashion. Championing is a voluntary act by an individual to promote a particular project. In the act of championing, individuals rarely refer to themselves as champions; rather, they describe themselves as trying to do the right thing for the company. A champion rarely makes a single decision to champion a project. Instead, he or she begins in a simple fashion and develops increasing enthusiasm for the project. A champion becomes passionate about a project and ultimately engages others based on personal conviction that the project is the right thing for the entire organization. The champion affects the way other people think of the project by spreading positive information across the organization. Without official power or responsibility, a champion contributes to NPD by moving projects forward. Thus, champions are informal leaders who (1) adopt projects as their own in a personal way, (2) take on risk by promoting the projects beyond what is expected of people in their position, and (3) promote the project by getting other individuals to support it (Markham 2000).

Championing skills include more than enthusiasm and foresight. A champion must possess the skills necessary to cross a developing product's Valley of Death. What is that valley and what does it have to do with championing a product? More importantly, how do you, as a champion, cross it?

THE VALLEY OF DEATH: WHERE CHAMPIONS RUSH IN

How many times have you seen good ideas, even obviously great ideas, lose approval while lesser ideas get adopted? The chance for a good idea to succeed diminishes in product development's Valley of Death. The Valley of Death is the gap between an idea's technical invention or market recognition and the efforts to commercialize it (see Figure 5-1). Many companies have resources, personnel, and company structures for technology development. These components appear on the left side of the valley in Figure 5-1. Similarly, most companies have resources for commercialization activities such as marketing, sales, promotion, production, and distribution. These appear on the right side of the valley in Figure 5-1. The resulting gap between discovery and commercialization indicates lack of structure, resources, and expertise. This lack of resources and expertise between technology development and commercialization activities is the Valley of Death. The champion's role is to drive projects across the valley.

Many reasons exist for the Valley of Death. Technical personnel (left side) often do not understand the concerns of commercialization personnel (right side) and vice versa. The cultural gap between these types of personnel manifests itself in the results prized by one side and devalued by the other. Networking and contact management may be important to salespeople but seen as shallow and self-aggrandizing by technical people. Also, both sides often have different objectives and reward structures. Technical people find value in discovery and pushing the frontiers of knowledge. Commercialization people need a product that will sell in the market and often consider the value of discovery as merely theoretical and therefore useless. Both technical and commercialization people need help translating research findings into superior product offerings.

Crossing the Valley of Death requires champions, resources, and formal development processes. Figure 5-2 illustrates the activities that take place within the Valley of Death. Often the champion's role and the need for resources are unclear and interact in an ad hoc fashion. A champion typically approaches

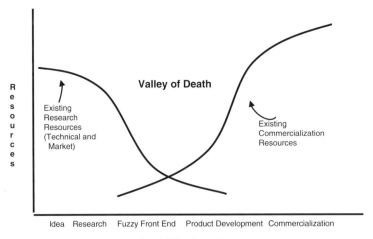

FIGURE 5-1. Valley of Death.

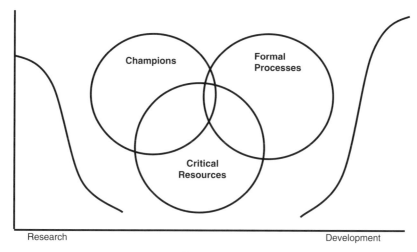

FIGURE 5-2. Factors in the Valley of Death.

individuals who have a needed resource, presents a vision of the opportunity to them, and informally asks for help. For example, a new diagnostic test to predict the onset of Type 2 diabetes may require a specific chemical reaction to be demonstrated before anyone will take the project seriously. The champion seeks the necessary lab time, chemicals, and personnel to demonstrate the chemical reaction. The champion then uses these resources to demonstrate the idea's potential. If the demonstration is successful, the champion seeks to have the project adopted into the formal development process.

Often the champion must demonstrate potential on more than one dimension, such as technical and market potential. For example, not only must the diagnostic test demonstrate the chemical reaction, but also the champion needs resources to demonstrate the medical need for the test to be ordered in large numbers. This informal process of presenting a vision and seeking needed resources repeats itself until the champion accepts defeat or the company formally accepts the project for further commercialization.

For a single project, the champion may repeatedly undergo the process of presenting a vision to gain access to resources. Yet the champion must follow a necessary progression of activities (see Figure 5-3). Even though any given project is likely to repeat previous steps one or more times, Figure 5-3 captures the general flow of activities in the Valley of Death. This figure shows that the champion first connects the research and needed resources to show a product concept's potential. He or she accomplishes this connection by developing a compelling business case (see also Chapters 3 and 4 in this volume). The business case presents the champion's vision by manifesting technical capabilities as product features and benefits that fit with demonstrated market needs.

After the champion creates a compelling business case and demonstrates

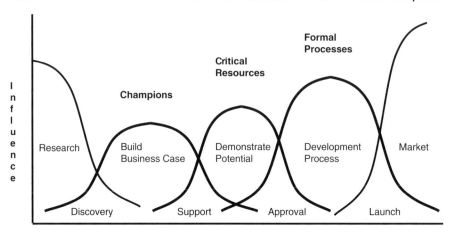

Level of Development

FIGURE 5-3. Linking research with markets through the Valley of Death.

the potential of the idea, he or she must translate the project into a proposal for the formal development process. Between each of these steps are transition steps. The three major steps of (1) creating a business case, (2) demonstrating potential, and (3) formal development, along with six transitional activities, represent nine activities a champion must accomplish to drive a project across the Valley of Death (see Table 5-1). It must be stated that most individual champions are uncomfortable with any linear sequence of the activities represented in the Valley of Death because they must often repeat or modify steps. Although this nine-step pattern does not emerge in a linear fashion, the champion must accomplish each step at some time to promote the project most effectively.

TABLE 5-1.
Nine Activities in the Valley of Death

1. Discovering that research has commercial value
2. Manifesting the discovery as a product
3. Communicating the potential through a compelling case
4. Seeking resources to establish potential
5. Using resources to reduce risk
6. Seeking approval of the project for formal development
7. Translating the project into the criteria used for approval
8. Deciding to approve or not approve project (not done by champion)
9. Developing and launching the product (not done exclusively by the champion)

CROSSING THE VALLEY OF DEATH

To cross the Valley of Death, the champion and his or her enlisted personnel must complete nine activities in any order (see Table 5-1). These activities aid in focusing energy on the common goal of creating a marketable product.

1. Discovering That Research Has Commercial Value

The first thing the champion must do is to recognize that a certain technical or market discovery has commercial implications. This insight is sometimes referred to as an entrepreneurial flash or as a techno-market insight (Jolly 1997). Looking beyond these singular events, the champion discovers that what looks to be a brilliant insight is actually the culmination of a great deal of work, often requiring a career's worth of insight.

The ability to recognize a technical capability that can meet a customer need and then be able to express that technical capability as a product is a challenging one and not always recognized as a workable task. All too often, a champion sees one connection between the technology and the market and pushes that one idea. Management often rejects any idea presented with this "technology push" approach.

Technologies typically have multiple capabilities. Devising product ideas that utilize those capabilities in a way that provides clear product superiority is the champion's goal. The champion must often investigate a number of product and market combinations that utilize different technical capabilities. Figure 5-4 reflects this combinatorial process. The logic of the technology-to-product-

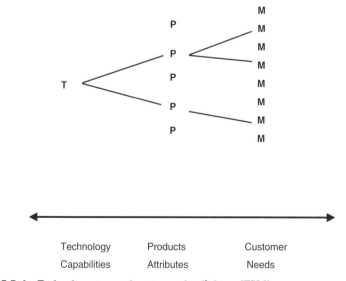

FIGURE 5-4. Technology-to-product-to-market linkage (TPM).

HINTS FOR RECOGNIZING COMMERCIALLY VIABLE TECHNOLOGIES

◆ Explore technologies with:
  ➤ Technical superiority, such as increased speed, lower cost, new needed function
  ➤ Ability to be protected by patents, trademarks, copyright, or secrecy
  ➤ Acceptable development time—the shorter the better
  ➤ Promise of large profit margins
  ➤ Inventors that will support commercialization
◆ Recognize that these technologies may originate in your organization with:
  ➤ Young, talented, ambitious researchers
  ➤ Older, experienced researchers
  ➤ Communities of practice or research groups
  ➤ Formal or informal programs
◆ Reduce your risk by finding people with a track record of successful inventions and commercialization

---

to-market (TPM) linkage must be clear to the individuals the champion approaches for needed resources. For example, the chemical reaction necessary for the Type 2 diabetes test may support a number of products, such as in-home tests, doctor's office test kits, or high-volume tests conducted by large clinical laboratories. Finding that a technology can develop into a number of products is uninteresting until the champion links one or more of the products to large market need. The Type 2 diabetes market can appeal in a variety of ways to patient, doctor, and health care payer. In this case, we have one technology, three products, and three market segments. Each segment represents a different value proposition for each product idea. Therefore, the ability to recognize commercially viable links is the essence of building a vision. The ability to express commercially viable links is the essence of communicating that vision.

2. Manifesting the Discovery as a Product

Rather than relying just on flashes of intuitive inspiration, successful champions learn to elaborate on TPM connections. They match a number of product ideas with a disciplined process of market segmentation for best commercial use of the technical capabilities. Champions recognize logical requirements. They develop the links between technical capabilities and enduring customer needs by means of product attributes. Product attributes unite technologies and markets. To develop the logical links between technology and markets, champions could use a Product Attributes Worksheet (see Figure 5-5). This worksheet features three steps: (1) listing the capabilities of a new technology, (2) listing multiple customer needs addressed by the technology capabilities, and (3) listing multiple product attributes or features that could be developed based on the technical capabilities and customer needs.

To draw on another example for variety, the cell phone example in Figure

This worksheet translates technology capabilities into specific product features. In combination with other components of product specification (customer, markets, commercial path, value chain), this worksheet helps develop product ideas.

Instructions
Step 1: Identify the unique capabilities of the technology.
Step 2: Describe what customers need in terms of the unique capabilities.
Step 3: Specify exact product features.

An idea may start in any column and then be extended to the other columns. If the idea cannot be extended to the other columns, the technology may not support a market manifestation or the technology may not address customer requirements.

Example
Technology: *High Performance Rectifier*
Product Idea: *Power semiconductor chip for power management in cellular phones*

Technology Capabilities	Customer Requirements	Product Attributes
Lower on-state resistance	Longer battery life	Cellular phone with 3X talk time
Lower on-state resistance	Lower-temperature operations	Lower temp 25°
Smaller die size	Lower cost	30% device price reduction
Lower defect rate	Lower cost	30% device price reduction

FIGURE 5-5. Product features worksheet.

5-5 lists "lower on-state resistance" as a technical capability for a new high-performance rectifier. This capability speaks to two customer needs: first, longer battery life for cell phones, and second, lower operating temperature. These technical capabilities and customer requirements allow the champion to promote a vision where cell phone users can talk three times longer with the existing batteries and operate in conditions that are twenty-five degrees cooler than conditions for other cell phones. Tying technical capabilities and customer needs together into a set of product attributes is a critical skill. One of the main reasons the Valley of Death exists is that few people explicitly make these connections for the individuals from whom they are trying to enlist support.

Going back to our diabetes test example, the capabilities of our Type 2 diabetes test include chemical methods that are much faster, of a lower cost, and more precise than typical methods. Yet it may give a number of false positive readings. The champion must match these capabilities and limitations with customer needs or requirements. For example, large laboratories must examine the high-volume and cost capabilities from the commercial perspective of how much faster and cheaper the test must be for a large laboratory to adopt it. Those capabilities will help define the product features of price and performance. Note that high-volume capability by itself is not of interest to doctors or patients, so the champion must use a separate worksheet for each segment.

HINTS FOR TURNING SUPERIOR TECHNOLOGIES INTO SUPERIOR PRODUCTS

◆ Know where your technology has the biggest advantage.
◆ Know your technology's disadvantages and limitations.
◆ Describe how your intended user will benefit from the technical capabilities.
◆ Describe how a user might actually use the product.
◆ Make sure the market recognizes the technical advantage as a solution to a large opportunity or problem.
◆ Clarify the difference between technical capabilities, product attributes and customer needs (see Figure 5-5).
◆ Develop a number of product ideas to chose from.
◆ Check your initial product ideas with industry experts.
◆ Check your revised product ideas with knowledgeable people.

3. Communicating Potential Through a Compelling Business Case

A proposed plan becomes a business case when the champion can tell a compelling story about how the company will benefit from engaging in the plan. The plan must provide enough promise to enable the champion to secure additional resources from key individuals. The proposed plan is not the ultimate business plan—it just has to be good enough to get other people to support the next step.

The fundamental thing a champion must do is encourage others to see the idea's value. Communicating the idea's worth through a business case accomplishes this task. Writing a compelling case is difficult since it requires the author to include technology, marketing, operations, finances, legal concerns, and company structure. Writing the business plan is by far the most complex championing activity because it involves several components.

Business Case Outline

The business case outline (see Figure 5-6) helps the champion develop compelling business cases. By following it, the champion addresses most major questions supporters are likely to ask. Even if the champion never makes formal presentations or written plans, developing the case for personal use makes a champion much more effective at presenting his or her vision to others.

The essence of the business case is profit. The champion must be able to clearly state how the company will make a large enough profit to justify the risk of adopting the plan. Thus the business case must present both risk and potential return. Many champions fail to adequately address risk (see also Chapter 8 in the volume) and present what must look like pie-in-the-sky fantasies. Ultimately, management can make a decision about adopting the champion's plan only if it includes an examination of risk. Organizations regularly accept high levels of risk, but only if a reasonable assessment has been done.

Usually five to ten pages in length, the business case contains the essential business issues. These issues include the core asset analysis and strategic factors presented in an order similar to that shown below. Business case development is a refining process that requires a number of iterations. Write a rough draft early and pass it around, because the quality of the case depends on the number of revisions.

1. Title Page: Identify the project name and include contact information.

2. Executive Summary: Include a summary of the major points, including bottom-line financial numbers and a clear statement about investment needs.

3. Problem and Solutions: Set up the issue by describing the problem and how the core idea solves the problem. Explain how the core idea provides an advantage. Include the size and severity of the problem you intend to solve in quantitative terms.

4. Technology Description: Describe what the technology does rather than the science behind it. Describe the product in terms of product attributes.

5. Market Analysis: Describe the market as it now stands without your product.

6. Marketing Plan: Explain how you will address the market described in the market analysis. Include highlights of the sales plan, unit volumes, and pricing.

7. Operations: In general terms, describe the availability of delivery mechanisms, cost of delivery, expected capacity, and quality of output. Also indicate the amount of time needed to start operations.

8. Management Structure: Identify who is involved with the project and in what capacity. Describe the type of start-up team you seek to assemble. Present plans for attracting the necessary individuals to the opportunity.

9. Financials: Include a five-year income statement and a break-even analysis.

10. Risk Analysis: Describe major risk components with when and how they will be addressed.

11. Appendix: Include articles, patent descriptions, papers, patent information, or other important information.

FIGURE 5-6. Business case outline.

For example, assessing the diabetes test potential would include examining new equipment requirements, FDA approval, and distribution to laboratories across the world. This plan includes many risks and potential delays. By failing to take these factors into account, the champion's business case lacks the credibility to offer a real opinion about the true value of the new test.

The key to preparing a compelling case is to tell a story rather than to describe the technology or present endless marketing data. Most of the time a technical person's attempt to write a business case is really a technology description. The way to address this issue is to write a comprehensive technical description and attach it to the business case as an appendix. This arrangement has two advantages: (1) it helps the champion keep the technology separate from

> **HINTS FOR PREPARING THE BUSINESS CASE**
>
> ◆ Include a two-to-three-page executive summary.
> ◆ Keep the entire text short (five to ten pages).
> ◆ Tell a compelling story that communicates the business potential. Remember that the business case is not an operating plan.
> ◆ Be dramatic, but do not overdo the drama.
> ◆ Write the case with your specific target market in mind.
> ◆ On the first draft, allow yourself to put in all the details you have. Edit details on the second draft.
> ◆ Rewrite the case until it is compelling and concise. The quality of the case is a function of how many times you rewrite it.
> ◆ Have other people review the case.

the business aspects, and (2) it completely describes the technology for patent disclosure processes.

Product Market Matrix

Substantiating revenue figures is one of the most important and difficult components of a business case. Taking a percentage of a percentage of a total market is never a convincing method to establish potential revenue. In fact, managers often refer to this method as "spreadsheet abuse." The most simpleminded form of spreadsheet abuse is saying, "The market is huge, so it's okay if we only get a quarter of a percent." Large markets do not by their nature bestow property rights for any percentage. A less obvious form of abuse is to present ranges of sizes and penetration rates, choose some intermediate point for the forecast (for example, the lower one-third), and then call it conservative. This approach is nonsense and sure to fail. A much more convincing method is a bottom-up approach to revenue forecasting, which relies on market segmentation and counting customers one by one to achieve a more accurate forecast. If you can personally name the first one hundred customers, you are on the right track.

Remember that product attributes derive from the recognition that certain technical capabilities address certain customer needs. The tie between capabilities and needs is the product's attributes (see Figure 5-5). The tie between product attributes and market segments is established with the product market matrix. The most effective way to fill in the matrix with ratings is to conduct telephone interviews with twenty people from each segment and ask them the importance of each of the product attributes. By having interviewees judge an attribute's strength of importance on a scale of 1 to 5, the champion can determine the product's overall attributes.

To complete the product market matrix, put the product attributes on one dimension and the market segments on the other. In the created cells, rate how receptive each segment is to each attribute. For example, as displayed in Figure 5-7, the diabetes test has three segments (patient, doctor, health care payer) and three product attributes (speed, cost, accuracy). On the product market matrix,

Instructions

Step 1: On the Y axis of a matrix, list all product attributes derived from the technology.

Step 2: On the X axis of the matrix, list all possible markets and segments for all attributes.

Step 3: Eliminate cells, rows, or columns based on obvious criteria, such as price, competitors, time to market, and low segment desirability.

Step 4: Identify cells that have high potential for a dominant product attribute in a market segment with a high need for those attributes.

Step 5: Develop a prioritized list of product attributes by market opportunities.

Markets and Segments

Product Attributes	Market A			Market B			Market C	
	Patient	Doctor	Payer	Seg X	Seg Y	Seg Z	Seg X	Seg Y
Faster Speed	5	3	1					
Lower Cost	4	2	5					
Higher Accuracy	5	5	4					
Product Attribute 4								
Product Attribute 5								
Product Attribute 6								
Product Attribute 7								
Product Attribute 8								

Instructions: Rate each product attribute from 1 (not important) to 5 (very important) for each segment. Ratings are derived from responses in interviews with customers.

FIGURE 5-7. Product market matrix.

the segments appear on one axis and the attributes appear on the other, which creates a three-by-three matrix. In each of the nine cells appears a rating of how important the attributes are to each segment.

By examining the columns in Figure 5-7, you can see that the customer segment is most responsive to the new test. Perhaps this sort of analysis is what led pharmaceutical companies and other medical technology companies to market directly to consumers. The ratings also suggest that overall accuracy is the best attribute to highlight in sales and marketing efforts.

Market Opportunity Worksheet

In addition to identifying a promising market segment, the champion must also quantify the opportunity in terms of revenue. The market opportunity worksheet (see Figure 5-8) assists the champion in quantifying and justifying revenue forecasts for the product idea. To develop a justifiable revenue figure, the market opportunity worksheet draws upon the results of the product attributes worksheet and the product market matrix to describe the market segments in terms of number of customers and drivers to adoption.

Market opportunity represents a critical part of the business case. If the revenue figures are unrealistic, the champion loses credibility and will be frus-

---

1. Describe the intended market segments and indicate number of customers.

	Segment Description	**# of Customers**
Segment 1		
Segment 2		
Segment 3		

2. Describe the customer/user/buyer in each segment.
   Segment 1
   Segment 2
   Segment 3

3. Explain why customers in each segment would adopt the product.
   Segment 1
   Segment 2
   Segment 3

4. Describe how each customer segment would use this product.
   Segment 1
   Segment 2
   Segment 3

5. What is the size of each segment?

	Total Number of Possible Sales	**Price Per Unit**	**Revenue**
Segment 1			
Segment 2			
Segment 3			
TOTALS			

Using the customer descriptions and reasons they would use this product, estimate and justify the number of unit sales by segment per year.

Number of Unit Sales:	**Year 1**	**Year 2**	**Year 3**	**Year 4**	**Year 5**
Segment 1					
Segment 2					
Segment 3					

FIGURE 5-8. Market opportunity worksheet.

HINTS FOR ESTIMATING MARKET OPPORTUNITY

◆ Find external data to support market size.
◆ Using real data for your product idea, establish the identity of the segment likely to adopt the product.
◆ Before picking your sample, ask experts to review your segments and questions. This step ensures that you proceed with the best information available to avoid making mistakes and leverage preexisting knowledge.
◆ Interview people to understand why they would use your product.
◆ Interview people to understand the likelihood of adoption.
◆ Do not conduct large market surveys. They are inappropriate at this stage of development.

---

trated by lack of support. At the earliest stages, the revenue figures do not have to be precise. Nevertheless, two issues must be very clear: (1) why the intended customers will buy the product, and (2) how many customers exist.

Using our sample business case as an example, the champion must establish how many people might use a test for Type 2 diabetes. Authorities estimate that more than fifteen million people have Type 2 diabetes. A recent consensus panel of diabetes experts estimate that seventy to eighty million people in the United States are at risk for insulin resistance, a condition that is a precursor to Type 2 diabetes. We also know that age, weight, and family history impact the likelihood of developing the disease. Therefore, the potential number of people who need testing is far greater than the number of people with the disease. Since the disease has grave health implications, we tentatively assume that there is strong reason for customers to use the test and that there are lots of potential customers (of course, a champion would not use the words "tentatively assume" or "potential").

Although an estimate may represent some maximum theoretical market, the champion must also determine market share. For this determination, the champion must understand why the customers will use the test. He or she then uses the strength of the reason to identify the number of early and late adopters. In the example, the champion might see people with a family history of diabetes who witnessed daily insulin injections as more motivated to avoid Type 2 diabetes than people who are just a bit overweight. He or she can use a small sample of each segment (thirty people from the customer base) to establish the adoption rate for a group. This method avoids spreadsheet abuse by (1) segmenting or defining the target markets and (2) gathering primary data on issues of concern directly from the intended customers.

Financials

The champion must prepare an income-and-expense statement and a break-even analysis for the project. After quantifying and justifying the revenue associated with a new product, the champion is almost ready to present a convincing financial story. All that is needed now is expense data. Use your company's administrative, marketing, and R&D percentages of revenue in your analysis. These are found in your company's latest quarterly report. Many companies maintain financial worksheets for estimating a project's income and expenses and analyzing its break-even point.

Strategy

Through the business case, the champion must provide a clear view of how the product will enter the market and then expand it. It is helpful to clearly tie the new product to stated company objectives. In addition, he or she must clarify the use of resources to achieve and maintain the idea as a successful product. The tactics to achieve the stated objectives must be clear and convincing.

4. Seeking Resources to Establish Potential

While building the business case, the champion often realizes that one or more parts of the plan need more work to be convincing. For example, the market opportunity may require a more detailed description, or the technology may not support the product attributes that the market actually wants. This realization naturally causes the champion to focus attention on these shortcomings. Often these failings become apparent when a champion prematurely seeks approval of the idea for formal development. When a shortcoming becomes apparent, the champion must seek support to develop and demonstrate that the project can perform on that critical dimension.

The source of support depends on the nature of the shortcoming. Often the source is different for each project, and it may be different for the same project at different times. For example, the champion may need to demonstrate a project's technical feasibility (the ability to develop from a prototype to a production model) or a project's quality. Similarly, from a commercial perspective, the champion may need to demonstrate the cost of production, the suitability of the market for competition, or that the drivers for adoption by a large number of people will generate the revenue being claimed. At this step within the Valley of Death, sponsors (the people who provide resources) can help by identifying other people in the company for the champion to approach for added support.

Seeking support can be a lonely time for champions because they must often persevere in the face of impending rejection. The organization is not responsible for making the champion happy by adopting the project. The champion must convince other people that the plan makes sense and promote it in a climate of organizational opposition or neutrality. Champions often do not seek solace from other people in the organization about the difficulty of gaining project support. They feel that seeking solace may result in a perception of weakness or lack of confidence. Yes, seeking support is a lonely and difficult process, but champions completely believe in their idea and that it's only a matter of time before support arrives.

Criteria for securing support include finding people who have access to the type of support needed at that time (from bench technicians doing a simple chemical analysis to an executive authorizing an extensive development project involving many people over months). Other criteria include timing the project to coincide with slack resources that the individuals control and finding individuals who are sympathetic to the project. In the diabetes example, the champion must find one or more persons or sponsors who can and will make people, supplies, and facilities available to test the proposed chemical reaction. The champion must find a person who has control over the needed equipment and who will either trust that the champion has a good idea or has some other reason to believe the proposed chemistry could work.

Given that a champion has a compelling business case prepared, how should he or she approach other people to support the project? A champion's effectiveness at gaining support does not heavily depend on polished presentation skills, bargaining techniques, political behavior, assertive requests, appeals to higher authority, clandestine activities, coercion, or rational argumentation. Rather, the champion's ability to generate support depends on his or her personal relationships with people.

The more credibility the champion has, the more easily he or she secures resources. Credibility derives from two sources. The first source is the champion's track record. Past success is often a prime determinant of obtaining support. The second credibility source is the plan's quality. A plan's quality includes both the quality of the science or market information upon which the idea is founded and the quality of the business case. The business case must clearly demonstrate how the company makes money from the idea.

New champions should focus on generating a high-quality business case and then develop a relationship with a key influencer. The high-quality business case is the lever to begin the relationship. Developing a relationship consists of informing the other person of what you want and then increasing the trust the other person has in you by being thorough, open, and eager to address concerns and rework the plan until the person is comfortable with supporting the plan. Building this relationship takes tenacity and willingness to do extra work. In the end, this person is risking his or her reputation to supply you with what you want. Make this person confident you can do what you say, and then overperform. He or she will likely be a source of continued support.

5. Using Resources to Reduce Risk

The champion must focus his or her resource request on the dimension of the business case that, if shown to be viable, will result in a large advantage to the company. The champion must determine how to reduce the company's risk. Having sought and received resources, the champion must use them in a way that validates his or her project vision. Obtaining resources without a clear idea of the objective results in using the resources without achieving the desired effects and ultimately hinders the project.

For example, if the chemical reaction necessary to show technical feasibility is accomplished, the champion will likely receive more support. If the champion then uses the resources to continue technical development to the exclusion of validating the market size and drivers, then the champion will not likely receive support in the next round even though the technology works.

Providing support to a champion's project is an unplanned use of resources by a department. The champion makes the special case that he or she needs the resources to demonstrate the value of the project idea. The resource provider generally provides the resource as an exception rather than as a standing offer. For each project, needed resources vary according to the critical dimension that requires validation.

The availability of these resources occurs in response to the size of the commercial opportunity. Therefore, it is helpful for the champion to specify the number of customers, the customers' identities, the customers' drivers for adoption of the new product or service, the unit volume, and the price. Given the nature of the product or services contemplated, the champion must secure the needed resources to effect the plan.

As the champion seeks and receives support to accomplish a specific goal, he or she relinquishes some control over the project and becomes answerable to project sponsors. The sponsor relationship is critical for the champion. Failure to accommodate a sponsor often ends direct support and assistance in finding other company resources. In fact, a negative sponsor experience becomes a negative reference for anyone else contemplating assistance to the champion.

6. Seeking Approval of the Project for Formal Development

The champion accomplishes each step to this point in preparation for seeking formal approval. Sometimes just having a good idea is sufficient to obtain support. Most of the time, however, finding support requires extensive preparation. Naive champions skip the planning and development work and immediately seek approval. When seeking formal approval, the champion must ensure there is no meeting to decide the fate of the project until all decision makers have embraced it. Without informal approval before a meeting, a project becomes subject to the vagaries of meeting processes, including lack of time, conflict between people, and emotional appeals.

For example, a diabetes test decision maker does not want to find out in the formal approval meeting that the sales and marketing people have never heard of the test before and that it will take months for them to add the test to existing doctor requisition forms. Similarly, development people could ask questions about development for which the champion does not have the answers. In a public meeting, not answering these questions could damage the champion's credibility. Usually there is little chance of short-term recovery after a formal negative decision.

Wise champions get their projects informally approved before the approval meeting. It is a mistake to face the decision makers for the first time in a meeting where they make a decision about the project. A champion finds out the identities of the decision makers and the positions they are likely to take. He or she devotes a great deal of preparation time to anticipating objections. After the champion is ready to respond to objections, he or she meets with each decision maker to learn about the process and gain approval. If decision makers voice serious concerns with the project, the champion has a chance to address each concern with a solution. The champion does not seek final approval until he or she has met with and addressed all the concerns the decision makers have about the project.

7. Translating the Project into the Criteria Used for Approval

Having addressed the largest risk factors, the champion and project sponsors must now prepare the project for the formal development process. Formal product development processes contain standardized decision criteria and guidelines for what constitutes strong or acceptable candidates for process admission. Even if a company does not have a formal development process, it likely makes informal decisions about what projects move forward with major resource commitments. In either the formal or informal case, the champion and sponsors must translate the project into the decision makers' language, objectives, goals, processes, timelines, and expectations.

Often a high-potential project does not receive adequate attention merely because the champion did not properly prepare for the transition from ad hoc project to formal project. At this stage, more people must coordinate their efforts to move a project along. The champion gives up more control over the project, and managers may diffuse responsibility to a number of individuals in order to accomplish many tasks simultaneously. The champion must begin to let other people make decisions about the project. Sharing control over a project must not be a roadblock to the approval process.

8. Deciding to Approve or Not Approve the Project

The approval phase is often the most frustrating phase for the champion, as during this period the champion has less control over the project's outcome than at any other time. After all the work, the champion must now rely on other people for moving the project ahead. Yet champions do have important tasks at this stage. They often directly try to influence the decision makers for their projects. They reiterate points, answer questions, provide additional information, and continue to support the project.

As explained in the previous section, champions should attempt to influence the processes and decision makers that determine the project's outcome. Nevertheless, not all processes or decision makers are amenable to such attempts. In these cases, the champion must be even more careful to anticipate challenges and know the preferences of the decision makers. For example, many decisions result in less-than-definitive outcomes. The decision makers may ask for more information or want to see more information before they will fund the project. Also, many decision makers are reluctant to reject ideas since rejection may discourage people from contributing other ideas. Champions must respond quickly and completely to all concerns before reapplying for approval.

If rejected, the champion should ask for an explanation without being defensive or confrontational. In the process, he or she learns more about the decision process and how to navigate in that setting. Finally, the content of the proposal and rejection needs to be fully understood in order to respond effectively in the next round.

9. Developing and Launching the Product

The role of the champion changes when the project enters formal development because the champion no longer directly controls the project's progress. In the formal development stage, resources come from established budgets. Managers establish timelines and project evaluation measures. Although the champion no longer has as much direct control over the project, his or her role does not end. Maintaining resources, generating enthusiasm, working on difficult parts, and continuing to expound on the vision are all critical. As a project moves through the formal development process, it leaves the Valley of Death and becomes part of the company's regularly funded commercial activities.

Many champions continue to use their informal influence to promote their projects in formal development. The champion may or may not be visible at this stage. Even though a project may be officially approved, it still faces many challenges, including continued funding and actually getting the needed resources to make the plan progress through its stages. Because a champion has a high level of understanding of and commitment to the project, he or she usually adds value by continuing to influence people to support it. The champion must often reinvigorate and encourage others through restating the project vision. Finally, when the company introduces the product into the market with production, distribution, sales, marketing, and revenue somewhere in the vicinity of the projections, the champion successfully crosses the Valley of Death.

BUILDING CHAMPIONING SKILLS

Crossing the Valley of Death requires champions to successfully negotiate all nine steps. Since there is often no reward and little encouragement for championing a project, a champion must possess passion for a project and the willingness to do extra work for that project. In addition to willingness, a champion needs knowledge and skill about how to promote projects. While some people with raw enthusiasm get a project approved and implemented, champions with well-developed skills, a strong track record, and years of experience shepherding projects succeed far more often.

Champions must devote special attention to developing their skills. Championing skills are not technical or content-based. Nor are they simply project management skills. Championing requires upward influence and peer-to-peer influence on critical decisions. Most companies discourage individuals from going outside their direct lines of supervision to secure resources or influence decisions. In fact, some companies see these actions as insubordinate and unacceptable. Seeking project support can be a delicate process and requires a unique skill set. This skill set includes the ability to seek project approval, an ability that helps differentiate between champions and project leaders. Champion skill building includes influence tactics, relationship building, business case writing, and project management skills.

Influence Tactics

Research has not shown the use of well-known influence tactics (presentation skills, bargaining techniques, political behavior, assertive requests, appeals to higher authority, clandestine activities, coercive actions, and rational argumentation) to be effective for champions. Overall, the use of these tactics to influence people to support projects results in a net reduction in willingness to back champion-led projects. Some of the tactics are particularly damaging to the champion's cause, namely, coercive actions and appeals to higher authority (going over someone's head).

Relationship Building

Although direct attempts to influence targets result in a negative outcome, the effectiveness of relationship building is clear (Markham 1998). A long list of contacts and friends is the result of a lengthy and diligent career in which one cultivates working relationships. While common for some marketing personnel, this skill is often more challenging for technical people. Nevertheless, the most effective way for champions to secure needed resources is through their personal relationships. Engineers may not have extensive networks due to personality characteristics and lack of exposure to other parts of the company. Networking and contact management are skills that a potential champion can learn in training sessions. While one cannot change personalities quickly, knowing how to mix at gatherings and how to keep track of people who express interest is a matter of utilizing simple management tools.

Many people do not naturally possess sales skills. Nevertheless, individuals who have a vision for what they are doing can generally express their ideas without much encouragement. Sponsors and others wanting to promote championing activities should address the acquisition of networking skills and relationship building for champions. Helping the champions expand their contacts and working relationships is a key role for sponsors.

Business Case Writing

The content of championing revolves around developing convincing business cases. Although seemingly foreign to many technical personnel, business case writing, like relationship building, is actually a learnable skill. Business cases have form and substance that the champion can learn in standard training courses. Building a business case does not replace the techno-market insight with which a champion begins. Writing an effective case can help refine and improve the idea. The case comprehensively presents the idea so that decision makers can evaluate the idea fairly.

Sponsors and companies that want to promote championing can assist business case preparation in a number of ways. First, they can provide direct training for champions to learn to prepare the cases. Second, they can provide access to others in the company who have the skills and information necessary to write the cases. Third, they can provide time for personnel in various parts of the company to assist the champion in preparing the business case. Access to information, both internal and external, is usually one of the most difficult challenges champions face.

Project Management

The ability to articulate a vision for a project does not guarantee that the champion can actually manage the project. In the early project stages, the champion still performs most of the work and demonstrates the idea's potential. This stage requires the effective and efficient use of limited resources. Project management skills do not ensure that a project has high potential; these skills only help the idea reach its potential. Many champions benefit from detailed instruction on how to run a project.

SUMMARY

Champions are great forces for change and innovation within companies. To be effective, champions must know what actions they need to accomplish in terms of getting projects approved. This knowledge includes understanding what developmental work needs to be done before formal approval for a project can be given. The Valley of Death is a model that helps champions understand the nature of the challenge and what must be done to meet it. Each step in the valley may or may not occur in the order presented here, but all must be accomplished. The level of development for each step depends on the nature of the project and credibility of the champion. The more difficult the project and the less credible the champion, the better the business case must be.

Championing projects is a rewarding and exciting activity, but it carries with it the burden of responsibility. When you ask someone else to support your project, not only do you risk your credibility, but you ask the other person to take a risk, too. To some extent, you also risk the success of the company when you divert resources from existing projects to your project. Champions, therefore, assume heavy responsibility. Fortunately, most champions do not consider these challenges as overwhelming. They persist in doing great things with or without permission.

REFERENCES

Jolly, V. 1997. *Commercializing New Technologies: Getting from Mind to Market.* Boston: Harvard Business School Press.

Markham, S. K. 1998. "A Longitudinal Examination of How Champions Influence Others to Support Their Projects." *Journal of Product Innovation Management* 15, 5, 490–504.

Markham, S. K. 2000. "Corporate Championing and Antagonism as Forms of Political Behavior: An R&D Perspective." *Organization Science* 11, 4: 429–47.

Schon, D. A. 1963. "Champions for Radical New Inventions." *Harvard Business Review* 41: 77–86.

BIBLIOGRAPHY

Badawy, M. K. 1988. "What We've Learned: Managing Human Resources." *Research Technology Management* 31, 5: 19–35.

Day, D. L. 1994. "Raising Radicals: Different Processes for Championing Innovative Corporate Ventures." *Organization Science* 5, 2: 148–72.

Frey, D. 1991. "Learning the Ropes: My Life as a Product Champion." *Harvard Business Review,* September-October, 4–10.

Frost, P. J., and C. P. Egri. 1991. "The Political Process of Innovation." *Research in Organizational Behavior* 13: 229–95.

Green, S. G. 1995. "Top Management Support of R&D Projects: A Strategic Leadership Perspective." *IEEE Transaction on Engineering Management* 42, 3: 223–32.

Green, S. G., M. B. Gavin, and L. Aiman-Smith. 1995. "Assessing a Multidimensional Measure of Radical Technological Innovation." *IEEE Transaction on Engineering Management* 42, 3: 203–14.

Howell, J. M., and C. A. Higgins. 1990. "Champions of Technological Innovation." *Administrative Science Quarterly* 35, 2: 317–41.

Lawless, M. W., and L. L. Price. 1992. "An Agency Perspective on New Technology Champions." *Organization Science* 3, 3, 342–54.

Madique, M. 1980. "Entrepreneurs, Champions and Technological Innovation." *Sloan Management Review,* winter, 59–76.

Markham, S. K., S. G. Green, and R. Basu. 1991. "Champions and Antagonists: Relationships with R&D Project Characteristics and Management." *Journal of Engineering and Technology Management* 8, 3–4: 217–42.

Reinertsen, D. G. 1999. "Taking the Fuzziness out of the Fuzzy Front End." *Research Technology Management,* 42 (6) November-December, 25–31.

Shane, S. A., S. Venkataraman, and I. C. MacMillan. 1994. "The Effects of Cultural Differences on New Technology Championing Behavior Within Firms." *The Journal of High Technology Management Research,* fall, 745–72.

Smith, J. J., J. E. McKeon, K. L. Hoy, R. L. Boysen, L. Schechter, and E. B. Roberts. 1984. "Lessons from 10 Case Studies in Innovation." *Research Management,* September-October, 23–27.

Van de Ven, A. H., and D. N. Grazman. 1995. "Technical Innovation, Learning, and Leadership." In R. Garud, P. Nayyar, and Z. Shapira, eds., *Technological Oversights and Foresight.* Cambridge: Cambridge University Press.

6

Managing Product Development Teams Effectively

Roger Th. A. J. Leenders, Jan Kratzer,
Jan Hollander, and Jo M. L. van Engelen

Multifunctional teams have become commonplace. Most companies today routinely form multifunctional teams, whether they call them that or not. Especially in more innovative projects, four out of five use multifunctional teams (Griffin 1997). Over the last two decades teams have become part of our managerial vocabulary and are now viewed as a central organizational building block. The results are often reported to be astounding. Unfortunately, the stories to the contrary are equally numerous. Who hasn't seen NPD projects gone sour because the marketing and R&D professionals couldn't—or wouldn't—work together well? How many NPD projects have you been part of in which information didn't travel smoothly and swiftly among the team members? Chances are it's more than you care to remember. In fact, every story of team success is contrasted by an equally profound story of failure. Managing NPD teams isn't easy; managing multifunctional NPD teams is even harder. Not only do they change the NPD process and call for different instruments for evaluation, support, and control, but they also require different management skills. No longer is a team manager's effectiveness judged by his or her ability to pilot the bureaucratic labyrinth of formal channels and vertical lines of authority. Rather, effectiveness is now judged by his or her ability to put together and run individual teams and networks of teams.

Of all the tasks required of a team manager, perhaps the most intricate challenge is in the coordination of multifunctional teams. Daunting as the task may be, there are some simple principles that have proved highly effective in managing NPD teams to higher performance. High-performing teams aren't enough to create high overall NPD results in your company. But good teams have been shown to be a precondition, both directly and indirectly, to the success of your NPD efforts. Moreover, a failing team is a surefire way to kill any project, regardless of how promising it might have been. The tools we present in this chapter can increase team performance and spot and predict poor performance; when they are used in conjunction with the many other tools discussed in this volume, you have available a comprehensive palette of tools that

141

can assist you in increasing NPD performance. It isn't easy, but you can improve virtually any NPD team. This chapter shows you how.

PRINCIPLES OF TEAM PERFORMANCE

How do you build and lead high-performing teams? The approach consists of two basic elements. Any gardener knows the trick. For a flower to flourish, one first needs to sow the right seeds at the right time of the year. Then the growing flower needs nourishment, care, and protection. For multifunctional teams to flourish, you first need to provide the necessary organizational preconditions. Recently, Patricia Holahan and Stephen Markham (1996) provided a useful overview of the factors that need to be taken into account in this effort. These factors include the way the NPD process and organizational support for the teams are structured, and the manner in which teams are selected. We recommend you take their suggestions to heart. The second step is nurturing the processes that take place within the team. In the present chapter, we focus on this second step and guide you along the principles that are most vital to the successful management of NPD teams and show you how to implement them into current and future teams.

The effective management of NPD teams revolves around two basic principles: team cooperation and team integration. We will first outline the essence behind these principles and then explain how you can convert these principles into practical, effective action. The brief overview of the two main principles gives you a framework to keep in mind as you think, reflect, and act upon the facts and advice we offer in this chapter or when you develop your own managing method. First the theory, then the action.

First Principle of NPD Team Management: Team Cooperation

Cooperation refers to the extent to which team members feel they are doing a good job together. When cooperation is high, team members turn to each other for advice. A cooperative team is characterized by a harmonious working atmosphere; team members feel comfortable with one another.

Since familiarity breeds trust and positive relationships, repeated interaction between persons in a team (including interaction in other teams in the past) tends to foster cooperation. Repeated interaction also helps team members develop the skills necessary to cooperate with other professionals with different backgrounds, languages, and personalities. Interaction thus breeds cooperation.

Conversely, the absence of a history of interaction is often a cause for lack of cooperation in teams. The good news is that, on average, repeated interaction will encourage cooperation. Give the team a common goal that induces them to rely on each other for success.

So far so good. You now know that cooperation has a tendency to come to those who have patience, and that once team members learn how to cooperate, this process will go much faster the next time they operate in teams. Does this mean that in the end everything will be fine? Unfortunately, no. Cooperation is good for quick decision making. Cooperation is good for a pleasant atmosphere. It even boosts development speed. But it does not necessarily boost the quality and market success of the resulting product. In fact, cooperation can be one of the greatest threats to NPD's goal of delivering competitive products to the market in a timely manner. Why? Because cooperation also breeds compromise. After team members have become used to each other's peculiarities, they start working out interaction mechanisms that diminish conflict and discussion. They start keeping information from others with different functions. This does, in fact, speed up the tasks performed in the team, but decreases their quality and integration and diminishes the quality of considered product alternatives.

How Much Cooperation Should There Be?

Cooperation is a two-headed animal. A high level of cooperation makes a team run smoothly, increases decision-making speed, and makes a team's operations efficient. Low levels of cooperation stimulate professional discussion and improve a product's fit with the market—it makes a team's operations effective. This contradictory character of cooperation is a blessing in disguise.

The early stages of NPD are characterized by high uncertainty regarding product and process. This demands much of the creativity and flexibility of a team and its members. Most knowledge necessary for the project still needs to be generated, and team members need to be open-minded and willing to work together and exploit the different kinds of knowledge presented to them. In the later stages of the project, uncertainty regarding the product is largely reduced—specifications and technical solutions have been decided upon, budgets and planning are clear, and a date for market introduction has been set. While in the early stages of the project the focus is on generating as many alternatives as possible, in the final stages alternatives have converged to decisions and specifications are not to be changed any further. A high level of cooperation is therefore good for teams in the later stages of a development project. Cooperation makes the team productive, as it is able to quickly make remaining development decisions without seriously questioning what has been decided before. On the other hand, high levels of cooperation are not desirable in the early NPD stages. Here, team members need to maintain sensible levels of disagreement—this keeps team members sharp and creative, and eventually results in products that better fit market demands. In other words, cooperation should increase over the lifetime of the project. It is a task of the team manager to ensure the team displays a trend of increasing cooperation over the span of the development cycle. This requires proactive management of interaction patterns in the team. A little later in this chapter we will show you how to do this.

Second Principle of NPD Team Management: Integration

Birds of a feather flock together. People tend to associate with others like themselves: those with similar interests, functional backgrounds, expertise, and history. In multifunctional teams, marketers will flock together, as will engineers. This occurs because they prefer to be among other marketers or engineers, and because they perceive the others as, well, "other." The preference for similarity creates subgroups, with most interaction occurring within these groups instead of between them.

The principle of integration refers to an even distribution of communication over the team members, rather than members creating subgroups. Members of subgroups tend to conform more to others within the same subgroup. Also, consensus is likely to arise within the subgroups, but not in the team as a whole. Low integration also restrains the flow of knowledge and information within the team. The good side of subgroups is that they create safe havens for functional discussion. Even in multifunctional teams, people from the same functional group in a firm need time and opportunity to discuss among themselves to find solutions to technical or commercial issues.

How Much Integration Should There Be?

There is no need to strive for 100 percent integration—overall, integration should be high and increase over the life span of the project. When decisions have to be made on the basis of mutual agreement, disagreements by single subgroups jeopardize the success of the product development process. As a manager, you need to ensure that these decisions are not controlled by subgroups within the team. Persistence of subgroups effectively kills the multifunctionality in a multifunctional team.

One caveat is in order. Innovation projects can be quite large; it is not uncommon for projects to include fifteen hundred members. A single team of fifteen hundred people makes no sense, cannot fruitfully be managed, and, simply put, is not a team. Large numbers of people have trouble interacting constructively, much less doing any real work together. Besides that, the project leader will be hard pressed to find solutions to logistical problems, such as finding enough physical space and time to meet. The common solution is to break up the group (after first breaking up the projects into distinguishable, coherent, and separable parts—use common project breakdown techniques for this) into smaller groups and, if needed, to further split up those into even smaller teams. For instance, the Anglo-Dutch information technology company CMG continues to break up groups until they consist of no more than fifty members. These groups are then expected to operate as teams and should have high integration.

HOW TO MAKE NPD TEAMS PERFORM BETTER

Above we discussed the two principles that underlie the performance of every NPD team. Now you are ready to take action. Based on these principles, in this section we will show you how to assess how your teams operate and how to improve their performance. You cannot sit and wait to see what happens. You need to be proactive in managing your teams toward optimal cooperation and integration and, thus, toward increased performance. Some steps can be taken today. Others can follow soon after. When applied with care, you will reap results you never deemed, or dreamed, possible.

What You Can Do Today

Today you can candidly assess the way your team functions. How cooperative are the team members? Do they work together well across functions or do they flock together in subgroups? You answer these questions by using our Team Spotter's Guide. The Team Spotter's Guide will provide you with a snapshot of how well your team functions. It consists of fifteen statements and can be filled out by you and your team members in several minutes. Instead of grouping the statements together along the principles discussed above, they are shuffled. To apply the Team Spotter's Guide, take the following three steps.

Step 1. Distribute copies of the NPD Team Spotter's Guide to the members of your team. The team members can choose to answer the statements anonymously or with their names on the form. Every team member is required to answer all the statements. Respondents answer by providing a score between 1 ("I completely disagree with this statement") to 10 ("I completely agree with this statement"). The tipping point between agreeing and disagreeing is between scores 5 and 6.

Step 2. After all statements have been answered, collect the forms and transfer the data to the Data Entry Sheet. The first column is reserved for your own answers (TM = team manager) and the others are for those of the team members. In the column labeled "Avg." you enter the team's average score for each statement. Not every statement has the same impact on the final score. Therefore, multiply these averages by the relevant weight (listed in the column "Wgt."). In the column "A × W" write the resulting scores.

Step 3. In the last step, copy the statement scores onto the Team Scoring Sheet. The scores on cooperation and integration follow after you add the appropriate weighted statement scores. For your convenience, the sheets allow space for three evaluations. Feel free to make copies of the sheet for additional evaluations.

You can transfer the results to the Team Trends form for later reference and your personal remarks.

You now have available a simple and direct overview of how your NPD team functions. In order to solidly improve your team, you need to know whether it is on the right track. Applying the Team Spotter's Guide provides you with a snapshot of the team. However, in order to effectively improve and sustain performance, a video of the team is better. Good teams can lose track along the way—a single positive outcome may induce a team manager to assume that all is well and no further monitoring is necessary, when in fact he or she needs to follow how the team's performance evolves over the course of

		[Disagree . . . Agree]
	Project name:	
	Date:	
	Respondent:	
	The Coordination Statements	[1 . . . 10]
1	I have frequent interaction with other team members about work-related matters.	
2	I consider myself to be a natural bridge builder between team members with different backgrounds, skills, and expertise.	
3	All our team members are focused on actively collecting knowledge for our project.	
4	There are no obvious subgroups in our development project.	
5	Our various working areas are located close to each other.	
6	Our team is characterized by a positive working atmosphere.	
7	On most topics, little discussion is required for teamwide agreement.	
8	All of my day-to-day interaction takes place with members of other subgroups rather than with members of my subgroup.	
9	If I have a problem or need advice, I will go to the other team members.	
10	Interaction with stakeholders outside the project team does not take place primarily within particular subgroups.	
11	The team manager plays a central role in the day-to-day functioning of the team.	
12	I cannot do my work without frequent communication with members of other subgroups.	
13	When I think my knowledge is beneficial to other members of my team, I do not hesitate to volunteer that knowledge to them.	
14	There are enough social events to allow team members to exchange their thoughts on the project in a casual atmosphere.	
15	In the future I want to work with these team members again.	

Every team member is required to answer all the statements. Answer by providing a score between 1 ("I completely disagree with this statement") to 10 ("I completely agree with this statement"). The tipping point between agreeing and disagreeing is between scores 5 and 6.

FIGURE 6-1. NPD Team Spotter's Guide.

													Date:		
Project name:													Project leader:		
Statements/Team Members	TM	2	3	4	5	6	7	8	9	10	11	12	Avg.	Wgt.	A × W
1 Frequent team interaction														0.1	
2 Bridge builder														0.2	
3 Collecting knowledge														0.1	
4 Divided in subgroups														0.2	
5 Working places separated														0.1	
6 Positive working atmosphere														0.1	
7 Little discussion necessary														0.1	
8 Interaction with subgroups														0.1	
9 Need advice														0.2	
10 External stakeholders														0.2	
11 Central leader														0.1	
12 Do work without communicating														0.1	
13 Offer knowledge														0.2	
14 Social events														0.1	
15 Work with members again														0.1	

FIGURE 6-2. Data entry sheet.

the project. For example, recently a large electronics company developed a new electric shaver that was packed with the latest technology. The shaver, for instance, studied the user's shaving behavior and adapted its energy consumption and position of the shaving heads to its owner's manner of shaving. From an engineer's point of view, it represented the Holy Grail of shaving technology. And that would become its kiss of death. When evaluating the team's functioning in the beginning of the project, the team manager was happy to see high levels of cooperation and integration. Indeed, at that time the team was doing well. He then stopped monitoring the team and focused his attention on other teams and ideas. Had he applied the Team Spotter's Guide over the life of the project, he would have noticed early and unmistakably that the team was starting to falter. The technical issues were an engineer's dream, so the development engineers cooperated closely and intensely on solving the many challenging technical problems. The marketers thus drifted to the team's periphery, as did the representatives of manufacturing, service, and accounting. When the product was finished—that is, when development finally deemed it

	Evaluation 1		Evaluation 2		Evaluation 3	
	Date:		Date:		Date:	
Cooperation	No.	Cooperation	No.	Cooperation	No.	Cooperation
Frequent team interaction	1		1		1	
Collecting knowledge	3		3		3	
Positive working atmosphere	6		6		6	
Little discussion necessary	7		7		7	
Need advice	9		9		9	
Central leader	11		11		11	
Offer knowledge	13		13		13	
Work with members again	15		15		15	
COOPERATION SCORE	SUM		SUM		SUM	
Integration	No.	Integration	No.	Integration	No.	Integration
Bridge builder	2		2		2	
Divided in subgroups	4		4		4	
Working places separated	5		5		5	
Interaction with subgroups	8		8		8	
External stakeholders	10		10		10	
Do work without communicating	12		12		12	
Social events	14		14		14	
INTEGRATION SCORE	SUM		SUM		SUM	

FIGURE 6-3. Team scoring sheet (maximum overall sum scores are 10).

ready—there turned out to be no market for the shaver. In addition, its high production price made it impossible to market. Timely and frequent application of the Team Spotter's Guide would have revealed these problems early and would have indicated how to manage the team toward a successful joint market-guided effort. With frequent application of the Team Spotter's Guide this disaster would have been readily preventable. Even the best teams can stumble along the way; it is your task to spot these hurdles and act on them in time.

By the same token, poor teams can be put back on track by effective man-

Symbol	Score	Evaluation 1	Evaluation 2	Evaluation 3
CP	Team Cooperation			
IG	Team Integration			
	Date:			
Remarks regarding Evaluation 1:				
Remarks regarding Evaluation 2.				
Remarks regarding Evaluation 3:				

FIGURE 6-4. Team trends.

agement action. Therefore, we urge you to apply the Team Spotter's Guide several times over the lifetime of the project. You will then start to see trends. You will be able to quickly measure the effect of any management action you take. You can assess the effects of changes in team composition. Some companies try to keep team composition stable over time. Ford, however, uses an American-football-style approach to its projects, adding and discarding members as the project progresses (Deschamps and Nayak 1995). Microsoft does a similar thing. After any significant event or change in the team, you can learn a great deal about the effects on the team by applying the Team's Spotter's Guide. It is a highly effective tool for the proactive management of your team toward greatness.

What You Can Do Tomorrow

Now that you have diagnosed your team's functioning, what you can do tomorrow (and in the days after) is take corrective action. The results on the Team Spotter's Guide offer hints and clues about what you can and need to do. For example, if answers show that team members tend not to consult others for advice on how to solve a problem (question 9), one of two causes typically is at the root. The first is that members simply do not want to ask for help because they think they know best themselves or they are afraid that asking for help will undermine their authority as an expert. In this case you can employ prac-

IN PRACTICE: WHEN TO APPLY THE TEAM SPOTTER'S GUIDE

Evaluating the performance of your team shows you how the team is doing. Multiple evaluations provide you with a chance to assess whether the team's performance is improving or faltering. Therefore, we encourage you to apply the Team Spotter's Guide as often as possible. Depending on the type (and length) of project you are involved in, try to apply the guide at least twice during each stage.

However, we strongly advise you *against* applying any team evaluation technique shortly before or after a gate meeting. Here is why: It is imperative that team evaluations not be used as part of the appraisal process at gate meetings. These meetings should focus on the product itself and consider the technical, commercial, and strategic issues associated with the development project. The Team Spotter's Guide is aimed at improving the team's functioning by exposing the team's weak points. When the Team Spotter's Guide is applied shortly before a gate meeting, you may run the risk of that information falling into the hands of the gate decision makers. If you do apply the guide before a gate meeting, keep the results to yourself. A second reason for not applying the Team Spotter's Guide right before a gate meeting is that gates typically spur team members to cooperate and integrate more; after all, the life of their project is on the line. Also, team members then tend to answer the guide's questions in a more strategic manner. The way in which teams operate just before such a critical and hot moment is very different from the manner in which teams function in quieter times. Similarly, excitement about the start of the next phase and adjustment to new goals and, sometimes, the presence of new team members prevent teams from showing their true colors shortly after a gate meeting. So after a meeting allow the team processes some time to settle before you measure them again.

In general, with the exception of the times just before and after gate meetings, we encourage you to use the guide often. The more often you use it, the better your understanding of the team's functioning will become and the more quickly you will spot potential problems. And you will become more effective in judging and accurately predicting the impact of any intervention you may consider.

tically any common team-building technique or invite an expert to train the team in joint problem solving. Another important technique is to set clear and strong team-level performance objectives; this reinforces the need to work together and opens up team members for each other's help.

A second common cause is lack of knowing who has the required expertise or skill. Some years ago, Philips's shaving division introduced multifunctional teams and focused strongly on high levels of cooperation within the team. Team members from various backgrounds were so strongly involved with their team that they lost the interaction with outside colleagues. As a consequence, these colleagues were no longer consulted when (technical) problems in the team needed to be solved. As a solution, Philips introduced glass walls on which every team graphically showed its progress and, most importantly, problems. Consequently, a specialist in team 1 who noted that his or her expertise might be useful to team 2 would then offer to help the other team; it was not necessary for team 2 to be aware that this expertise existed. A more powerful and structural approach you can employ is to assign (additional) people as liaisons or bridges with the specific objective of keeping groups and teams informed about

OVERVIEW OF ISSUES FOR WHICH THE INTERVENTION TOOLS ARE MOST EFFECTIVE

Proximity
- Stimulates and increases frequency of interaction; especially useful for interaction between subgroups
- Increases speed of decision-making processes
- Impedes subgroups from arising; is highly effective in breaking up established subgroups
- Stimulates a team feeling

Bridges and Liaisons
- Intensifies contacts between members across the team
- Stimulates and organizes interaction with stakeholders outside the team
- Enhances the effective flow of knowledge

Mode of Interaction
- Organizes knowledge transfer
- Affects speed of decision making
- Affects team members' creativity
- Creates new impulses and a new working atmosphere (changing between primary modes of communication is perhaps the cheapest and most subtle intervention technique—it can have a profound impact without negatively disturbing individual and teamwork processes)

Information Systems (Know-how)
- Affects the effective flow of knowledge (in terms of both outcomes and members' attitudes)
- Assists in writing the formal part of the project's postmortem

Information Systems (Know-when)
- Improves bridge-building activity, both within the team and with outside stakeholders
- Assists in writing both the formal and informal parts of the project's postmortem
- Allows the team manager to keep the team on schedule without having to be too central

Objective Setting
- Aligns the goals of subgroups and team members with the overall project goals
- Stimulates interaction and collaboration between subgroups
- Allows the team manager to keep the team focused and energized without having to be too central in the early stages of the project

Leader Centrality
- Strongly affects speed of decision making
- Very important to team atmosphere
- Can prevent the emergence of subgroups
- Affects the active flow of knowledge
- Affects the amount of interaction between team members
- Influences the tendency of individuals to act as bridges

Each intervention option has an effect on many different issues. In this list we summarize the issues an intervention option is generally most useful for. We stress that most interventions can also successfully be applied to the solution of other problems.

their progress, problems, and solutions. In this way, the liaison will actively look for (and know of) people with specific knowledge and expertise. For example, let a production engineer spend part of his or her time in production and the other part with R&D. The engineer will make an excellent bridge between the two functions.

Several intervention options, formal and informal, are available for you to improve your team's functioning: You can make adjustments to the team itself (organizational design), you can facilitate better the way team members do their joint work (support structure), and you can alter your own interaction with the team (management style). Below we will discuss seven intervention options for you to consider and help you build your portfolio of effective techniques. The box summarizes which problems each intervention option generally is most effective for.

Organizational Design

Proximity and Facility Design

If teams are separated in space, unite them. It is often remarkably simple to let team members work in the same room or in rooms close to each other. In many companies all it takes is to (re)move dividers. Alternatively, you can trade space with another manager who wants to do the same thing with his divided group. If space restrictions require you to spread out the team over several rooms, make sure to combine opposite types of functions together, rather than combining similar types. For instance, do not put marketing and sales in the same room, but combine marketing with R&D in one room and join sales and manufacturing in another. This will enforce cooperation—or, for starters, interaction—and will prevent the formation of subgroups along functional lines. Philips is transforming its R&D facility in Eindhoven (The Netherlands) into an open campus, hosting employees from all functions. In addition, although to a limited extent, individuals from other companies and from universities are invited to participate in the business creation processes.

Other practical ways in which you can support cooperation and integration are to rent a (temporary) facility to house the entire group, encourage integration of multiple locations through job rotation and temporary reassignments (for example, put a marketing person on temporary assignment in R&D or production), or adopt an open office design (get rid of dividers and private offices). A good way to temporarily increase cooperation and joint problem solving—especially useful for issues of teamwide importance such as the selection of design alternatives and the setting of time schedules—is to convene a group off-site. When Microsoft organizes an off-site (several times over the course of a project), it uses a few standard locations in beautiful areas. The attractive surroundings are enough to make team members look forward to an off-site with joyful anticipation. The presence and contribution of most team members make these gatherings very productive for Microsoft.

Especially when the team is working on a radical innovation, consider creating a joint facility where the team can reside. Creating something completely new requires the ability to try new procedures and processes, to break out of the company's culture, to experiment and be original. For this task, house all team members together in a facility away from your company's main location. In this fashion, you will proactively create the conditions conducive to team cooperation and integration, which will stimulate high levels of creativity and innovativeness, out-of-the-box problem solving, and a true team effort.

For example, Control Data Corporation (CDC) housed its team working on the development of the 5.25-inch hard disk drive in Oklahoma City, away from the company's principal Minneapolis facility (Christensen 1997). Many other companies are doing the same thing. It is vital, however, to return the project physically to the principal facility as soon as possible in order to incorporate and engrain it in the rest of the company's project and product portfolio.

Bridges and Liaisons

Above we mentioned that the appointment of liaisons and bridges pays off handsomely in NPD projects. Besides formally appointed liaisons, every team has informal ones; the trick is recognizing them. Although most team managers are convinced they accurately know most of the interaction patterns of their team members, both within the team and with other teams, research suggests that managers in fact get the lion's share wrong. If you get 30 percent right, you are among the absolute top! As numerous managers have sadly found out, chances are you won't even make double digits. Unless you invite a network researcher to perform a formal network analysis of the communication pattern of your team (or provide you with thorough training in network observation), there is only one way to find informal liaisons: ask! Talk regularly with the members of your team about the problems they face and the way they look for solutions. Ask them whom they involve in this process—not just who provides them with required knowledge but, more importantly, who helped them locate this knowledge in the first place. Ask them if they are involved in problem solving for other teams or other functions. In virtually any team, you will find one or more liaisons. In most cases, management did not expect these people to be liaisons. Often they are not even considered experts in a particular field. The only surefire way to find them is to ask. Also, ask other team managers; informal liaisons on one project are often informal liaisons on the next one, too.

Another way in which you can encourage bridges between subgroups and teams is to establish regular meetings of groups that must coordinate with each other. With larger teams, introduce formal relationships between subteams. For instance, a new copier requires development efforts on quite remote technological aspects such as optical imaging and electrochemical processes. It is not useful to force the subteams to work closely together on a daily basis; instead you should coordinate the development efforts on the overall team level. Such networked team structures have proven to be both effective and efficient in technologically complex environments.

Also assign dual reporting relationships so that team members must report to different functions. For example, you can have a development engineer report to both his or her functional manager and a marketing representative. Another development engineer may report to manufacturing and to representatives of an outside client. In any case, make sure that no one becomes the sole bridge between two groups or parties. The existence of multiple bridges strengthens project performance by providing parallel channels. Also, when your sole bridge between two pools of knowledge or functions falls ill or is taken off your project, you still have at least one bridge remaining. And people who bridge on this project will make natural bridges on future NPD projects. More is better!

Finally, you will find it fruitful to create work-related interdependencies. Break down the walls of subgroups by giving them assignments that make them depend on each other to some extent. One way to do this is to provide subobjectives that cross the boundaries of specific subgroups. For example, a project manager in an airline company provided marketing and R&D with the objective of developing a newly shaped aircraft that would carry more passengers, offer each more space, have a weight equal to conventional aircraft, and be commercially viable. The original idea was to "rotate" the standard aircraft body by 90 degrees, making it wider while reducing its height. The resulting design was never actually introduced to the market, but it did bring marketing and R&D closer together, created liaisons that turned out to be effective in many subsequent projects, and produced knowledge beneficial for future aircraft development in the company. In effect, it both broke down the walls between marketing and R&D and produced the cement for productive future collaboration between the two.

Bridges and liaisons do arise spontaneously. Some people are naturals at bridging; others need explicit encouragement to do so. But anyone can become a proficient bridge. Practice makes perfect. Since we cannot overstate the importance of bridges, formal or informal, it is good practice to actively and consistently build and encourage them. Bridges will help disseminate knowledge and increase the quality of decision making. They will pull together the team and make it run more efficiently and effectively. Build bridges.

Support Structure

Mode of Interaction

The most effective way in which you can support cooperation and prevent subgroups from forming is to support face-to-face communication. However, people spend large chunks of their professional life in unproductive meetings, and many NPD professionals have developed a psychological allergy to them. Your team members are most likely tempted to run most communication over e-mail. Increased use of e-mail decreases cooperation as compared to face-to-face meetings.

There is much to be said about the pros and cons of electronic communication versus face-to-face communication, but the bottom line is this: In the early stages of the project, meet face-to-face regularly. In fact, discourage electronic communication. Have team members talk and talk, and add your share of face-to-face interaction too. Face-to-face team interaction is an effective way to stimulate active and sincere cooperation and prevent functional silos from forming. Especially when you have created the situation in which team members are in close physical proximity to one another, you can easily encourage them to talk, rather than send e-mail. Also, meet weekly with the whole group. A classic mistake is to think that final decisions need to be made in meetings like these. The attempt to do so is what renders most meetings unproductive. So don't. Instead of deciding on a color, for example, decide on the requirements that need to be taken into account and determine who will make the final color decision. Their decision can then be discussed in the next meeting. Most NPD managers are experts who are admired for their technical skill and knowledge but often lack the training to lead meetings. It is far more productive to admit this to yourself and have your company provide you and your colleagues with proper training in how to lead productive meetings. These weekly sessions will then be the backbone of your successful effort to stimulate cooperation and integration.

As the project progresses, increasingly but slowly stimulate the use of e-mail by setting the example yourself. The lower the uncertainty involved in the project, the fewer and shorter plenary sessions should become. Over time you slowly start communicating more via e-mail. By carefully shaping the way in which your team members communicate, you craft the levels of cooperation and integration needed for the next phase of the project.

Information Systems: Know-how

Information systems relate to computer support for two sets of problems: those related to know-how and those related to know-when. When it comes to matters of know-how, computers are often proclaimed to be the cure for all of our problems. It is true that electronic bulletin boards allow team members to query hundreds of people at once. But there are serious limitations. Despite their promise and great advantages, computers cannot become the cornerstone of your information system, especially not in the early stages of the NPD project. Many systems are at your and your team's disposal, but all have the same problems: They provide only aggregated information and old news. They are also necessarily incomplete. When a team chooses a particular technical solution, for example, one would prefer to also include in the information system all the alternatives that were considered but discarded. No NPD professional has the time or taste to insert all of this into a computer system. Besides, skills and experience are virtually impossible to type into any system.

As a team manager, do support the use of databases and information systems. But especially stimulate face-to-face interaction. You can do this through liaisons (see above). Reward team members for actively sharing their knowledge with members inside and outside the team. A group meeting is an excellent time to let someone pose to the group a request for help with a particular problem. It is good to show team members that face-to-face interaction provides a rich source for finding important knowledge and experience, so let them offer their help in the meeting. However, do not let them actually start helping during the meeting—there will be ample time later.

You can also organize weekly sessions with randomly chosen participants across teams and functions; limit yourself to ten to fifteen participants per meeting. The goal of the meeting: informal information exchange. These meetings are used to hear everybody's concerns, get and give information, and keep everyone in touch. Trust is key; you need to apply the sacred rule that there will be no retaliation, no one is going to be hurt by what they say, and upper management will not be told. Without trust these meetings don't work. (And add doughnuts and soda too.) Such meetings (and variations on them) have been very successful in organizations as diverse as MCI, AT&T, Federal Express, the U.S. State Department, Mitsubishi, and many Silicon Valley firms.

Also, make sure that team members are not assigned to a team full time. Most managers shiver at this thought, but it is an important one. In many companies a good distribution of time is 80-10-10. The team member spends 80 percent of his or her time on your team; this is the person's main task. Ten percent of his or her time is spent with functional colleagues. By designating time for discussion among functional specialists (say, one day every other week), functional knowledge and experience generated in one project can be transferred to the specialist's colleagues on other projects. In this fashion, teams benefit from what goes on in other teams. In addition, a group of functional specialists together can solve problems that specialists in isolation cannot.

The final 10 percent of the time is spent on another team. The reason is that this helps to disseminate knowledge between teams and it assists in increasing organizational support for teams involved in radical innovation. For example, KPN Telecom, a large European company, used to have several departments housing product development efforts. Each department could start NPD projects of its own and recruit specialists from across the company on its teams. The teams were multifunctional but used the NPD methodology that had been developed by the department at hand. At one point, one of the departments started a project that involved a radically new product to be sold to groups of companies in industrial areas. None of the other departments supported it, so the project lacked access to vital pools of knowledge (and funding). We then suggested that the project manager "lend" his team members to other teams for limited amounts of their time (10 percent). Although it slightly decreased manpower, it worked miracles for the project. By virtue of their co-assignments, the team members had access to all the important sources of knowledge. In addition, they could personally and enthusiastically tell about the new product they were developing and quickly gained support for the project throughout the entire company.

Information Systems: Know-when

Support issues relate to "know-when" with the help of workflow management (WFM) systems. Such systems can be successfully employed when your team members need to follow many formal procedures or when decision-making moments have been specified in advance. Many companies routinely run WFM systems to control project teams' formal operations. However, it will be clear by now that these systems can also be used to assist you in effectively managing a team's levels of cooperation and integration. When developing a new car, the engine designers and the designers of the bodywork must work together. By nature, they won't. After all, their tasks and fields of expertise are quite different. But the dimensions of the engine and the points of attachment to the body cannot be decided upon without due regard to the space available under the hood of the car and the available connection points.

Use the WFM not only to remind two groups of designers to work together but to actually schedule meetings. Also, use the system to have them meet with other functions and subgroups. By using a WFM cleverly, you can boost knowledge flow—both knowledge related to know-how and knowledge related to know-when. And the best thing is that you can create these bridges and interactions yourself ahead of time. When the team reconvenes after a successful gate meeting to start working on the next stage, you can proactively insert moments of interaction, such as joint cross-functional tasks, giving those individuals unlikely to become a bridge naturally in the group a chance to participate in bridging activities.

In general, WFM systems become more beneficial with increasing complexity of design tasks. Compare, for instance, the use of task-structuring QFD (quality function deployment) techniques in NPD.

When having to work with a WFM for the first time, most people initially resist. However, while you should be sensitive to their reservations, your team members will soon discover that the system helps them get rid of annoying and bothersome administrative tasks. When you intelligently and proactively apply the system to boost team cooperation and integration, your team will notice that their work progresses more quickly, increases in quality, and becomes more fun. Quite a generous payoff!

Management Style

Setting Objectives

Performance goals are compelling. They motivate and energize. The combination of purpose and specific goals is essential to performance. Clear performance goals help a team keep track of progress and hold itself accountable (Katzenbach and Smith 1993). First, set clear objectives against which the performance of the team can be measured. To this add subobjectives or milestones to be reached along the way. The type of project and the uncertainty surrounding it will determine what kinds of objectives you can set. The most stimulating (sub)objectives are those set by the team itself. It is your task as team manager to ensure that these objectives fit within the overall objective of the project.

Make sure the objectives require a team effort. Nothing is more detrimental to a team's functioning than goals that mainly depend on or can be reached by the effort of only a subset of its members or functions. Set compelling goals, the type of goals that may seem impossible to achieve at first sight. The goals must be high enough to inspire extraordinary effort, but can't appear so unattainable as to discourage people from reaching for them. But by all mean, shoot a high as you can imagine and then add a bit to that. At GE, Jack Welch calls this type of goal setting "stretch." Stretch goals create we're-all-in-this-together teamwork. Where single individuals may despair of accomplishing a monumental task, teams nurture, support, inspire each other, and stimulate impressive joint results.

Over the course of the project it is imperative that you check whether every team member has the same objectives in mind. Depending on the way the objectives are formulated, the tenure of members on the team, and the team members' personal and functional background, various team members may interpret the objectives differently. Hold regular plenary meetings to establish a common perception of the team's objectives, but also frequently ask various team members to describe the team's objectives to you. You may be stunned by the variation in stories you will get. When you get a too large a variety of perceptions of the team's objectives, an off-site is called for. Famous are the discussions in European dairy farming about the so-called milk–value, representing the quality of milk and thus the amount of money the farmer receives. For over two hundred years now there have been quite different perceptions about this milk–value, even though the product, milk, hasn't changed at all in this time period.

Being Participatory

As a participatory manager, you support your team members and assist them. You are more of a facilitator, rather than the "boss." The opposite is an authoritarian manager, who occupies a central and highly visible position in the team. The more central the manager, the more he or she regulates and controls decisions, discussions, and information transfer. Especially in the high-uncertainty world of NPD, central managers tend to become overloaded with data and decision needs. As a result, the quality of your decisions deteriorates.

A well-run NPD process calls for the involvement of professionals with varying functional backgrounds and responsibilities. No matter how knowledgeable you are as a team leader, the quality of proposed alternatives and solutions is always improved when different specialists are involved. No single manager can solve every issue or even completely know whom to route the problem/knowledge/solutions to.

It gets worse. The more you regulate and control the flow of information, knowledge, and problems yourself, the more you hamper your team's creative capability. Intense communication and interaction between colleagues is indispensable in the process of formulating product ideas and solving problems. The more central the team's leader, the less creative the team.

If you are highly central in your team, this is not necessarily a bad thing, however. The good thing is that it speeds up decision making. In effect, it makes the team process more efficient. This is especially profitable in the later stages of the NPD project. Central leaders can ensure that past decisions are followed through on. Projects with central leaders go faster through the phases of test runs, engineering, production, and market introduction. You need to be more central at the later stages and less central at the beginning.

You yourself determine your centrality as a leader. If the matter is left to chance, most leaders are too central in the early stages of the project and not central enough at the end. Fortunately, making changes is easy to do. Being less central means taking a backseat at plenary meetings. You can chair them, but it is better to have others do this (so long as they have the ability to effectively run meetings). You take note of what is said and speak up only if decisions are made that lead away from the project's main objectives. When you do chair meetings, you make sure the different functions all get similar amounts of airtime, and you do not take a formal position in the discussion. Being more central means controlling discussions more in plenary sessions. You decide and overshadow the discussion. Such meetings are shorter. You make most decisions outside of meetings and then announce them to the rest.

A participatory manager can manage multiple teams; a central manager usually cannot. The most difficult thing about working with your style of leadership is to let it gradually change from being at the periphery to being central. Most of us have one style we are most comfortable with, and we tend to move to that style too soon or stay with that style too long. The Team Spotter's Guide will warn you when your style needs to change: If functions do not involve each other in decisions, old discussions are reiterated, or upper management starts worrying about the project's time schedule, you have a strong clue that you need to become more central. The best team leader is a true chameleon, and also has the ability to project him- or herself into a situation. Rapid and reliable assessment of the situation is an important condition for style adaptation, and the Team Spotter's Guide is designed for this task. If you want a role model for chameleonlike managers, look at politicians. Take them as examples, study their behavior, leave out the baloney, and there you are: the participatory manager.

What You Can Do in the Long Run

What you can do in the long run is manage conditions so as to create the context that will help you and your (future) teams consistently function better. In multifunctional teams, technical skill is not enough. Team members and team managers need to be trained in their networking skill. The better you are at it, the more effectively you can teach others the skill as well. You will reap immediate results. Better networking enhances both the coherence between the various functions within the team and the speed at which team members can find knowledge from sources outside the team. Seize these opportunities and arrange for training and education. Provide education and training in "soft skills," such as observation, listening, giving and receiving feedback, group dynamics, and team building. Although there are books with exercises that will help you achieve this, there is no substitute for the hands-on training that consulting firms can provide. Several of them specialize in teaching networking skills to professionals. So go out and engage in network management training. Unfortunately, and strangely enough, most companies spend most of their training budgets on cognitive skills (know-what and know-how) and lack significant funding for training in interpersonal skills (know-who). Far too often the budget for soft skills is one of the first items slashed when a company goes on a cost-cutting crusade. Given the importance of relationships, this training should be the *last* budget item to go. If you have the authority to assign budgets, guard the lines for training in interpersonal skills. If you do not have this authority, talk with those who do, and do not rest until you get the funding. Then spend this money on providing your team members with such training. Regardless of your tremendous management skills, wouldn't you agree that life would be easier—and, most likely, more successful—if your team members could take care of the team management themselves? Ideal would be a team in which each application of the Team Spotter's Guide showed a team on the fast track to success. You turn your team into a high-velocity train by training your team members well in networking skills. There is no substitute for starting right. Prevention is better than cure. Although most NPD team managers initially fear that such training money will simply be "lost," all of them who have committed to such training have experienced a multifold return on the investment. Expect the same.

When recruiting members for future teams, always assess their networks. By actively keeping track of the persons who tend to be informal bridges (see "Bridges and Liaisons," above), you can make sure your teams have ample natural liaisons present. Having a dedicated notebook for jotting down names of liaisons—both the formal and informal ones—and their networking performance will enable you to quickly add and build bridges by selecting appropriate team members. Keep such a journal in collaboration with other managers and you have a powerful management information system that no other system can ever match.

Make sure you consistently use the management tools we have described above. They go a long way and will improve not only the performance of your present teams but also the functioning of future ones. Together they cover the gamut of problems related to the cooperation and integration of your teams. Make sure you also actively devise your own intervention tools. Many more intervention options exist, some of which are idiosyncratic to your organization and the type of projects you are involved with. With these seven intervention options, we have provided you with a taste of options and how and when to apply them. Studying and applying them will supply you with the finesse necessary for you to develop a portfolio of intervention tools most effective in your situation.

For you to learn the effect of each intervention option, it is absolutely imperative that you apply only one at a time. Guided by our list, you may be tempted to cast off and stimulate the flow of knowledge throughout the team by redesigning their working areas, creating bridges, stimulating more face-to-face interaction, and introducing a computerized information system. This may well work. However, by applying multiple interventions simultaneously, you will create a host of changes. It will be not be possible for you to gauge the effect of the new workplace design because its effect will be moderated by the effects of the other interventions. Even though the results may be beneficial, you still haven't learned what to do next time when you want to change the flow of knowledge in a team. Therefore, be wise and patient: Apply only one intervention at the time, document it well, and allow some time for the effects to show up. Then feel free to add a second intervention if necessary.

In the box "In Practice: When to Apply the Team Spotter's Guide," we warned you that the results of team evaluations should be kept out of the hands of upper management. However, do involve upper management in your interventions. When you want to change the workspace arrangement of your team, make sure you get the (financial) backing from upper management. When you have created effective bridges and liaisons in your team, make sure they are not moved to another team by upper management. Resist! These bridges are instrumental to effective NPD team performance, so you cannot afford to lose them. Involve upper management when you want to introduce or adapt a WFM system. Make sure you have an agreement with upper management that they will not modify the challenging and energizing goals you have set for (and with) the team. Each of these intervention options is powerful and can create profound and lasting positive change. But every beneficial effect is immediately lost if upper management cancels your measures because you did not involve them in the decision. You then also lose the chance to advantageously apply that intervention option next time, because your team members will not believe and trust it will last. So do not involve upper management when you routinely evaluate the quality of your team. But always, always, always involve upper management when you take drastic interventions.

Vital to the continued high functioning of teams is to carry out a post-mortem, regardless of whether you celebrate success or mourn failure. Document the lessons learned and incorporate them into revisions of project methods and procedures. In addition to common topics included in a postmortem—such as an evaluation of the extent to which objectives were met, the correctness of risk analyses, and the accuracy of planning—it is imperative that you include a thorough evaluation of the cooperation and integration achieved by the team over the life of project. Whom did you use as liaisons and bridges and how well did they do? Were there liaisons you hadn't expected? Who? Why did you miss that? Who bridged between teams and who between functions within the team? When doing this, you cannot be gentle about yourself, either: Candidly evaluate how your management style and interventions affected the team's cooperation and integration along the way. In addition to your project post-mortem, keep an active logbook over the course of the project. Write down which actions you took, why you took them, what the effects were and why. In the next paragraph write down how you could have done better and what you will do in a similar situation next time. You absolutely need to sincerely evaluate, admit your own mistakes, and allow yourself to do better next time. Only then will you truly learn the lessons put forward in this chapter and become a better manager consistently.

And last but not least, read. Read anything you can lay your hands on that is relevant to your line of business, from daily newspapers and *Business Week* (to follow your competitors, partners, and customers) to scientific publications on the fundamentals and theory of your business. Even though most managers do not handle scientific theory well, it is mandatory reading because a profound understanding of the principles of business provides you with the best information and knowledge you can build your strategy on. You will then become a manager who knows how things react and why, instead of a manager who only knows how things are (see Baker 1994; Drongelen 1999; Hollander 2002; Kratzer 2001; Leenders and Gabbay 1999; Muller 1999). As the saying goes, knowing improves doing, and doing enhances knowing.

IN PRACTICE: KEYS TO SUCCESSFUL INTERVENTION

◆ Apply the Team Spotter's Guide frequently and diligently so that you learn how to spot problems before they occur. In time, you will be able to immediately spot the fractures in your team's cooperation and integration levels.

◆ Before using any intervention, clearly formulate the goal you want to achieve. A well-formulated problem is more than half the solution.

◆ Set team-level objectives for your interventions. Reward the team for a successful intervention.

◆ Learn from other/previous interventions, both those performed by you and those performed by others in your company.

◆ Try a different perspective; build a repertoire. Do not become myopic; there is always more than one solution to any problem.

◆ Apply only one intervention technique at a time. Never apply several techniques simultaneously, or you will never know what intervention caused what effect.

◆ Involve team members actively in the search for solutions to team problems.

◆ Protect your flanks by making sure you have the support and commitment of upper management. Also, engage external stakeholders—manage your interdependencies actively.

◆ Report results to the team, especially the positive ones. Most people consider change unpleasant, so by advertising the success to the team, you will keep team members motivated.

◆ Always perform a postmortem and document your interventions. This is useful both for you and for others who can learn from your experience.

◆ Be consistent in your approach to a problem. Do not change your approach along the way because you become impatient and don't want to wait for the results of an intervention.

CONCLUSION

This chapter provides you with the tools necessary to act immediately. By frequently using the Team Spotter's Guide, you keep track of the functioning of your team. Look for changes in answer patterns. Apply the intervention tools and use the Team Spotter's Guide to keep track of the effects. This is where the rubber meets the road.

The process of learning how to manage NPD teams well never ends, however. What you can do for long-run success is to continue to read and apply the principles and tools described in this chapter. Paths are made by walking, and improved team management requires that you apply the tools consistently and frequently.

The tools presented in this chapter are based on intensive scientific study and extensive practical experience. They are very powerful and will, when applied conscientiously, considerably increase project performance today, tomorrow, and in the long run.

BIBLIOGRAPHY

Baker, Wayne E. 1994. *Networking Smart*. New York: McGraw-Hill.

Christensen, Clayton M. 1997. *The Innovator's Dilemma*. Boston: Harvard Business School Press.

Deschamps, J., and P. R. Nayak. 1995. Product Juggernauts: How Companies Mobilize to Generate a Stream of Market Winners. Boston: Harvard Business School Press.

Drongelen, I. C. van. 1999. *Systematic Design of R&D Performance Measurement System*. Enschede (The Netherlands): Print Partners Ipskamp.

Griffin, Abbie. 1997. "PDMA Research on New Product Development Practices: Updating Trends and Benchmarking Best Practices." *Journal of Product Innovation Management* 14: 429–58.

Holahan, Patricia J., and Stephen K. Markham. 1996. "Factors Affecting Multifunctional Team Effectiveness." In Milton D. Rosenau Jr., Abbie Griffin, George Castellion, and Ned Anschuetz (eds.), *The PDMA Handbook of New Product Development*. New York: John Wiley and Sons.

Hollander, J. 2002. "Improving Performance in Business Development." Ph.D. dissertation, University of Groningen, School of Management and Organization, Cluster of Business Development (The Netherlands).

Katzenbach, Jon R., and Douglas K. Smith. 1993. "The Discipline of Teams." *Harvard Business Review* 17: 111–20.

Kratzer, J. 2001. Communication and Performance: An Empirical Study in Innovation Teams. Amsterdam: Tesla Thesis.

Leenders, R. Th. A. J., and S. M. Gabbay. 1999. *Corporate Social Capital and Liability*. Norwell, MA: Kluwer Academic Publishers.

Muller, P. C. 1999. "Teambased Conceptualization of New Products." Ph.D. dissertation, University of Groningen, School of Management and Organization, Cluster of Business Development (The Netherlands).

7

Decision Making: The Overlooked Competency in Product Development

Mark J. Deck

Phase-based go/no-go decision making is the glue that holds the new product development (NPD) process together as projects enter the pipeline and move from phase to phase toward completion. Defects in that glue are almost always the key causes of product development performance problems. Ill-formed decisions, late decisions, ignored decisions, no decisions—all those failures stem from very correctable defects in the NPD decision-making process. Good phase-review decisions have six specific characteristics. They are timely, complete, appropriate, high-quality, operational, and expeditious. This chapter explains what those characteristics mean in concrete terms, and outlines the surprisingly straightforward elements of an effective NPD decision-making process. It concludes with a simple self-test for evaluating your company's NPD decision-making effectiveness.

Most companies rely for their growth on the flow of new products and services to market. If your company obtains 40 to 60 percent of its revenue from products you have introduced to market within the past three years, then product development is one of your vital competencies. The many processes through which you deliver more—and more valuable—products to market, and with greater speed, are your keys to success. But the one process that most critically determines your success as a product developer is your NPD decision-making process. Yet the very role of decision making in the development process is poorly understood and little appreciated. That's actually an understatement. Many companies don't even perceive decision making as an explicit element of the product development process.

This borders on the inexplicable. After all, what is product development but a series of decisions? Key decisions, such as when to start, stop, or change the course of developing a new product, service, platform, technology, or solution, can determine success or failure. The earliest or precursor decisions are of particular strategic importance. Is there a demand, whether active or latent, for the product you have in mind? In what way will this product be better than your competition's product? Is the product's creation within your capabilities? Will it consume more of your capabilities than you have the capacity to cover?

Does the product fit your business strategy? Can you complete and introduce the product while the market timing is right?

These decisions are often referred to as phase reviews, phase gates, or business decision points. Making these decisions also entails decisions about the proper priorities of the various opportunities in your development portfolio. These decisions about priorities—about the sequencing of your NPD efforts—can make or break a company whose lifeblood is product development. The fact is this: Product development *is* decision making.

We have diagnosed the causes of more than 350 companies' product development performance problems, and deficient decision making is consistently a basic cause. Not lack of talent, technology, resources, or experience. Not any of the limitations that outsiders might assume are the stumbling blocks to strong NPD performance. To reframe this in the affirmative, the beauty of NPD decision-making problems (as problems go) is that companies are not resource-constrained from solving them.

THE DAMAGES FROM DEFECTIVE DECISION MAKING

To equate an ineffective NPD decision-making process with incorrect decisions per se is to underestimate the harm it does. Consider the following:

◆ Ineffective decision making slows product development speed and responsiveness. It routinely results in new products being late to market and missing their windows of opportunity.

◆ Poor decision making leads to the inefficient use of resources and to an accumulation of development projects in excess of the development resources available.

◆ Poor decisions increase costs and hurt profits. The direct cost of an unsuccessful project is compounded by the lost opportunity to have spent the time and resources on a better project.

◆ Informal decision processes fail to give senior management an explicit role in the product development process, undermining most efforts to improve performance.

Decision-making speed is impeded for various reasons. For example, decisions that could be made expeditiously by a designated team of decision makers have to be made sequentially because there is no such team. The plodding sign-off procedures that attend sequential decision making are, in turn, a symptom of a more invasive malady: decision politics. Project teams that need a decision from management are often bounced between decision makers whose isolation from one another makes a consensus decision practically a contradiction in terms. But the most insidious danger of such splintered and slow-motion decision making is that the organization itself may become frozen by confusion over technological and market discontinuities: "What's going to happen? Should we go after this opportunity or not? Better wait and see what happens.

THE PIPELINE INDEX

Jon McKay, a former director at PRTM, developed a formula for maintaining an NPD pipeline at optimal volume. His work stemmed from the realization that managers had no quantitative indicator of pipeline underloading (rare) or overloading (typical). The formula, called the Pipeline Index (PI), is disarmingly simple. It compares the present volume of projects in a company's product development pipeline to the flow of development projects that the company has demonstrated it can handle. The PI is based on fundamental factory principles relating to cycle time, inventory, and throughput—principles that were learned the hard way by U.S. and European manufacturers in the 1970s and early 1980s. They learned, in a phrase, that "more on the floor doesn't mean more out the door." Overloading the factory simply created problems, ranging from high work-in-process inventories to excessive scrap.

Theoretically, a company with a PI of 1.0 has committed to precisely the volume of NPD projects it has proved it can handle. Our experience shows that actual PI values fall into four ranges:
◆ PI less than 0.9: pipeline underload, wasting capacity
◆ PI 0.9 to 1.3: optimal pipeline load, maximizing productivity
◆ PI 1.3 to 1.6: probable overload, somewhat degrading productivity
◆ PI greater than 1.6: definite overload, significantly degrading productivity

---

Wouldn't want to make a mistake." The opportunity roars by, and the company dimly realizes that its decision to wait and see was its final decision on the matter. Management tries to console itself by saying that it didn't have enough facts to make an informed decision, but could it have had the facts? In many cases, it *did* have sufficient facts. What it didn't have was the decision-making process discipline to make a high-quality decision under some pressure.

Phase-review decision making is, in a very real sense, the throttle on the NPD engine. Throttle back to slow the engine down, throttle forward to speed the engine up, while being careful not to flood the engine with too many projects. A steady hand on the throttle is essential. Making decisions to pursue development projects without the resources to complete them is irrational, but companies do it all the time. Our benchmarking research shows that this kind of head-in-the-sand pipeline decision making can cut a company's NPD pipeline throughput by half. (See sidebar on the Pipeline Index [McKay 1998].)

A sample performance comparison:

Industry	Best NPD Performers	Other Companies
Computers and electronics	1.0	1.6
Medical devices	1.4	2.2
Telecom equipment	1.4	1.7

To summarize, pipeline underloads are rare, and high loads correlate directly with NPD underperformance. The leading technology producers have come to view their R&D organizations as development "factories." The Pipeline Index provides a means of calibrating the NPD "factory's" utilization rate to its capacity.

The costs of ineffective decision making are often paid in hard dollars. If you decide to create a major new product and it doesn't sell, you've probably thrown away millions in development and launch expenses. But the damage extends beyond direct monetary measurement. Ineffective decision making slows new products' time to market and thus imposes lost-opportunity costs. How so? Every failure to make a complete, timely, and final development decision decelerates the entire NPD cycle. If you start a project and then decide to cancel it, you lose whatever you invested in the project (other than the project learning), but at least the waste stops there. Procrastinate in canceling the project, and you effectively decide to go on wasting resources for as long as you can.

The impact of decision making on product development speed, throughput, and cost is generally apparent. Less apparent is the fact that decision making is the glue that binds the various elements of the NPD process, such as effective cross-functional teams, a consistent and flexible work process, and robust resource-pipeline management. If we've learned one thing from hundreds of product development interventions, it's that if you don't establish and maintain an effective phase-review decision making process, none of the other NPD processes and practices you try to paste into place will stick.

When setting up the NPD process for the first time, or when making major changes to improve NPD performance, it is essential for senior management to drive the initiative. Establishing a world-class NPD process often involves making some deep and organizationally stressful changes. Raw nerves in the company's development culture may be exposed, and long-simmering frustrations may come to a boil. The longer your organization has been accustomed to slow, political, and incomplete decisions, the harder it will be to establish an effective decision process. Change of sufficient magnitude to make a real difference is driven forward by top management, or not at all. Consequently, managerial decision making at the highest level needs to be an *explicit* element of any product development system. If it's simply an implicit element, nothing is likely to change because there is no explicit role for management in the decision-making process.

What are the crucial qualities of good NPD decision making, the primary focus of this chapter? Experience has taught us that good decisions have six vital characteristics.

1. *Good decisions are timely.* The time to plot a ship's route is before it sets sail, not after it is hundreds of miles out to sea. The earlier top management becomes involved with a development project, the more it can do to help the project succeed. Don't procrastinate until the project has gained its own momentum and direction.

2. *Good decisions are complete.* A complete decision regarding a development project addresses all the issues that need to be addressed. Every-

one who needs to participate in a decision in order for the decision to be complete, clear, and credible must participate.

3. *Good decisions are appropriate.* A good decision is one that is strictly about the relevant issues. Decision makers must know and observe the difference between the many issues related to a development project and the specific issues that currently demand resolution.

4. *Good decisions are of high quality.* Consistently high-quality decisions are the next best thing to decisions that are always right. A quality decision is a decision about the right issues, based on the right information.

5. *Good decisions are operational.* A project decision doesn't mean much if everyone involved in the project doesn't know what to do in response. A good decision generates specific expectations and leads to specific actions by the project team and its supporters.

6. *Good decisions are expeditious.* Decisions, like products, need to be produced efficiently. Every member of the decision-making team must have done his or her homework before the team meets, and the decision session needs to follow a preset agenda and protocols. Preparation and organization are the keys to making expeditious decisions about development projects.

DECISIONS MUST BE TIMELY

Fact: The early project decisions have the most impact on final outcomes. With early decisions, you're steering the project. Late decisions are often efforts to countersteer, which can be very difficult and costly. The companies that consistently move products from concept to market without enduring crises en route are the companies that do their homework early. The more work you do at the outset to understand the market, understand the need, and make sure you have a product that will address the need, the better the course you can plot for a given development project. The same is true for the decisions that shape the product's charter: "What are we doing? Why are we doing it? How are we going to do it?"

One reason why NPD decisions are typically made so much later than they should be is top management's inclination to restrict its decisions to investment decisions. Executive management traditionally weighs in when it's time to invest, not before. The perceived "time to invest" in a project is usually at the beginning of the pilot production stage. But by that time, some very important strategic project decisions have already been made, a very substantial amount of money has been spent, the costs of project cancellation have gone up, and redirecting the project has become much more difficult. Would the project's resources have been better spent on a different NPD opportunity? It's a bit late to be asking.

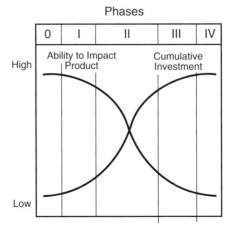

FIGURE 7-1. **Management impact and resource involvement by phase.**

Figure 7-1 makes the paradox plain. Senior management's ability to guide a development project declines at almost exactly the same rate that the cumulative investment in the project rises (McGrath, Anthony, and Shapiro 1992). Wheelwright and Clark (1992) found that the attention paid by senior management to development projects follows a curious pattern: Attention is minimal through the knowledge acquisition, concept investigation, and basic design phases; abruptly rises throughout the prototype-building phase; peaks at the beginning of the pilot production phase; and drops sharply as pilot production moves forward, only to spike again midway through manufacturing ramp-up.

Top management should be making frequent business decisions throughout the entire NPD process, from start to finish. Why? Because it's along the way to the "big" decision points that the enterprise must make certain that the project remains aligned with strategic intentions and business priorities. Early and frequent decisions also enable organizations to clear any roadblocks in the overall development process.

Just as the early decisions have the most impact on the fate of an NPD project, each and every project decision must be made on time. Sometimes there's decisional lethargy at the rough idea stage, when opportunities are still indistinct objects on the horizon. Decisions at this stage are understandably difficult. After all, companies are weighing different NPD opportunities under a great deal of uncertainty. That's one of the reasons why decisions and investments are phased with increasing depth of information and risk mitigation as projects mature and project investments increase.

Uncertainty is no reason to abdicate the responsibility to make those early decisions, however. Sitting back and waiting for development opportunities to sail into full view can take an awful toll on your development process speed. Don't let your project decisions slide until you reach the midway point between the design and development phases of a project. A good NPD strategy process includes both scrutinizing opportunities at a distance and compelling your organization to make decisions on those opportunities early in the development cycle.

Another critical aspect of making decisions on time is making them on an as-needed basis. Some companies are married to calendar-driven decision making. For example, management reviews each project every few months, whether a decision is necessary at the time or not. Or every project team is forced to prepare for a review every few months, whether a review is needed or not. Why? Simply because the project is up for review that month. Yet if a project needs a decision in the interim between fixed review dates in order to resolve a problem that's seriously impeding its progress, it languishes for another month or two until its review date rolls around. Would someone with a broken arm be expected to wait until his or her next scheduled checkup to have the bone X-rayed and set? High-performance NPD decision making is event-driven, not date-driven. Midphase decisions are made whenever project teams feel compelled to call an interim decision meeting. Those who make project decisions are available when events demand decisions, thereby avoiding unnecessary, low-value, and time-consuming project reviews.

DECISIONS MUST BE COMPLETE

Why do decisions get revisited—and frequently redecided? We see three reasons why decisions don't stick: the right people weren't on board to make a decision that the organization would find credible, the decision makers weren't clear about whether they were saying yes or no, or they didn't think through the implications of their decision. In all cases, the decision is viewed as incomplete, and thus it has no credibility.

One of the fundamental principles of effective product development is collaboration among all the functions that must contribute to bringing a project from idea to reality. This principle of cross-functionality applies to decision teams as well. A decision team, like a project team, must have representatives from the complete array of requisite business functions. It's equally critical to have every key business function represented during decision-making sessions. Otherwise, the decision won't be complete and final, either because some aspect of the decision could not be properly or expertly considered or because the functional resources could not be committed to the project. In that event, the decision effectively remains on hold until the absentee can be brought up to speed on all the issues.

Countless companies have solved the problem of how to create a balanced, credible, and responsible NPD decision-making body by forming what is typically called a product approval committee, or PAC (McGrath 1996). The PAC can be thought of as a company's NPD executive committee. It is explicitly charged with approving and prioritizing NPD investments from the portfolio of NPD projects for a given business area or unit. Because the PAC's work is strictly to make decisions, the group should remain small; six to eight senior executives is an appropriate size. A PAC typically includes the general manager

(GM), along with top managers from marketing, R&D, finance, operations, and customer care.

The PAC has the authority and responsibility to do the following:

- Initiate new product development projects
- Cancel or reprioritize projects as they move from concept to market
- Ensure that products under development fit the company's strategy
- Allocate resources to projects across functions from phase to phase as they progress

Product development is a horizontal business process that cuts across vertical functions. Many times, the functions are structured as separate organizations, which can make it difficult to bring together all the right functional executives. In some companies, these organizations don't come together until you get to the CEO level. Many companies make the mistake of setting up their decision team or PAC to include only the GM and his or her direct reports. That may be sufficient in some cases, but what happens when one or more functions are corporate functions spanning a number of GMs? What happens when a complete solution requires resources from the product organization and from a service organization under another GM or another product organization?

Since product development is a horizontal business process, it needs a horizontal governance and team structure. Project teams should be organized across functions, regardless of organizational boundaries. They should be drawn from the organizations that are needed to rapidly create the new product or solution, in just the way that a start-up operates. That enables the team to move with quickness and agility, because it has all the resources and functional representation it needs to succeed. Decision teams need to be cross-functional and cross-organizational as well. The PAC structure should be aligned with project structure. Just as different NPD projects may have different structures, they may also have different decisional needs. The PAC structure needs to flex in order to accommodate those differences.

In practice, that ability to flex is achieved by having a core set of PAC members who represent all key business functions on every project decision. When necessary, that core membership is augmented by others, such as a PAC member from a partner's business unit or a functional manager who controls a different set of resources being tapped by a certain project.

The product approval committee makes complete decisions because all its members are heavily involved, in real time. An effective PAC recognizes that there are really only three possible decision outcomes that can stick: go, no go, or redirect.

Go means "Proceed to the next stage. You'll have the resources and backing you need." It does not mean "We like what you're doing, so keep going for now, but we're not ready to fully back you just yet." *No go* means stop. Many companies don't know how to hold up a stop sign in a way that shows they really mean it. Instead of pulling the plug on a project, they allow it to burrow down and continue life as a sort of covert operation. Someone conveniently neglects to reassign the project's personnel or reallocate its resources. One company with zero tolerance for such "underground" projects is W. L. Gore & Associates. In an interview, Bob Henn, Gore's global leader of R&D, explained that the company classifies projects as either "above the line" or "below the line" (*PRTM's Insight* 2000). A below-the-line project might be raised above the line at some point in the future, but until or unless that happens, the project receives *nothing:* no work, no resources, no exceptions.

No-go decisions are often based on throughput or resource issues, which are facts of life in the NPD game. What's surprising is how many companies simply can't bring themselves to say no. They're afraid of demoralizing their project teams, not recognizing that nothing is more demoralizing than working on a dead-end project. When you're stuck on such a project, you pine for somebody with the courage to walk in and put the project out of its misery. Good NPD decision makers encourage project teams to recommend a no-go if the teams regard that as the best business decision. Early, crisp, and clean project cancellations are actually celebrated, to set a positive example for other project teams that may be keeping their misgivings to themselves.

A *redirect* decision is one that states, "We can't make a go/no-go decision at this time. We either don't have the information to make the decision or don't have the appropriate people to make it. We're going to defer the decision until the decision team is better informed and prepared. At the same time, we're directing the decision team [and/or the project team] to get itself better informed and prepared."

Eight times out of ten, a redirect decision results from a poor decision-making process or poor execution of the process. If all the decision makers are clear about what information they need in order to make a decision, if they have that information on hand, and if they're clear about the decision criteria, then they shouldn't have to make a redirect decision. When redirect decisions are made, it's usually because the information needed to make the decision wasn't detailed enough and/or because the decision makers didn't understand what it was going to take to make a decision. They hadn't "worked the details" in advance.

A legitimate redirect decision could result from something new happening in the market. Or it could result from something (hopefully nothing obvious) coming up at the decision-making meeting that no one had thought of before. Or the assessed level of risk, after due diligence, might simply be too high, requiring further risk reduction before proceeding. Such situations should be rare, however. A good product development process produces go and no-go decisions, with rare exceptions.

The third aspect of a complete decision—a decision that sticks—is awareness of the decision's wider implications. What are the interdependencies involved? How is the development project in question related to other development projects? A telecommunications equipment company, for instance, may be working on a network solution that involves a combination of components. If there is a delay in the development of one component or an element of that component, then the development of the entire solution is delayed ("for want of a nail . . .").

Beyond project interdependencies, there are almost always resource inter-dependencies. What happens when a phase review reveals that a stage needs to continue for an extra three months? For a complete decision, the PAC needs to assess the resource-availability effects on other NPD projects. No project is an island. Each lives in the context of others. Most companies have a whole pipe-line of projects they're managing, so how could they fail to grasp this resource-based interdependency of projects? Some companies establish phase gate systems where teams push their projects individually through decision gates based on self-assessment rather than management's decisions. This state of affairs encourages the assumption that decisions are made about single projects at each phase, in isolation. Decisions must be made about each project in rela-tion to other decisions about other projects and the overall availability of proj-ect resources. It's a juggling act. Unfortunately, product development managers have been known to keep their pet plate in the air while letting the other plates in the set fall all around them.

A complete development decision is a fully resourced decision, which is why resource management is so important. Without the requisite resources, you can review a project or even approve it, but you can't move on it. That's the distinction between a decision and a review. Think of each phase gate in a project's development cycle as a business decision point in a portfolio context, not simply as a review of the project's progress.

DECISIONS MUST BE APPROPRIATE

NPD decision makers cannot afford to become distracted by issues that are not immediately germane to the decision at hand. One of the most common mis-takes is to allow the business decision to become intertwined with a tactical review of the technical approach to the project. While the technical merits of a project may be relevant to certain go/no-go decisions, product approval com-mittee members should beware of becoming sidetracked by matters such as which technical approach to use, what the specific design of the product should be, or how to construct an appropriate channel partnership. Such decisions are appropriate for the project team to make. The PAC is not the project team. If it starts making all sorts of detail-level decisions, it undermines the project team's charter and discourages the team from thinking for itself.

A good way to keep go/no-go decisions appropriate is to clearly define and communicate the mechanisms for addressing different kinds of decisions. For example, technical reviews should be included as part of the standard integrated project template. Just as with a business decision made during a phase review, there must be clarity about who needs to be involved in the technical review, when it should take place, what its purpose is, and what preparation is expected. This helps clarify the distinction between a technical review and a business decision meeting, making both types of sessions more effective. It also helps keep decision makers exclusively focused on the directly relevant issues.

Another aspect of appropriate decisions has to do with matching the decision to the project phase. At early decision points, there is a tendency to ask for decision inputs more appropriate to later decision points. This is quite understandable, since there is so much more uncertainty early on. To deal with this issue, the decision makers need to agree that a first-cut business case is all that is needed for the first phase review and then stick to that agreement. If they don't, then their decision becomes a premature decision regarding the next project phase, effectively pushing the project into the next phase criteria. The next section of this chapter will deal more explicitly with the information and the decision criteria that should go with each phase.

DECISIONS MUST BE OF HIGH QUALITY

Decisions should always be right, but it's an imperfect world, so they need to be of consistently high quality. To make a high-quality decision, you need to have the right decision criteria and the right information relevant to those criteria. You don't want to overlook a key question or fail to consider a key issue. You'll never be guaranteed the right decision, but a good checklist can help you avoid making a poor-quality decision that will almost certainly wind up being a wrong decision.

Arriving at High-Quality Decisions

High-quality decisions result from timely and high-quality answers to the questions that matter. There are four types of indispensable considerations in the NPD decision-making process: market/customer considerations, technology/capability considerations, business considerations, and project management considerations. These considerations vary from phase to phase and need to be tailored to a company's context. Here is an illustrative set of questions that might be used to evaluate a project after the initial concept definition and feasibility assessment phase has been completed:

Market/Customer Considerations—Does the Customer Care?

- ◆ Are the planned product's critical functionality and performance characteristics fully and clearly defined?
- ◆ Do we understand who the target customers are and how well the product will meet their requirements?
- ◆ Is the product's planned differentiation from our competitors' products understood, and is it adequate? Will it provide enough value to lead customers? To end users?

◆ Do we have a solid understanding of our target market's maturity and growth, and of the best timing for market introduction?

◆ Is there a compelling win strategy? That is, does our market-attack strategy address how aggressive but achievable dollar- and sales-volume targets will be met?

◆ Have the product's intended lead customers been engaged to provide direct input regarding the product's design?

Technology/Capability Considerations—Can We Do It?

◆ Are the requirements for developing, testing, producing, marketing, distributing, and supporting the product well defined and acceptable?

◆ Are key technologies and capabilities identified and sufficiently understood to feasibly develop and produce the product?

◆ Does the architectural solution provide an adequate competitive advantage, and have lead customers been sufficiently involved in the architecture's selection?

Business Considerations—Do We Care?

◆ Does the proposed product support our product strategy?

◆ Is the product aligned with our market/product/technology roadmaps?

◆ Do the project financials justify the investment required? Do we understand the risks associated with achieving the cost and revenue targets? Do we have contingency plans that adequately address those risks?

◆ Are the project's program budget costs understood, achievable, and acceptable?

◆ Have we adequately considered all potential patenting, licensing, and other legal/regulatory issues?

Project Management Considerations—Are We Ready?

◆ Are the deliverables and milestones identified in the project plan achievable?

◆ Do we sufficiently understand the resource requirements (people and skills) and costs? Specifically, do we understand those requirements in detail for the next phase of the project?

◆ Does the plan for the current and following year provide sufficient funding for the project's staffing, capital, and expense requirements? If not, is the recommended plan for overcoming those insufficiencies acceptable?

◆ Does the proposed core team for the project have the right set of skills and experiences?

By explicitly asking and answering the right questions at the right decision points, the risk of making bad development decisions is greatly reduced. If you haven't asked the right questions at the right decision points, your decision process is apt to devolve into what Mike Anthony, a PRTM colleague, describes as "the rock game" (Anthony and McKay 1992). The rock game is a metaphor for management-dominated product development, based on whim. Mike gleaned the term from a fed-up project manager at a client site: "It's as if management has told us to go out into the field and bring back a rock. When my team does bring back a rock, management tells us that it's the wrong rock, and to try again. This has happened over and over. If they will just tell us what kind of rock they want, we'll bring it back!"

As my colleague observed, "Management in companies playing the rock game is waiting for the next brilliant product concept to be handed to them on a silver platter." The rock game drives product development teams to distraction, ties up critical development resources, and—unless someone happens to stumble across the perfect rock—reduces time-to-market speed to a glacial creep.

To avoid the rock game, the product development decision criteria need to be clearly defined, and agreed to by all concerned, *before* any NPD decision is made. Senior management has a vital role to play in defining and communicating these decision criteria.

Good decision-making processes use the four types of decision considerations—market/customer, technology/capability, business, and project management—to define different sets of specific questions that apply to different phases in the development cycle. For instance, early-phase business considerations may entail questions such as "Do the financials appear attractive?" At a later phase, the question might be "Are the sales projections by month sufficiently robust to commit our manufacturing capacity?" Figure 7-2 shows how the business focus of each decision changes by phase. The specific decision criteria need to change in order to reflect these differences.

Note that all the aforementioned decision criteria are phrased as questions. That's intentional. Many companies fall into the trap of establishing gate criteria that are simply deliverables. Crucial business-decision questions, such as "Is the marketing plan complete?" or "Is the product spec signed off?" are thus implicit rather than explicit. Granted, there are key deliverables that should be expected at certain junctures, as prerequisites for the project's advancement, but high-quality decisions are evaluative. They ask the gut-check questions that need to be answered, such as "Do we really believe that this development project will give us the lead in the market?" That's why all the key decision makers need to be present at every decision-making meeting. If a good decision simply required a sign-off on some gate deliverable, what would be the point of holding the meeting? Because these decisions can be tough, they require intense discussion of the facts by those who are accountable for making the decision.

Another aspect of a high-quality decision is an open and honest dialogue among all of the decision makers. This dialogue can be very difficult to generate if past decision making has been less than collaborative or highly politicized. An expectation needs to be set that all functions will be represented and all perspectives will be heard. How can management expect a team to go off and be successful after a go decision if the team has hidden disagreements or reservations about the project? Sooner or later such unresolved conflicts or ten-

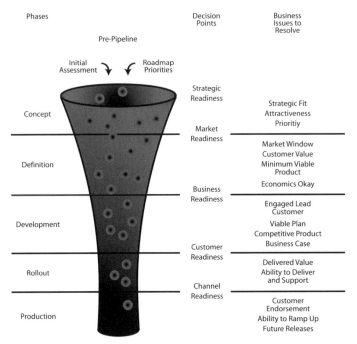

Phases
Pre-Pipeline
Initial Assessment   Roadmap Priorities
Concept
Definition
Development
Rollout
Production

Decision Points
Strategic Readiness
Market Readiness
Business Readiness
Customer Readiness
Channel Readiness

Business Issues to Resolve
Strategic Fit
Attractiveness
Prioritiy

Market Window
Customer Value
Minimum Viable Product
Economics Okay

Engaged Lead Customer
Viable Plan
Competitive Product
Business Case

Delivered Value
Ability to Deliver and Support

Customer Endorsement
Ability to Ramp Up
Future Releases

FIGURE 7-2. Example of decision focus changing by phase.

sions are bound to destabilize support for the project or slow its progress. A good practice to follow is to have only the PAC and the project team attend the decision-making meetings. The fewer "outsiders," the greater the chance of open dialogue that resolves the issues at hand.

It is also a good idea to have a set element of the meeting agenda during which the product approval committee confers in private to discuss any sensitive issues. For example, suppose that the PAC wants to discuss the possibility of canceling another project (Project X) to make room for the project in question. If the Project X team learns of the discussion, it is likely to draw its own conclusions and lose motivation. A better plan is to arrange to discuss such issues in private.

DECISIONS MUST BE OPERATIONAL

A decision is only as good as its outcome. If that seems obvious, then why do companies so often make decisions without being clear about what they expect to be done in response? An operational decision generates specific expectations. Who is accountable for meeting those expectations? What are the deadlines? What happens if an expectation is not likely to be met? Sometimes no one has a clue. If that's the case, decision makers can expect to have a hard time making good on their decisions. They are likely to end up micromanaging the personnel charged with following through on ill-defined expectations, tormenting them with monthly or even weekly status reviews that clarify nothing but the rising level of stress. Senior executives should concern themselves with operational business decisions. Their involvement in project progress reviews only squanders the project team's competency and undercuts its effectiveness. If it makes a go decision, senior management needs to unleash the project team, not supervise it to death. The team simply needs to know what is expected of it in terms of time/schedule, cost/margins, performance/quality, and budget/resources.

Within the boundaries of its authority, the project team is now empowered to do whatever it takes to drive the project to success until the next project decision point. (This is how fast-moving product development teams operate at start-ups.) If the project team goes "out of bounds," or needs to go out of bounds in order to achieve the results expected of it, then it needs to reconvene with the decision makers. This is an important point, because project teams sometimes go off in directions that are not aligned with business goals, or are forced to improvise beyond the bounds that were originally set in order to make progress. If a project team doesn't have a mechanism for calling an unscheduled meeting with senior management to explain its situation, or if senior management simply waits until its next scheduled meeting with the team to get it back on the right business course, it may be too late to deal with the problem. The sorts of issues that should trigger an unplanned, or interim, phase review of an NPD project include a significant schedule overrun, a product functionality shortfall, a major cost overrun, or a serious resource-availability problem.

An especially helpful construct to put in place is what is often called a project contract. The contract lays out the critical few parameters that define what is tolerably "within bounds." For instance, the contract might specify a launch date plus or minus four weeks. Now everyone is clear: "If it looks like we can make the date, plus or minus a month, we're 'in bounds.' If not, we need to reconsider the business decision, including its interdependencies."

If the ability to remain within the boundaries of a project contract is in doubt, it is an important responsibility of the project team leader to understand what action to take and when to take it. When a project is definitively out of bounds, the project team must assess the situation and the alternatives and then make a clear and honest recommendation on whether to continue or stop the project. The PAC must then consider that recommendation and make its go, no-go, or redirect decision.

A good project contract typically covers four areas: time, performance/quality, financials, and resources. It's essential for the contract to be limited to the critical factors that are deal breakers. Its purpose is not to rehash the project plan. It's also critical for the tolerances to be looser early on and tighter later on, as more becomes (or should become) known about the likely outcome. It is also helpful to establish a project contract template. Table 7-1 contains an example.

For a decision to be operational, everyone should agree to an explicitly defined NPD project contract. The contract works in two ways: It says that the project team is empowered to do whatever it deems necessary within the bounds of the contract but that the team is also accountable for initiating an interim

TABLE 7-1.
Project Contract Example

Project: Mustang
Phase 1 Review   Phase 1 Review Date: 01/19/99

	Targets	Allowed Variance
Next Decision Point (Phase 2 Review)	2/28/99	30 Days
Launch Date	9/1/99	60 Days
Customer Requirements	Per Plan	No Major Changes
Functional Requirements	Per Plan	No Major Changes
Product Cost	200K	5%
Development Cost—Phase 2	14.5M	10%
Capital Requirements-Phase 2	10.5M	10%
Total Project Budget	116M	20%
Personnel Requirements	Per Plan	20%
ROA	5.5M	20%
Gross Margin	45%	10%
New Unit Sales		
Year 1	350M	20%
Year 2	450M	20%
Year 3	550M	20%
Year 4	700M	20%
Core Team		

Team Member: _____

Team Member: _____

Team Member: _____

business decision if it is headed out of bounds. At the same time, executive management agrees under the contract not to meddle in the team's affairs and to provide the agreed staffing, funding, contacts, and so on.

DECISIONS MUST BE EXPEDITIOUS

Good decision making doesn't occur by happenstance. It's orchestrated. It involves a lot of advance work behind the scenes to ensure that the decision-making team shows up prepared to do its job. The little things that are done to prepare the team to make a decision add up. They include a preset agenda, some protocols regarding how the decision session will be organized and run, and the setting of a time frame for making the decision. The decision meeting itself need not be longer than two hours in most cases. Figure 7-3 depicts a typical agenda.

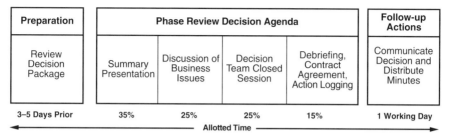

Preparation	Phase Review Decision Agenda				Follow-up Actions
Review Decision Package	Summary Presentation	Discussion of Business Issues	Decision Team Closed Session	Debriefing, Contract Agreement, Action Logging	Communicate Decision and Distribute Minutes
3–5 Days Prior	35%	25%	25%	15%	1 Working Day

Allotted Time

FIGURE 7-3. Example of meeting mechanics.

Typically, the participants agree to set aside about forty-five minutes for the presentation of facts, a half hour for discussion, another half hour for making the decision, and fifteen minutes for communicating it back to the project team. On that schedule, the PAC is clearly not going to try to make six decisions in four hours. By agreement, material is sent out for review by the decision-making team three to five days ahead of the meeting. That gives PAC members enough time to review the material and to anticipate the issues that will need to be worked through in the course of the meeting. Without this advance review, foreseeable issues that could have been resolved in advance are still unresolved when the meeting convenes. What's the outcome when this happens? Everyone takes the time and trouble to attend a decision-making meeting that fails to produce a firm decision. Instead, there is a redirect, and two to four weeks get added to the project's cycle time.

When a decision session is effective, the project team is informed of the decision and everyone agrees to the contract. Decision outcomes are recorded along with related actions and are communicated to all the parties involved within twenty-four hours. None of this happens by itself. All the participants in the decision session need to understand that the sole purpose of the meeting is to reach a timely, complete, and high-quality decision within the time allotted. They also need to understand their respective roles and responsibilities in making this happen.

It is helpful to establish clear protocols or operating rules for the entire decision-making procedure, and then to enforce them. Seeing to it that the necessary materials are sent to the decision team members ahead of time can sometimes be as difficult as it is for the team members to find the time to review the materials prior to the meeting. But difficult or not, it has to be done. If necessary, reschedule the decision meeting to a date and time when everyone can be ready. The same discipline applies to attendance. If a member of the decision team cannot make a meeting, he or she needs to choose between sending an alternative representative and opting out of participation in the decision. In either case, the agreed protocol needs to be that an executive cannot recall or second-guess the decision that is made. Table 7-2 provides a list of situations in which a specific protocol should be established, along with an example of a suitable response to each situation.

One great way to call attention to the level of preparedness is to track

TABLE 7-2.
Example of Decision Protocols

Situation	Potential Response
Core team leader unavailable	Find a different date/time
Core team member unavailable	Core team makes a call on whether or not to proceed
Documentation is distributed one day before meeting	Decision team manager makes the call on whether or not to reschedule and notifies participants with urgent voicemail
Presentation has a major hole	Decision team attempts to draw out the information and redirects team if unsuccessful
Presentation is unfocused or dragging	Decision team chair pushes presentation back on track
Team does not present a clear recommendation	Decision team chair requests recommendation and rationale; continued lack of clarity is cause for redirect

decision outcomes. By tracking the number of go, no-go, and redirect decisions, the number of decisions that are later recalled, and the length of and attendance at decision meetings, NPD decision team members obtain a better understanding of their own behavior and effectiveness. In our experience, the number of redirect decisions usually starts high but declines steadily as decision-making preparation improves. By the same token, decision meetings tend to run longer than planned for the first few months, then become increasingly time-efficient as the process matures.

YOUR DECISION-MAKING EFFECTIVENESS: A SELF-ASSESSMENT

How can you assess your company's decision-making maturity—that is, your level of proficiency at decision making? One way is to go back and review the most recent decisions you've made on development projects. Look at a sampling of a dozen decisions. Were they made in a timely manner? Were they made early enough? Were they complete decisions? Were there a lot of redirect decisions? You can get a useful sense of your decision-making aptitude, so to speak, by scoring your decisions against these criteria.

Below is a helpful diagnostic aid. Give yourself a 0 if the practice is not evident, 1 if the practice is understood but not always practiced, and 2 if the practice is in place and well used. A score of 36 to 40 is world-class. A score under 20 probably means that you have a major opportunity for cutting your costs and cycle time and improving your productivity.

Major Decision Process Elements	Subelements	Score
Decision Timeliness	Decisions are made early enough.	_____
	Decisions are event-driven.	_____
	Decisions are made quickly and without delay.	_____
Decision Completeness	Full cross-functional decision authority is consistently present.	_____
	Decision outcomes are clear (go, no-go, redirect).	_____
	Decisions include project and resource interdependencies.	_____
Decision Appropriateness	Decisions are focused on business issues.	_____
	Technical and design reviews are conducted separately.	_____
	Decision milestones and deliverables match the decision at hand.	_____
Decision Quality	Decision-making information is specific, complete, and relevant.	_____
	Dialogue is open and honest.	_____
	Evaluative and comprehensive decision criteria are used.	_____
	Projects are canceled when appropriate.	_____
Decision Efficacy	Specific targets and contract out-of-bounds parameters are set.	_____
	All functions support and follow through on commitments.	_____
	Action items and decisions are quickly communicated to all.	_____
	Out-of-bounds triggering accountability is clear.	_____
Decision Expeditiousness	Clear decision process logistics and protocols are observed.	_____
	Decision makers properly prepare themselves in advance.	_____
	Decision outcome metrics are in place.	_____
Total		_____

IMPLEMENTING NEW DECISION-MAKING PRACTICES

If your NPD decision making is not as effective as you want it to be, here are a few basic improvement steps:

1. Make sure that you have clarified who the decision makers are. Which individuals make up your product approval committee? Make sure that the GM of the business unit, or the senior executive for the portfolio of projects in question, is personally involved and chairs the PAC. Spend time with the prospective PAC chair in order to help the individual understand the extreme importance of this process element, and establish an objective assessment of the current level of decision-making effectiveness.

2. Conduct a training session with all of the decision makers in the PAC who need to be involved, in order to teach them what the determinants of effective decision making are. What has your organizational self-assessment revealed to be the weak points of your current decision-making process?

3. Work with the PAC to set up protocols for a more effective process. Deal with the details. Get the PAC members to discuss how they will operate and how they will enforce the needed discipline.

4. Work with the PAC to arrive at a clear set of decision evaluation criteria. There should be a set of questions and key deliverables for each business decision point that covers the four types of considerations described above.

5. Define the content of your project contract and have the PAC discuss and approve it. Make sure that the expectations and responsibilities in connection with out-of-bounds decisions are clear to everyone.

6. Establish a typical PAC meeting agenda and set rules for advance preparation.

7. Set up a PAC calendar, making sure that the decision team has regular time set aside for decision meetings as necessary.

MAKING PRODUCT DEVELOPMENT DECISIONS STICK

♦ *Identify the people in your organization who have the power to make or break the decision's credibility,* and involve them as necessary in the decision process. Every organization has both opinion leaders and opinion followers. If the leaders are behind the decision, the followers will follow.

♦ *Take control.* All too often, project teams and project champions have learned to ignore disagreeable product development decisions until they go away. Don't tolerate this behavior. Make it absolutely clear that go means go and stop means stop.

♦ *Examine the wider implications of every decision.* How will this development decision affect other development projects and decisions? What are the interdependencies? Decisions made in isolation tend to get redecided once their wider effects become obvious.

8. Broadly communicate the new decision-making protocol. Make it an explicit part of your product development system. Ensure that project teams and their leaders understand the new decision expectations, especially the decision criteria and contract parameters.

9. Coach your decision makers over the next 10 to 15 decision-making sessions to be sure that they're implementing what they're learning. Work behind the scenes with the decision makers and the project teams involved to orchestrate world-class decision making the first few times. After that, it will get much easier as everyone realizes how valuable these meetings are.

10. Track your decision-making outcomes. Every quarter, provide an update on your "outcome tally" and on your decision process maturity. Get that score above 38 and you'll see an enormous difference.

Effective decision making is a core skill in successfully developing new products from concept to market. It's also a learnable skill and an endlessly refinable one. If your decision effectiveness score is less than 30, you should consider spending more time on this element of your process than on other process elements that might not afford as much leverage. Effective decision making is incredibly visible. Done right, it demonstrates that management is "walking the walk." Done right, it will make all the other elements of your NPD system work better. Done right, it will bring you big payoffs in speed, throughput, and productivity. It's a critical competency that's worth persistent management.

BIBLIOGRAPHY

Anthony, Michael T., and Jonathan McKay. 1992. "Balancing the Product Development Process: Achieving Product and Cycle-Time Excellence in High-Technology Industries." *The Journal of Product Innovation Management* 9, 2: 140–147.

McGrath, Michael. 1996. *Setting the PACE in Product Development*. Boston: Butterworth/Heinemann.

McGrath, Michael, Michael Anthony, and Amram Shapiro. 1992. *Product Development: Success Through Product and Cycle-time Excellence*. Boston: Butterworth/Heinemann.

McKay, Jon. 1998. "Twin Indices of Portfolio Success: The Pipeline Index." *PRTM's Insight* 10, 2: 9–13.

PRTM's Insight. 2000. "Sustaining a Culture of Innovation: An Interview with Robert Lyon Henn, Global Leader of R&D, W. L. Gore & Associates." *PRTM's Insight* 12, 2: 10–17.

Wheelwright, S. C., and K. B. Clark. 1992. *Revolutionizing Product Development*. New York: The Free Press.

8

How to Assess and Manage Risk in NPD Programs: A Team-Based Risk Approach

Gregory D. Githens

All good NPD program teams have a collaborative process for risk identification, analysis, and response. Figuratively, these projects have a radar that looks out over the horizon for incoming threats. The radar allows the project to anticipate threats, make decisions to prioritize the threats, and apply appropriate countermeasures to effectively avoid or mitigate the risk event. Teams that can anticipate risks and respond effectively are the ones that develop agility and are more likely to succeed. The sidebar "Risks in the Firebird Project" shows how an NPD program manager was able to tactfully get her boss to recognize risks in a new product launch strategy.

TEN STEPS FOR TEAM-BASED RISK MANAGEMENT

This chapter describes a compact and practical team-based risk management approach for projects and programs. Figure 8-1 summarizes the ten steps at the project level.

I am using the term "program manager" for the person who leads an NPD launch effort and is ultimately responsible and accountable for success. In many NPD organizations, the project manager is the person who works with scheduling software and is nothing more than a coordinator. I want to ensure that I get and hold the attention of people who are thinking strategically about NPD.

Why a team-based approach to NPD project risk management? The NPD program is like a tightrope-walking team. If one person loses his or her balance, the entire team may tumble into failure. Each NPD team member needs to understand how his or her contributions fit into the balance, how to use the tools of his or her function or discipline, and how to execute his or her roles according to plan.

STEP 1: PREPARE

Perform this step before assembling the NPD team in a risk management meeting.

Step	Name	Activities
1	Prepare	Conduct initial research and agenda setting.
2	Build communications with common language	Build common language and vocabulary.
3	Generate a list of the team's concerns	Capture and record individual and group risks and issues.
4	Classify	Group risks and issues into categories for further action.
5	Analyze the risk	Assess the probability and impact of risk events, and evaluate issues.
6	Prioritize risks and issues	Quantify stakeholder risk tolerances and develop a ranking of risks and issues.
7	Plan risk responses and manage issues	Determine practices used for risks, if they occur, and issues. Issues can be typically captured and documented as tasks in the work breakdown structure.
8	Integrate risk responses in program strategy and document project baseline commitments	Identify/accept buffers and contingency in the final program plan. Issues management flows into the work breakdown structure.
9	Execute and control the risk response strategy	Implement planned actions.
10	Learning from risk management activities	Capture learning and build a culture for effective risk taking and risk management.

FIGURE 8-1. Summary of NPD project risk management steps.

NPD performance depends upon cross-functional contributions. Thus it is essential to perform risk identification collaboratively and systematically in a focused session, *not* with the program manager shuttling from individual to individual in one-on-one meetings.

A certain amount of homework is necessary before any team meeting. Here are six recommended tasks for the program manager and the meeting facilitator:

♦ *Confirm the "right." Consider these questions:*
  ➤ *Do you have a real project?* Many projects are launched as half-baked ideas.
  ➤ *Do you have access to the right people?* Many people do not participate in early stages of project planning because their experience is that their assignment will change. Thus they are reluctant to participate in anything that is not product design or product launch.

RISKS IN THE FIREBIRD PROJECT

Here is an example of how one individual, by asking a few simple questions, was able to create an awareness of risk that turned a possible disaster into an excellent result.

Wendy Chou was project leader for the Firebird project for Spectrum Corporation, a global information technology company (all names are disguised). Firebird was a software application development project that was integral to the technical and commercial performance of Spectrum's product strategy.

Early one week, Wendy excused herself from a training class to attend a meeting with her functional manager, Elise MacDonald. Elise was a brilliant, sometimes egotistical engineer, and often stubborn about her point of view. Elise wanted to explore adding a new set of features, which would offer the opportunity to attain proof of concept for another emerging technology.

As the week progressed, Elise's interest in the Firebird project increased. Wendy became concerned that the new features and technology would cause a complete rethinking of the product architecture, but she kept her thoughts to herself. Elise determined it was time for a preliminary technical review meeting with the team. As the issues unfolded it became clear to all that the project was undergoing a major rethinking. Still, Elise confidently pushed forward.

Murphy's Law (anything that can go wrong will go wrong) had struck. Wendy got an awful feeling in the pit of her stomach. If she defied Elise, she would raise an unpleasant conflict. If she acquiesced, she would be ensuring a difficult, time-pressured scramble that would require heroic effort. Wendy looked at her team members, saw their trepidation, and knew she would have to summon the courage to confront the dilemma.

Remembering a technique that she had just learned; Wendy stood and walked to a nearby flip chart. "Okay, assuming this new product architecture is our direction, let's jot down a few risks that we must accept or manage." Quickly her team members chimed in: "It will take more specialized resources." "Users will need added training." "It creates a dependency on the work product of overseas units." "There might be other unidentified stakeholders." "Any delay opens a window for competitors."

Listening as Wendy and her team brainstormed risks, Elise's eyes opened and she exclaimed, "Maybe we need to reconsider." She even added a few risk events of her own. Quickly everyone remembered that the project had started with a deadline and a just-do-it attitude. The team had skipped planning and was working from a list of dates and wishful thinking.

Soon the group realized that they lacked good market and customer data. They invited their marketing representative's participation. He validated some of the risks but also helped clarify the demands of the market (and opportunities). While the risk identification stimulated a lot of additional work for the team, they agreed that it led to one of the most successful project introductions in Spectrum's history.

The Firebird project offers many lessons to other NPD projects. I congratulate Wendy for taking the initiative by asking some simple and basic questions that stimulated risk identification. She included her technical team, her director, and the marketing organization. She engaged the cross-functional team—one of the principles of NPD performance.

> *Have you invited key suppliers?* In my experience, my biggest frustrations occurred when I did not get suppliers involved in projects early enough.

> *Do you have cross-functional representation?* A common mistake is delaying the involvement of the downstream functions such as procurement, testing, sales, and manufacturing.

> *How realistic is the value proposition of the proposed product?* The quickest way to create cynicism is to champion an idea that does not make sense for an existing or potential customer. If you are creating a radical innovation or trying to disrupt a market, make sure that people understand that purpose.

If you are the program manager and have answered no to most of these questions, I would sit down with the executive sponsor and get commitments before making a major commitment to the risk analysis session.

◆ *Identify stakeholders.* There are always multiple project stakeholders. Examples of commonly overlooked NPD stakeholders include product service and support, legal, purchasing, advertising, logistics, safety/environmental, and insurance. Each has a different perspective on risk and a different risk tolerance. Many stakeholders are not initially considered, and their eventual outspokenness sparks many problems. Their input is necessary to do a good job in Steps 5, 6, and 7, which have to do with analysis, prioritization, and response planning.

◆ *Establish your own sense of the gaps.* Identify what you expect to be the major gaps in resources, skills, and other enablers of success. How might others' evaluations of gaps differ from yours? Work by David Perkins at the Harvard Graduate School of Education suggests that something like 90 percent of all errors in thinking are errors of perception. Discipline yourself to examine a wide range of stakeholder perspectives; you will be rewarded with powerful insights.

◆ *Research past projects.* NPD organizations tend to make the same mistakes repeatedly. The goal is to avoid repeating past mistakes so that the project team can focus on innovating. From what was learned in past projects there should be a body of knowledge of common mistakes and problems, which typically include unstable or changing requirements, resource flux, unclear roles and responsibilities, and problems with interfaces. The preceding items are frequent frustrations, and you need to establish a vision that the current project will be different. You should also try to identify opportunities to leverage other organizational capabilities and to create new technologies, strategies, and intellectual property.

◆ *Conduct a sanity check on project planning.* Avoid using risk analysis as a substitute for project planning, because the standard project planning tools do a good job of identifying known tasks. You will get the best results from risk analysis and management if you have a rudimentary plan in place that documents phases and tasks, resources, roles and responsibilities, task timing, and preliminary definition of requirements. Also, spend some time to confirm that the project plan is aligned with the product launch and positioning strategy. Give thought to this question: How is success defined and measured? I have seen many NPD organizations, accustomed to launching derivative products, fail at more radical innovations because they did not anticipate or recognize critical distinctions. This sanity check should include evaluating product concept, product requirements, and the proposition of competitive value. Save risk analysis for focusing on what is unknown. Risk management is much more than the completion of checklists.

◆ *Communicate the purpose.* You need to ensure that the participants understand the purpose of the meeting. Figure 8-2 provides a sample agenda. Send out your announcements well in advance. Allocate sufficient time for the session; it is easier to end the session early than it is to schedule the additional time needed to finish response planning. While you will hear "I don't have the time," remember that the adverse impact from one risk event can easily cause weeks of schedule delay and overrun. Recognize that there is a learning curve: If people are new to the risk management process, plan for *at least* a day. They will become more efficient as they develop fluency with the terms and tools. These decisions happen best in face-to-face environments.

In today's fast-paced global environment, people often meet virtually. While I have seen some successes, most virtual meetings are poorly facilitated and harvest only 5 percent of the potential benefits. Meet face-to-face at one location until you have built an experienced project cadre. Risk management is fundamentally a process of communications and group decision making.

Topic	Duration	Who Leads
Welcome/executive kickoff, including strategic significance/introductions	½ hour	Program manager, executive sponsor, facilitator
Icebreaker	½ hour	Facilitator
Gaps in performance	½ hour	Program manager and/or facilitator
Level set (common language)	½ hour	Facilitator
Listing and classifying concerns	2 hours	Facilitator
Risks and issues analysis and prioritization	3 hours	Facilitator
Planning risk responses and issues management	3 hours	Facilitator
Documentation and wrap-up	1 hour	Facilitator

Note that time durations are typical of a NPD team that has experience and has a good facilitator. Expect it to take much longer if this is the first time. Consider separating the work into two or three half-day sessions.

FIGURE 8-2. Sample agenda for the risk analysis and response planning meeting.

**VALIDATE TARGETS AND PROCESS: SOME QUESTIONS
FOR EXECUTIVES**

Neither executive exhortation nor the team's hard work will cause an unrealistic goal to happen. Pressure to achieve results causes all kinds of people to make mistakes. Watch out for the excesses of wishful thinking, bottom-line fixation, and macho images. Poor early decisions are especially significant because many commitments are difficult to reverse. Keep the focus on good decision making.

Accept for the moment the possibility that the NPD program has unrealistic expectations. Instruct the project team to validate all parameters of the target (e.g., time, product performance, and expenses). As senior managers learn to be better project sponsors, they probe deeper for the enablers of good project results. Ask the team to perform a risk analysis. Questions to ask: "What is the probability you could complete the project earlier? What is the probability you complete it on time? What is the probability you will be late? When can you get me the answers to these questions?" Be aware that people often panic when given due dates. They will tend to ignore the organization's process.

Other questions to ask: "Are our processes going to get in the way? What changes to the process might be necessary? What is the probability you will return to me saying that you can't fulfill all aspects of the request? When can you get me the answers to these questions?"

There are quantitative risk assessment techniques (supported by software tools) available that can help generate data to answer these questions. Be patient if your organization wants to use these quantitative techniques. It takes time to assemble quality data inputs and to learn to use the tools.

Finally, everyone, but especially the program manager, should keep a strategic, holistic outlook. The sidebar "Validate Targets and Process: Some Questions for Executives" provides some perspective.

You are now ready to assemble the NPD team in a meeting. You have framed the relevant knowledge, identified team members, and established the rationale for applying risk management to your NPD program.

STEP 2: BUILD COMMUNICATIONS WITH COMMON LANGUAGE

Step 2 exposes the team to concepts that are helpful in establishing effective team responses to risk. Common language creates grounding for good communication and prompt action. It is vital to align the team's mental models and lexicon before diving into analysis. If you do not establish a common terminology for the team, you will undermine the benefits of the remaining steps.

I like to start with a risk identification icebreaker, asking the assembled team this question: "Consider your firm as well as others. What makes NPD projects perform poorly or even fail?" The answers foster a common mental model about the strategic aspects of NPD. Too, the question creates thoughtfulness and stimulates deeper probing into the complexities of the project. Record the output on flip charts and hang up the charts to remind the team of sources of potential concerns. This icebreaker should take only fifteen minutes.

Ask executives to participate in the icebreaker exercise, as it decriminalizes risk, helps to point out the differing paradigms that people apply to NPD performance, and sets the stage for effective risk allocation. The goal is to develop leverage: The team needs to impress upon the executive what the potential risks and issues are, that *some* of these risks and issues require executive ownership, and that the executive needs to support the decision-making process by indicating priorities.

This icebreaker is a top-down approach to thinking about failure. It widens the perspective of subject matter experts. Most NPD people, especially those with strong technical backgrounds, tend to approach projects from a narrow specialist perspective and to be unconcerned with the integrated functioning of the system. By directly addressing success and failure, you can get them to focus on the strategic issues and not insular details.

Case study analysis, personal experience, and review of empirical studies suggest that there are a number of predictable, recognizable red flags that signal poor NPD project performance. Based on my experience and research, I created a simple diagnostic instrument (Figure 8-3) to help you identify and list a number of common causes of NPD frustrations. You can use this tool to assess your project's likelihood of getting into trouble. Feel free to change the weighting and the factors to suit your own situation. Look deeper for the root causes of the frustrations.

You do not need to spend large amounts of time on this icebreaker exercise of project/product failure analysis. I find it best to lead the program team toward project-relevant activities as quickly as possible. At this early stage, you are trying to move the skeptics a little outside their individual comfort zones and into a basic shared understanding of recognizable sources of poor performance. So move on to establishing common definitions for risks and for issues.

A risk event is a discrete occurrence and is describable as a cause and its associated risk consequence. There is a probability that the risk event will occur. A practical criterion for recognizing a risk is "maybe it will happen and maybe

Red Flag	True	False	Weighting
Insufficient staff			10
Lack of market information			9
Dysfunctional teamwork			9
Insufficient financing			9
Unrealistic time frame			7
Lack of focus			6
Lack of management support			5
Technical limitations			4
A score of greater than 25 indicates a project heading for trouble.			

FIGURE 8-3. Red flags for NPD programs.

it won't." If that risk event occurs, then there will be a consequence. All three elements (event, probability, and consequence) are necessary to define a concern as a risk.

In contrast to the discrete and probabilistic definition of risk, issues are certainties and are evaluated and managed differently than risks. If something has a 100 percent probability of occurrence, it is not a risk but a certainty. Therefore, if you're sure it is going to happen, or if it has already happened, it is an issue and you must proactively resolve the issues. I will provide additional insight on this crucial distinction in Step 3.

Here is a helpful technique for orienting the team and clarifying terms: Draw a figure like the one in Figure 8-4 on a flip chart page. As you draw, explain how the team needs to identify general concerns and subsequently resolve them into lists of risks and issues. This process removes sources of anxiety and creates actionable work items.

In clarifying the project team's lexicon, use familiar layperson-type examples before getting into more complex business and technical examples. Illustrate with some commonplace examples such as flat tires, computer crashes, or lost luggage. For an example of risk, consider the following: It is the start of the day and it is partly cloudy. You check the weather forecast and see there is a 50 percent chance of thunderstorms today. You define the risk event as a thunderstorm and assess the probability of occurrence and the potential consequences. Prudently, you decide to include a contingency provision in the day's plan: You take an umbrella with you in case the risk event occurs.

For an example of an issue, assume that you look out the window and see that it is raining heavily. Rainfall is no longer a risk but a certainty that you must manage. As a certainty, the issue demands resolution by accepting ownership and actively working the issue. As a rational and prudent person, you do not foolishly deny the presence of rain or think that you can superhumanly dodge the raindrops. You put on your raincoat and open your umbrella!

The presentation of common language should also include a brief introduction of the four generic approaches to risk response planning: avoidance, mitigation, transference, and acceptance. I define and discuss these terms in Step 7.

Good leaders establish an effective culture for the team, and risk management can be a catalyst for that. There are many characteristics of effective team

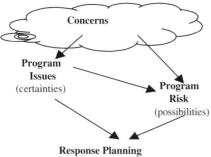

FIGURE 8-4. Decompose concerns into risks and issues for subsequent response planning.

culture, and a few characteristics would include open-mindedness, honesty, authenticity, a balance of inquiry and advocacy, patience and consideration for others, and passion and willingness to commit. The Texas Instruments example at the end of the chapter provides more insight on the relationship of risk management to effective culture. Paradoxically, organizational culture in many U.S. organizations (for example, be optimistic, confident, cheerful, competent, and cooperative) is a source of risk because many people are wearing a happy-face mask and suppressing their reservations. Leaders have to help individuals overcome their reluctance to raise NPD concerns that stem from their fear of being labeled as "not being a team player" or "not being positive." A supportive team culture helps ensure that people focus on the success factors rather than developing defensive routines that result in finger-pointing.

To summarize, in Step 2 you established the common language that is essential to good team development. These terms include issues, concerns, risks, probabilities, certainties, response planning, avoidance, mitigation, transference, and acceptance. You also know about tools such as a risk icebreaker and risk diagnostic questionnaire. All of these have the benefit of helping people articulate and grapple with project success factors.

STEP 3: GENERATE A LIST OF THE TEAM'S CONCERNS

The goal of Step 3 is simple: to anticipate threats to success by generating a list of risk events and a list of issues. By recognizing hazards, the team can take steps to avoid them, react more quickly and effectively, or turn the threats into opportunities.

Figure 8-5 presents a list of proven risk identification techniques. These tools help individuals and teams reframe their mental models and avoid potential decision-making blind spots. I also list a few advantages and disadvantages. Note that there are common disadvantages to each of the techniques: They take time to learn and time to perform, and they create psychological discomfort that may result in interpersonal conflict. Some of these techniques are common and some are not.

As the team generates concerns, they should segregate the concerns into risks and issues. Risk events are discrete and meet the criterion of "maybe it will happen and maybe it won't." Issues are statements, sometimes vague, of concerns that have happened, are happening, or are certain to happen.

Many people often regard the difference between risks and issues as an unimportant fine point. Jeff Tucker, a program manager with Acterna, a manufacturer of test equipment for the communications industry supports this: "At first, NPD team members don't see the relevance of separating risks from issues. No matter how many examples you give them to define risks and issues (e.g., a flat tire is a risk, and driving on a near-empty gas tank is an issue), people will still struggle with distinguishing between risks and issues." However, experience shows that the distinction is very helpful.

Try to make the concerns as specific as possible and you will save the team

Technique and Description	Advantages	Disadvantages
Assumptions analysis: Assumptions are factors that, for planning purposes, are true, real, or certain. These assumptions may be invalid during implementation. Assumptions are incrementally altered, or totally invalidated, and the resulting impact is assessed.	Can be supported with quantitative analysis (e.g., sensitivity analysis)	Vague and ambiguous
Brainstorming: The cross-functional team generates ideas about project risk, often done with the support of a facilitator.	Energizing and creative; easy, familiar	Results may be perceived as ambiguous and difficult to organize
Checklisting: Based on historical information and knowledge of the system, checklisting helps to make sure that past mistakes are not repeated.	Quick	Tedious, often overwhelming, and may result in uncritical "checklist mentality; not prospective
Delphi: This technique is best used for complex, messy problems where there are numerous perspectives on the nature of the problem and the solution. For example, a question could be "What is the future for electric-powered vehicles?"	Remotely extracts expert opinions and builds consensus	Time-consuming; results are often ambiguous
Document reviews: Assemble and analyze documents for content (literal meaning) and context (application). Include initial plans, assumptions, past experience, statements of strategy, requirements specifications, and review.	Establishes common basis for further analysis and decisions	Tedious
Diagramming: A variety of techniques to make models of stock, flows, causality, decisions, and assumptions more explicit. (Diagramming techniques include fishbone, flow charts, influence diagrams, relationship mapping.)	Analytical; complements other techniques	Difficult to get agreement on accuracy.
Independent assessment: External assessors use a variety of techniques to inspect the project.	More objective, comprehensive; domain expertise often required	Less ownership
Interviewing: An analyst elicits knowledge from others through guided questions and probes.	Depth of inquiry and follow-up questions possible	Bias
Triggers: A list of symptoms or warning signs that indicate a risk event has occurred or is likely to occur. Analogous to dummy lights on your automobile's dashboard.	Effective and efficient once implemented	Requires integrated understanding of system and tolerances

FIGURE 8-5. Identification techniques.

196

a lot of frustration in the later steps. The cause-effect structure, as illustrated in Figure 8-6, is helpful in understanding and recognizing a risk. This example shows the outcome or effect called "delayed launch." Note the two potential causes, which are typical of what emerge during a risk identification brainstorming. They are risk events because they meet the test of "maybe they will and maybe they won't." The "supplier late" factor results from deeper potential causes, and this is where the analysis gets interesting. The item titled "supplier given insufficient lead time" is not a risk but an issue that likely belongs to the buyer. Thus, by the buyer not providing sufficient lead time, it has created the potential for a delayed launch. Now the NPD team can step up and work, in a partnership, to ensure that the supplier can meet obligations. I could make a similar argument about design: If the supplier does not have a modular and robust design capability, then it can't respond to late changes in requirements. To minimize risk, the NPD team needs to look across the entire supply chain and work proactively to build capability, instead of finger-pointing and blaming suppliers.

Diagramming techniques can be helpful for modeling the chain of cause and effect. Too, diagramming techniques help to point out the systematic nature of product development and emphasize that responsibility for success is shared by many stakeholders.

A common question is "When do we stop the cause-effect decomposition?" The answer is to work backward through the chain of causes to a point where an individual has influence on the risk event's occurrence. When the team has judged that they have analyzed to a sufficient level of granularity, stop and

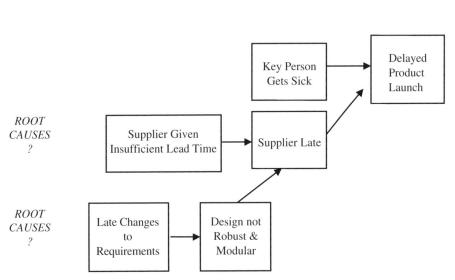

FIGURE 8-6. Example of a cause-and-effect diagram for a delayed product launch.

document the risk event. A list of discrete, specific risk events is the input to the quantification and assessment process.

Generally, early in the NPD program you will find a larger quantity of issues than risks. This is because the team usually has an incomplete understanding of the product development effort, so it is hard to develop specific cause-effect statements and natural language is ambiguous. With practice, the team will develop the skill to extract specific risk events from issues. Nevertheless, the real value of the process is getting the cross-functional NPD team to articulate their concerns in language acceptable to everyone else.

This ideation of concerns—and stating them specifically—does not have to take a long time and frequently can be done in less than an hour. Often the team is impatient and reluctant to do a comprehensive identification of risks. However, it is possible to sustain energy by reminding the team that if a risk is not identified, the team ends up with a workaround that will consume precious time, effort, and money. It's a judgment call on when to stop ideating.

The biggest dangers result from errors of omission rather than errors of commission. An example of an error of omission is lacking or ignoring essential technical or functional input, whereas an error of commission might be allowing a design review meeting to turn into a problem-solving meeting.

You now have a list of team-identified concerns for further analysis. The more specific you can make these risks and issues, the better it will be for analysis and response planning.

STEP 4: CLASSIFY

In Step 4 the team takes its list of risks and list of issues lists and evaluates the sources of them. The classification process helps the team to clarify what they know and what they don't know so that they can deepen their analysis and determine an action plan. I find the following labels work very well for classifying the sources of NPD program risk:

- ◆ *Technical risk sources:* those sources associated with the design or operation of the project's product and/or production processes
- ◆ *Logistic risk sources:* those sources associated with supply, procurement, inventory, maintenance, and support.
- ◆ *Programmatic risk sources:* those sources associated with obtaining and applying program and project resources such as technical experts, project-specific tools, and budgets for capital and project expenses
- ◆ *Commercial risk sources:* those sources of risk that change assumptions affecting revenues, costs, market share, profitability, and so forth

Avoid classifying risks by functional area (e.g., engineering risks, manufacturing risks, and marketing risks) because this practice perpetuates finger-pointing and blaming others. Most project and product-performance problems occur at interfaces.

Often it is worthwhile to deepen the analysis to determine strategic and operational factors. Operational risks are those that affect delivery time and development expenses. Technical problems are usually operational, not strategic. Most technical problems can be solved given sufficient time and resources. The team is developing products with time-functionality-cost trade-offs. While in technical development, an NPD team will spend most of its time with operational risks.

However, do not overlook the strategic concerns that affect project selection or the ability of the project to meet the business objectives, such as profitability or customer satisfaction. Periodically, the team should evaluate risk events against the customer value proposition. Project cancellation is a legitimate response to some risk events or issues. I return to this point in Step 9, "Execute and Control the Risk Response Strategy," where I discuss how risk is key to operating a stages-and-gates NPD process.

Step 4 is the start of converging on actionable responses. By classifying the concerns into common groupings, you have enhanced the team's ability to analyze, prioritize, and manage the risks and issues.

STEP 5: ANALYZE THE RISK

Recall that risk events are discrete events that meet the test of "maybe it will happen and maybe it won't." Risk analysis is the process where the NPD team estimates both the probability and the consequences of the risk event. The output of the calculation is an estimate of the overall risk exposure. The formula for a risk exposure is:

$$\text{Probability}_{event} \times \text{Consequences}_{event} = \text{Risk Exposure}$$

It is important to make an overall assessment of each risk event. This is done by multiplying the probability value by the impact value to yield an expected value. For example, a high probability times a low impact results in a medium overall risk event. This technique is suitable for qualitative analysis as well as quantitative analysis of the risk exposure. The quantitative approach usually attempts to accurately estimate probability and define all consequences in monetary terms. Most NPD teams, realizing the work and ambiguity involved, use the qualitative approach for analysis of project-level risks.

Figure 8-7 illustrates a risk map, showing six risk events arranged by probability and impact for overall risk. There is one high-risk event, three medium-risk events, and two low-risks.

Figure 8-8 shows the format for structuring an analysis of risk. I usually do this on a flip chart, but it can also be set up on an overhead transparency. Some people swear by projecting a spreadsheet with a computer, but I find that the computer stifles discussion of the subtle, important assumptions; flip charts are much better.

The quantification process helps the NPD team deepen its understanding of which risk events require more attention. Recall the words of Lord Kelvin:

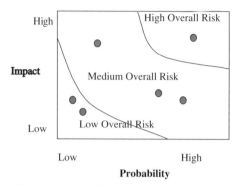

FIGURE 8-7. A probability-impact matrix for evaluating overall risk.
Each point is a risk event within a project, and they are plotted in regions of high, medium, and low overall risk.

"When you measure what you are speaking about and express it in numbers, you know something about it, but when you cannot express it in numbers your knowledge about it is of a meagre and unsatisfactory kind." I typically advise teams to scale the rating by anchoring the extremes to the definitions of project success and failure that they developed during the icebreakers in Step 2. Assessing the risk impact helps the project team to evaluate critical success factors.

In nearly every risk clinic I've facilitated, I've had at least one individual object to the large number of potential and trivial risk events that may be present. They fear the volume may overwhelm them with work and waste their time. I respond by introducing the concept of *residual uncertainty*, which is the uncertainty remaining in the project after a reasonable analysis. "True, it is not practical, and probably not possible, to identify everything," I tell them. A reasonable analysis would consider that the team has a limited capacity for attention to risk. The main work of development is designing and developing communications. Further, some events have extreme consequences; thus extra analysis is justified for events with low probability but high criticality, whereas other risk events, if accepted, have effective workarounds.

IBM has made project risk management techniques an important part of its NPD process. Says Nancy Whetstone of IBM in Rochester, Minnesota, "I tell teams to first establish your risk review cycle," for example, quarterly, monthly, or at each development phase. This helps to gain agreement on the visibility and benefits of risk management. After determining the risk review

Event	Probability$_{event}$	Consequences$_{event}$	Risk Exposure (Probability$_{event}$ × Consequences$_{event}$)	Rank

FIGURE 8-8. Chart for structuring risk analysis.

cycle, Nancy advises, "the team needs to clarify what high, medium, and low mean for probability and impact. For example, the team may determine that high probability means an 80 percent or higher chance the risk event will occur." She also advises examining time, expenses, and targets in the impact analysis.

I offer this caution when you see very high probabilities: You may be looking at an issue rather than a risk, and you will want to use the issue management technique rather than the risk management technique described in Step 7. In my experience, there are few technical risks in projects; rather, there are product performance issues that are resolved through design trade-offs. Almost any technical problem can be solved given enough time and money.

Recall that I define operational risks as those that affect development time and expenses. Figure 8-9 is a table for assessing NPD program operational risks in each of four potential impact areas: development expenses, target product cost, schedule, and product performance. It is important for the team to anchor impact ratings to their own business environment and scale them specifically to each project and its organizational context (e.g., a two-week delay may be unnoticed in some companies or projects but disastrous in others).

Radical innovations need larger risk tolerances, while incremental innovations typically have smaller tolerances. Incremental products have more known and knowable items and thus more predictability. For example, a radical innovation might tolerate a 50 percent slip in schedule or budget, whereas an incremental innovation would consider a 10 percent slip as high-impact.

Figure 8-10 lists several NPD project risk quantification techniques that the project team can use to understand the probability and impact at a deeper level.

Project teams often perform structured reviews (also called "red teams" or "walk-throughs") to enhance their risk analyses. The team invites an independent reviewer to probe the validity of estimates, verify boundaries and interfaces, confirm the categorization and specificity of risks, and confirm ownership. The team retains the responsibility for project performance but gets

Impact Area	Impact		
	Low	Medium	High
Development expenses	<15% overrun of target	15%–30% overrun	>30% overrun
Target product cost	Insignificant overrun	3–5% overrun	>5% overrun
Schedule	Insignificant delay	>6 weeks	>3 months
Product performance	Insignificant loss of functioning	Defects present but manageable	Significant performance issues

Operational risks are those that affect the project team's ability to meet targets.

FIGURE 8-9. Impact table for assessing operational risks.

Technique Description	Sample Application in NPD Programs
Decision Analysis/Decision Trees: Applicable to selected discrete events and limited risk outcomes	Investment in well-defined alternatives
Expected Monetary Value (EMV): When combined with decision analysis, can assess consequence of risk (Probability × Impact = EMV)	Selection of product configuration options
Failure Modes and Effects Analysis: Starting with a functional analysis of a product or process, the team identifies failure modes, assesses the impact of the failure mode on the customer, assesses probability, and assesses detection	Used to evaluate the reliability of products and processes; common in complex mechanical products
Monte Carlo: A computerized simulation of the project model for assessing the probability of meeting time and cost targets	Develop estimates and confidence intervals for completion dates and expenses
Program Evaluation Review Technique (PERT): Useful project technique for probabilistic evaluation of time and cost; requires some estimate of variance in the input data	Estimate confidence in time and expenses by estimating an average project duration and a variance (the variance is a quantitative measure of cost or schedule risk)
Real Options Analysis: Allows managers to identify risk components and decide which ones to hold, hedge, or transfer; preserves flexibility and allows for frequent adjustment; can lead to different conclusions than those from traditional discounted cash flows	Investment decisions as projects progress through development funnel
Scenario Analysis: Develop different alternative scenarios of the future	Strategic planning of different levels of competitive response
Sensitivity Analysis: Examine the degree of change of an output with the change in inputs; the more sensitive measurements are perceived as riskier	Common for testing assumptions in business case for pricing and profitability analysis; easily done with a spreadsheet.

FIGURE 8-10. Risk quantification techniques.

a more objective external perspective on their true status. The review forces the project team to reflect and develop self-awareness of priorities, and it keeps them from rushing to premature closure.

NCR has an excellent approach to understanding opportunities and threats, called the risk and opportunity assessment model (ROAM), to evaluate new business opportunities. ROAM gauges a project team's perception of the risk at the beginning of a project. The analyst examines threats and opportunities. Examples of threats include low customer commitment, schedule urgency, instability of requirements, the delivering organization's lack of capability, and technology immaturity. Examples of opportunities include alignment with strategic direction, good revenues, high profitability, high growth potential, new markets and technology, good resource utilization, good fit with needs,

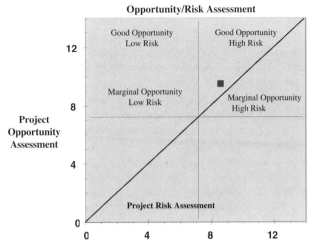

FIGURE 8-11: A risk and opportunity assessment model output.

and high value added. Then the analyst plots the project's ROAM score on a chart like that shown in Figure 8-11. (Please note that the underlying calculations are proprietary to NCR and cannot be shared in this chapter.)

Terry Murphy, NCR's risk manager, is a veteran of hundreds of project risk assessments. He notes, "The shortfall with this or any tool that quantifies risk is that people 'game' the numbers; they manipulate the input values to the outcome they want." If a salesperson wants to secure a new account, he or she has a natural incentive to discount the threats and highlight the potential gains. The value in the analytical tool is in surfacing assumptions and recognizing biases. Terry stresses the importance of involving all functions, sales as well as engineering.

A final and important point is to beware analysis paralysis. It is common for people to get lost in details and quibble over small, subjective numbers. I keep ready a plastic overhead foil labeled "The purpose of analysis is insight for improvement" for when I see that the team has become bogged down in trivia. I also remind them of the concept of residual uncertainty, defined as the uncertainty that remains after a complete risk analysis.

In Step 5 you analyzed risk events for probability and impact. You now know about many analytical tools. However, be on guard for analysis paralysis. The real value is in developing a common understanding of risks by the NPD program team.

STEP 6: PRIORITIZE RISKS AND ISSUES

A typical NPD project can easily have hundreds of risks and issues, which is more than the team can or should try to manage. The next step is for the team

---

DETERMINE STAKEHOLDER RISK TOLERANCES

It is essential to determine stakeholder risk tolerances as part of the prioritization process. As a decision-reaching process, the project team discusses the results and establishes a cutoff level to determine which risks the projects will accept and which the project will avoid. Stakeholders have to accept and avoid risks.

Jerry Groen of Abbott Laboratories says, "To evaluate the level of risk and the project's returns, we capture the opinions of experts from all parts of the organization on the probability of success of a project: the project team, in management, and even outside the organization. The program manager acts as the facilitator."

Jerry adds, "We then summarize the data. We request that the participants provide a rationale for their estimate so that executives can understand the underlying rationale. We think that the original numbers have value in understanding the range of plausible outcomes. The stakeholders can then consider their personal risk tolerances. We don't try to get the outliers to congregate to the middle numbers.

"The intent of the process is project selection—the go/no-go decision," says Jerry. "Project evaluators continue to use it to advise management of the riskiness of a project early during project selection so that they can make a determination whether to continue. We also use it in project reviews."

While it might seem logical to determine stakeholder risk tolerances *before* quantification and prioritization, I find teams usually wait to develop the analysis and then discuss the data with stakeholders. In examining the range of potential outcomes under various scenarios, the stakeholders look directly at the relevant data and make an intuitive decision on what they can and cannot accept. Here are some good questions for assessing the individual or organizational risk tolerances:

◆ What is the magnitude of the upsides and the downsides? When do the risks occur?
◆ Can we reverse or recover losses? How much control do we have?

---

to prioritize the classified risks and issues. There are a number of techniques available for prioritization; weighting is probably the most common technique.

One of the more common ways to present risk prioritization data is the risk matrix, as illustrated in Figure 8-7. This, or the risk exposure calculation in Step 5, provides a simple way for developing a common NPD best practice, the top ten risk list. In prioritizing, always keep stakeholder risk tolerances in mind, as discussed in the sidebar, and keep the list of items compact.

STEP 7: PLAN RISK RESPONSES AND MANAGE ISSUES

Risk response planning (Step 7) is the process of developing strategies for reducing threats to the intended project and product outcomes. The team takes their prioritized risk list and begins response planning for the top-ranked risk.

As a frequent flyer, I know that most commercial airlines insert schedule buffers so that flights usually arrive on time. I have experienced occasions when the plane departs the gate thirty minutes later than posted and still arrives at the scheduled time. In addition to schedule buffering, pilots use response options, such as flying faster, that allow them to meet their arrival times. Airlines manage customer expectations by setting conservative, buffered schedules and then empower their pilots (who are analogous to an NPD program manager) with resources and authority to develop and apply responses.

Before delving into the practice of risk response planning, I want to discuss the too-common approach of passive acceptance of risks, which is defined as accepting the consequences of a risk by ignoring it or practicing wishful thinking. Bruce Horwitz of TechRoadmap is an engineer and veteran of many NPD projects. Bruce is referring to passive acceptance when he says, "People tend to do the easy stuff on their projects first, and then they sweep problems under the rug. A good leader will focus first on the tough things." When NPD teams "sweep problems under the rug," they have passively accepted the risk event.

Organizational culture often dictates the risk response. Everyone is familiar with the shoot-the-messenger syndrome, for example. In cultures where shoot-the-messenger occurs, individuals are reluctant to ask challenging questions of themselves, because it will expose their lack of knowledge and raise the specter of conflict. Identifying risks will probably lead to more work for individuals who are already overwhelmed. Passive acceptance of risks permits teams to stay in their comfort zone, which is often functionally siloed. This comfort zone often creates psychological denial, weak self-responsibility, and excuses for fire-fighting and crisis management. The effective program manager leads and encourages participation in initiating and sustaining the effort to manage program risks.

There are four generic risk strategies that the NPD team should consider (see Figure 8-12), as revealed in the following series of questions:

♦ *Risk avoidance.* How could we avoid this risk event? Are the hazards associated with the risk event so significant as to justify canceling the project?

♦ *Risk mitigation.* How could we reduce the probability of the risk event happening? How could we reduce the impact of the risk event on the project's success?

Avoid: Change the project plan to eliminate the specific risk events or conditions. By avoiding the risk, the project team removes a source of poor product or project performance. Avoidance can also include project cancellation. However, by practicing avoidance the organization may miss opportunity.

Mitigate: Reduce the probability or impact to an acceptable threshold. Strengthening a product's reliability (e.g., reducing mean time between failures) is an example of reducing the probability, and robust design (building in redundant or backup systems) is an example of reducing the impact.

Transfer: Moving the responsibility for the risk to another party. NPD examples include joint ventures, subcontracting, and purchasing insurance and warranties. Transfer is often the most "out-of-the-box" concept for many NPD teams, which have traditionally thought of NPD work as technical and market development.

Accept: Active acceptance is to develop a contingency plan. Passive acceptance leaves the project team to work around risks as they occur.

FIGURE 8-12. Generic strategies for risk response.

RISK RESPONSE EXPERIENCE AT NCR

Terry Murphy

Avoidance

While working with a customer to fully define a new application development project, I noticed that the customer planned to develop the new application on its current legacy platform and then migrate it to its new platform at a later date. The risks from a development standpoint were obvious and significant. Rather then accept the risk, I went to the customer, explained the risks, and offered a good business case for them to purchase a developmental platform for the new application, thus avoiding the risks completely.

Mitigation

While developing a new systems solution for a customer, we could find no way of realistically predicting whether end users would find value in the proposed solution and use it. To mitigate the risks, I recommended a controlled implementation that called for recruiting a few end users on a prototype system, then using the data gathered to finalize the solution.

Transfer

Once we needed a technology that was not within our core business. We were concerned about its performance. We elected to transfer the risk by identifying a niche vendor with the appropriate expertise and developing a highly structured third-party teaming agreement.

Acceptance

The implementation phase of a project's schedule was very aggressive, and the business case for that aggressive schedule was compelling. I developed a contingency plan that included a 16 percent risk pool for use at the discretion of the project manager, a list of prevalidated third party vendors that could be used as needed, and an aggressive schedule risk review plan.

◆ *Risk transference.* Could another organization better handle the risk? How could we contract out the risky work or the risk event?

◆ *Risk acceptance.* If we have to accept the risk, what contingency strategies are available to us?

Instruct the team to work through each risk and apply each response, even if the response does not initially seem feasible. The sidebar "Risk Response Experience at NCR," by Terry Murphy of NCR, provides examples of how Terry has applied the risk response strategies to his programs.

You will consistently find that the NPD team adopts the proactive strategies of avoidance, mitigation, or transfer. If they decide to accept the risk, they will have contingency plans for it. Residual risks are managed through a buffer established for that purpose or by accepting lower profits.

In Step 4, I discussed classifying risks into categories. Often the largest category is technical risks. I find that most technical risks involve product functions that do not perform perfectly. Thus technical risks are really product reliability issues. They are risks only in the loosest sense of the word. Here is where a common product engineering technique—failure mode effects analysis (FMEA)—is helpful. FMEA involves identifying product functions; judging the performance of those functions; selecting failure modes for the function (for example, intermittent failure); and assessing the probability of a given cause leading to the failure, its effect on the customer, and the ability to detect and correct the failure mode. FMEA is commonly taught in engineering schools and continuing education programs, and there are numerous references to it on the World Wide Web, so many of the engineers on an NPD team are already familiar with FMEA. The process I describe in this chapter is for program- and project-level risk; FMEA is better suited for technical reliability analysis.

Ideally, the team should assess each and every listed risk, but in practice, they select a cutoff level and address the risks above the cutoff level. Recall that IBM's Nancy Whetstone advises establishing this cutoff level early. My experience in NPD projects is that teams will generally choose to consider thirty to fifty risk events but plan responses for only about ten of them.

Effective NPD risk management has its foundation in Step 2, "Build Communications with Common Language." Recall that a risk is a discrete event with a probability and an impact, whereas issues are concerns that have already happened or are certain to occur. The team will get frustrated if it did not do a good job of understanding the difference between risks and issues and distinguishing between them. Because an issue is a certainty, the NPD team is concerned not with whether the issue will happen but with how to resolve the issue to closure. Acterna's Jeff Tucker comments, "As the teams work through the analysis, the risk-issue partitioning helps the teams see that they *collectively own* the responsibility for the success of the project. The risks and issues are theirs, and they can't rely on wishful thinking." He goes on to add, "The process of listing risks and issues helps the program manager build an effective project plan that has buy-in." Often the team can resolve issues simply by adding a task to their work breakdown structure and schedule. Also, the team will be frustrated in planning effective responses if they did not do a good job in Step 5, "Analyze the Risk."

Management of project issues is straightforward. The team starts with the top-ranked issue and further clarifies it. They next determine an individual owner and establish closure criteria for the issue. I find it extremely helpful to assign ownership for investigating and making recommendations, often with a different person deciding on the resolution of the issue. The program manager can keep NPD team meetings productive by encouraging the off-line resolution of issues and then using the team meeting to confirm the decision that the issue is now closed or needs to remain open. My experience with NPD projects is that the team will typically choose to manage about twenty or so issues as a team and delegate the remainder to individuals.

Step 7 is one of the most significant steps in this process. In it, you developed risks and issues response plans. There are generic strategies for risk response planning: avoidance, mitigation, transference, and acceptance. Manage issues separately.

STEP 8: INTEGRATE RISK RESPONSES INTO PROGRAM STRATEGY AND DOCUMENT PROJECT BASELINE COMMITMENTS

Good NPD program managers make commitments to their management to achieve certain results for scope, time, and cost. The program team has to make some decisions, such as: What are the stakeholder's priorities? How much buffer should the team allocate to each project element (time, scope, and cost)? How does the team allocate ownership for the risks? Program managers are just like pilots who file their flight plan before starting, and they take into account risk events that might compromise their objectives for schedule, passenger comfort, budget, and safety. If the weather is severe, they may apply the response strategy of avoidance and cancel the flight, as both the pilot and passengers have zero tolerance for accidents. In the language of risk, a weather-related accident is a low-probability, high-impact event.

Recall that the notion of stakeholder risk tolerances (part of Step 6, "Prioritize Risks and Issues") means that any internal or external stakeholder has certain comfort levels. Competent program managers explicitly consider stake-

---

RISK MANAGEMENT CHARACTERIZES ORGANIZATIONAL MATURITY

Some firms have more capability to manage risk well, and these firms are the most consistent in their growth and profitability. Perhaps the simplest test for examining risk management maturity is to examine the level of authority given to the NPD program manager: If authority is high, then the organization is probably positioning itself well to manage risks, but if authority is low, then the blinders may be on. Another test is the use of checklists: If ticking off a checklist is the sole company response to risk, then organizational maturity is low. Risk management provides an excellent lens by which to evaluate a firm's ability to integrate and balance strategic intent with operations.

Many firms ignore risk management because they have not seen the need for it. They perceive their industry as stable and mostly focus on their competitive rivals and operational challenges. The risk management process described in this chapter is also appropriate for sensing and responding to industry-level threats and opportunities. By addressing risk at the project level, you encourage the organization to surface additional strategic concerns.

Top NPD firms have a sophisticated capability for risk management, and they will "book" a project plan, pay attention to the details of product scope and project scope, use risk management tools such as computer simulations and principle-based negotiation, and document their plans and assumptions. These more mature firms are the ones that will consider risk in establishing project baselines and contracts. For example, Nortel uses a concept called "out of bounds" that provides the NPD program managers with the freedom to make trade-offs in time, performance, cost, and other factors. Risk analysis and management is an important tool.

Less mature firms typically establish a due date and pay attention to little else (and in my experience, this is the majority of firms). Firms that use the decision rule "Hit the launch date" default to passive acceptance—hiding the risk instead of managing it. Firefighting and crisis management characterize their organizational culture, and their strategic performance is inconsistent. These firms are like the mythological character Icarus: They fly high but come crashing down because they ignored easily recognizable risk events.

holder risk tolerances and include them in their baseline project plans. For example, some stakeholders can tolerate reduced product performance but cannot tolerate delays. The plan identifies and quantifies the amount of buffer available for use by the program manager. Buffer is more accurately called management reserve; it is the calculated expected value of negative risk events. This management reserve is a resource belonging to the program manager for the purpose of absorbing the negative consequences of risks. Recalling the concept of residual uncertainty in Step 5, there are occasionally freak occurrences that the project must address, and that is one purpose of the management reserve. Baselining performance and the management reserve ensures ownership of risks and accountability for results. Of course, the program manager's authority to manage risk varies with the organization, as described in the sidebar "Risk Management Characterizes Organizational Maturity."

Finally, I note that contracts are risk allocation mechanisms. Therefore, all good program managers must have some practical knowledge of the different types of contracts used in projects so that they can understand the relationships among price, risk, and scope. The clarity of requirements is the starting point: More vagueness equals more risk. Firms set prices to cover the expected monetary value of the risks as well as to cover costs and meet market demand.

John Goodpasture describes an equation for strategic project risk, which I interpret this way:

$$\text{Resources Investment} + \text{Project Risk} = \text{Desired Results}$$

The job of the program manager is to manage invested resources so as to deliver a result, taking measured risks to do so. The contract formalizes the allocation of risk between the project and the team's customer.

Do not neglect to integrate the responses into the baseline plan of the program. Step 8 is ultimately the step that determines whether your planning effort will bear fruit or not.

STEP 9: EXECUTE AND CONTROL THE RISK RESPONSE STRATEGY

This step encompasses all of the implementation work that the project team performs after the risk response analysis and development, described in Steps 2 through 8.

In the execution of the project, the program manager makes sure that the team is following its own processes. The team should have a prioritized list of risks and issues, a strategy for how they will manage the risk events (if they occur), a sense of how they will investigate and resolve issues, and a plan detailing what information they will record and archive.

NPD teams cannot afford to ignore or dismiss critical NPD risks. Individuals generally make choices that are logical for their own mental model. Many can identify the trivial details but have huge blind spots that can lead to disaster.

A good practice is to revisit the risk analysis and response planning periodically in the NPD program. I suggest a weekly review of the risks and issues list, including asking the question "What new risks or issues does the NPD team need to be aware of?" Another commonsense practice is to formally reopen the risk analysis when team-defined triggers appear. To identify the triggers, the facilitator would ask the team, "What triggers would stimulate a complete relook at the project?"

Recall that in Step 4 I defined strategic risks and operational risks. Strategic risks are those that can lead to project cancellation, whereas operational risks affect the accomplishment of targets. In my opinion, a characteristic of world-class product development is a balanced approach of strategic and operational risk management. A good gating meeting focuses on strategic risk and recognizing the need for early project cancellation. A project review or technical review meeting focuses on operational risks. It is a mistake to focus exclusively on operations; keep a balance between strategic and operational risk.

Figure 8-13 provides a useful self-assessment instrument for ensuring that the team remains aware of its responsibilities for managing program and project risks.

While "flying your project" you should note these two lessons from pilots. The first lesson is loss of situational awareness and is often a team-based error. The flight team must know the location, altitude, and heading of the plane at all times. As an example, on December 20, 1995, pilots on American Airlines Flight 965 chatted away as their airplane descended on the wrong approach path into an airport in Colombia; the aircraft crashed into a mountain, killing 159 people. The NPD analogy is the 90 Percent Syndrome—projects appear to be on track until they are 90 percent complete, and then the last 10 percent of the scope takes the same amount of time as the first 90 percent. Complacency leads to loss of situational awareness and is a huge danger.

Team Practice	Never	Sometimes	Always
The team keeps track of the identified risk events and identifies residual and emerging risks.	☐	☐	☐
The team keeps a list of open and closed issues.	☐	☐	☐
For risks, the team assigns and verifies individual ownership of the cause and/or effect of the risk event. For issues, the team ensures that the issue owner has the needed knowledge and resources.	☐	☐	☐
The team collectively determines the content and quantity of risk information that is communicated to stakeholders (neither too much or too little) and the effectiveness of the modes of risk communications.	☐	☐	☐
The team assesses the efficacy of the response strategies, watches for the inevitable new risks that occur, and refines the risk response strategy.	☐	☐	☐
The team updates the risk database at each major review milestone or decision milestone.	☐	☐	☐

FIGURE 8-13. A self-assessment for executing the risk plan and for ensuring good communications.

The second lesson is target fixation, a phenomenon where military pilots become so narrowly focused on the target that they become oblivious to other things and crash their planes into an object. I have seen similar experiences in NPD projects when people become overly focused on the due date. Somewhat panicked, they declare, "We don't have time to plan!" and then they make unrealistic assumptions, make foolish compromises on quality, and burn themselves out with rework.

As program manager, you need to maintain your alertness and a sense of perspective. The project team will take their cues from your actions. When risks and issues emerge, get the facts and work on the responses. This is not a personal insult to you.

Step 9 is risk response execution and control. This step is straightforward if you have done a good job with the preceding steps. It is when you will get feedback on the quality of the up-front steps.

STEP 10: LEARNING FROM RISK MANAGEMENT ACTIVITIES

The final step in this team-based risk process is to review, reflect, and capture learning for the future.

Risk offers an excellent opportunity for creating the organizational learning that undergirds product development. In the postimplementation review, the team members should consider these questions to stimulate reflection:

◆ Did we implement risk responses as planned? Why or why not? Were they effective? What can we learn?

◆ How effective was our issue resolution process?

◆ Are we blaming individuals and taking positions, instead of looking deeper and more systematically into our organization and practices? What have we learned about our cultural preferences toward decision making?

Wisdom is an intangible that comes from experience and reflection. Often humor helps by revealing our weaknesses and pointing out the pathway to wisdom. My favorite Sunday comic strip is Bill Watterson's *Calvin and Hobbes,* the escapades of a six-year-old boy (Calvin) and his philosopher-cum-stuffed-tiger (Hobbes). I best like Calvin's soliloquies as they race down a steep hill in a wagon. I cannot help but consider the wagon as a metaphor for NPD program management, and I imagine Calvin as the program manager and Hobbes speaking for the program team. My favorite is one where, as they start down the hill in the wagon, Calvin states, "It's true Hobbes, ignorance *is* bliss." Shortly before going over the cliff, Calvin states, "If you're willfully stupid . . . you can keep doing whatever you want," and then covers his eyes. Calvin and Hobbes end up wrecked and bruised, and team member Hobbes moans, "I'm not sure I can stand so much bliss." Calvin responds, "Careful, we don't want to learn anything from this." Calvin exhibits hubris, the boastful pride that people claim for themselves and their capabilities. We may find hubris humorous in a cartoon character of a six-year-old boy, but it is downright dangerous in the high-stakes environment of NPD.

Good program managers are *not* like Calvin; they keep their eyes open, anticipate the hazards, and learn from their mistakes. Ultimately, decision making is cognitive and culturally influenced, so a good program manager knows that an attitude of humility improves decisions. I agree wholeheartedly with what Scott Adams writes in *The Dilbert Principle:* "People are idiots. Including me. Everyone is an idiot, not just the people with low SAT scores. The only difference is that we're idiots about different things at different times. No matter how smart you are, you spend much of your day being an idiot."

I have learned that a sense of humor is the pearl in the oyster of humility. Humor relieves tension and sparks creativity. Relax! We all make mistakes. Let's try not to make big ones.

While there are plenty of methods for capturing learning, from simple bulleted lists to complex systems-thinking diagrams, the essential practice is learning to think strategically. Take off your mask of competency and be willing to honestly analyze your own decisions and behaviors.

In summary, Step 10 involves learning from risk management. I know from experience that risk management is a practical, nonthreatening way to get people to learn profoundly. It has the potential to transform the NPD organization, as I describe in the final section of this chapter.

BUILDING A CULTURE FOR GOOD DECISION MAKING

You now have a seen a simple ten-step approach for building a team-based approach for NPD program risk management. In nearly every step, I linked perception and decision making to individual psychology and organizational culture. Risk is more than a nice-to-use "tool"; it is a Zen-like discipline of consciousness building.

The sidebar that I introduced earlier, "Risks in the Firebird Project," illustrates that a highly structured and formal approach is not necessary—just find the courage to ask questions. Risk management does not need to be complicated. Kelly Benjamin, director of NPD at Essilor of America, confirms the value of simplicity: "We have a rather rigorous risk method that we use worldwide. Because each project is different, we leave it to the project manager and team to match the intensity of risk identification effort to the product strategy. Simpler projects get a more streamlined response."

Kelly's comment emphasizes the importance of judgment and flexibility in matching technique to strategy. Just start with some simple inquiries: "How do you define project completion? How do you define success? What will keep you awake at night?" Because each project is different, its risks are different, and the response must be flexible and robust. A sound risk management approach scales up or down to the innovativeness of the marketing, technology, and production strategy. Too, the program manager should scale the amount of risk management process to the ability of the team to assimilate and use the process. Hence the use of standard checklists and templates can focus the NPD team's attention on the wrong areas and lead to a false sense of security.

Team ownership of the project's risks is essential. I recently heard an executive complaining that the NPD team would not step up and make decisions, no matter how much he encouraged them. It was one of his biggest frustrations. Risk management is one of the very best ways to cause people to confront the difficult issues in the NPD project and structure information so that the team can make good decisions. Communicating, team building, and decision making are processes, not events.

Kent Harmon of Texas Instruments notes that organizational culture sometimes conditions individuals to be distrustful. This distrust undermines good decisions. The fear that "I might be judged for something out of my control" often leads to avoidance of participation in risk, or deflecting the concerns to other team members. Conflict-avoidance behaviors often reinforce the culture of distrust. Conflict avoidance seems especially common in technical organizations. The pattern is this: People are busy and focus on their competency and their controllable area, which is a personal comfort zone. They avoid cross-functional issues because they fear appearing incompetent. Eventually they can no longer avoid a given issue, and people react with anger and urgency, often blaming other people or organizations. A strange feature of this situation is that competency illogically equates to trust: If an individual perceives someone as lacking competency, he or she illogically concludes that the other person can't be trusted. This gives the individual a reason to avoid further contact with the other person. Regardless, the urgent issue gets resolved, and people go back to their comfortable way of doing things. Because the root problem of avoidance and distrust doesn't get addressed, the cycle repeats. Risk management, because it is perceived as rational, detached, and objective, gives people a logical, practical method for breaking the cycle. Kent and his staff have helped Texas Instruments improve their NPD program management by encouraging active listening, by engaging people, and by coaching senior management on accepting the risk data.

When I get on an airplane, I expect that the pilot has assessed all risks and developed appropriate responses for his or her flight plan. I expect the crew to

FIGURE 8-14. The Chinese character for "risk."
The character for "risk" is a blending of the characters for "danger" and "opportunity."

keep their heads up and not get absorbed in their technology. NPD organizations should expect similar standards of professionalism from their program managers. Successful entrepreneurs and executives know that taking calculated risks is important to their success. No organization can make progress without taking risks. Risk management is arguably the single greatest competency of NPD program management.

I have repeatedly seen a no-longer-surprising side effect of proactively managing project risk: People get more creative. They take the lemons and make lemonade. Risk is not always bad—risk is a choice, not an outcome. Did you know that the Chinese character for "risk" (see Figure 8-14) is a blending of the characters for "danger" and "opportunity"? NPD program managers must thoughtfully manage risk and be willing to assume accountability for the performance of the project. A good decision is one that increases the probability of success and decreases the probability of failure.

Part 3

Process Owner Tools

The tools of Part 3 target improving the firm's process for developing new products. The individual or group in the firm that is charged with maintaining, improving, and deploying the NPD process, the NPD process owner, will find these tools useful in giving them new avenues to improve the firm's NPD process. While team leaders may derive benefit from learning and applying these tools to improve NPD for their particular project, the responsibility for seeing that these tools are available in a consistently applied format across project teams will fall to the NPD process owner. Part 3 presents tools in the order of their use through the stages of the NPD process.

Chapter 9, "Capturing Employee Ideas for New Products," reviews the history and relative effectiveness of different kinds of employee-driven idea-capturing programs, and provides an eleven-step method for designing and implementing a successful employee new product idea program. Having a formal idea-capture program allows business unit and program managers, as well as project leaders, to tap into a database of potential new product ideas, when they need a new product to close a forecasted revenue gap. Although teams and project leaders would likely benefit from an idea-capture program, it is the NPD process owner who needs to own, design, implement, and maintain the program as part of the NPD process. Virtually any firm should benefit from creating and maintaining a warehouse of NPD ideas, in some form.

In Chapter 10, "Lead User Research and Trend Mapping," two interrelated tools are presented. Lead user research is a method to find innovative product users on the leading edge of market change and use them to help construct new product concepts the firm can investigate commercializing. The chapter provides details on each of the four stages of the lead user process: framing the project, trend mapping, finding lead users at the front edge of the trend, and holding joint invention sessions to create potential solutions to trends-related needs. Within stage 2 of the lead user method, the chapter presents a detailed procedure for defining and mapping the leading edge of one or more trends that have the potential to change the user needs of a broader potential market. Trend mapping involves performing interviews to capture company and industry knowledge and then developing a roadmap. Firms whose markets or technology bases are changing rapidly and in unpredictable ways will benefit the most from implementing these tools.

Traditional Stage-Gate NPD processes provide a structured (yet somewhat flexible) mechanism for firms to step through the tasks and activities of product development. Firms that use formal, structured NPD processes have higher product development success rates. However, traditional Stage-Gate is only applicable when essentially all of the technology development for a product has been completed. Chapter 11, "Technology Stage Gate: A Structured Process for Managing High-Risk New Technology Projects," presents a tool to use to manage the technology invention process prior to moving into product development, when there is high technical uncertainty and risk. This process manages tasks for high risk projects within and at the transition between the Fuzzy Front End and formal new product development in the Stage-Gate sense. This chapter clearly delineates the differences between the two processes, and presents details of how to complete the six elements of the Technology Stage-Gate (TechSG) process: project charter, review process, review committee, structured planning, development team, and process owner. Chapter 5 in this ToolBook provided a people-based mechanism for crossing the valley of death between technology invention and formal product development. Another way to look at this chapter is as a process-based tool for first managing technology risk, and then additionally moving the technology development across the valley of death and into product development.

The final chapter of this part, Chapter 12, "Universal Design: Principles for Driving Growth into New Markets," presents principles and methods for making products, services and environments "usable by people of all ages and abilities, to the greatest extent possible." These are techniques and precepts that need to be incorporated into the NPD process at the specification development and physical development stages. The chapter presents enormous detail about the different aspects of potential human limitations (cognition, vision, hearing, speech, body function, arm function, hand function, and mobility) and their impacts on a person's ability to use a product. It then defines the seven Principles of Universal Design, which were developed to help firms overcome design limitations in a structured way. Universal Design Performance Measures are provided that can be used to evaluate how well your firm's products satisfy the seven Principles of Universal Design.

9

Capturing Employee Ideas for New Products

Christine Gorski and Eric J. Heinekamp

Whether you call it the Information Age or the new economy, the driving force of a successful enterprise has changed. We have seen the transition from automation to reengineering and now must move from continuous improvement to radical new product design to remain competitive. The challenge to today's winning firm is to outthink the competition and move nimbly as opportunities arise. This shift requires companies to rethink their approach to new product innovation and places a much greater emphasis on the ability to generate fresh new product ideas.

It is impossible for a company to anticipate when, where, or how a new product idea will be generated, or who will be involved. Every employee has something that he or she alone knows about the company because of work experience and detailed job knowledge.

Some new product ideas result from happy accidents, as was the case in 1965 when a research scientist working on a new antiulcer drug brought an unexpected gift to pharmaceutical company G. D. Searle and Co. After a series of mishaps caused the chemical he was heating to boil and spatter onto his bare hands, the scientist licked his finger without thinking and discovered an extremely sweet taste. After some convincing and more than sixteen years of testing, the drug company launched the new sweetener to the market. Today NutraSweet is a billion-dollar product.

Other ideas are unexpected, as was the case at Japan Railways. Constructing the new train line required tunneling through mountains, and water running in the tunnel through Mount Tanigawa caused problems. While the engineers were drawing up drainage plans for the water, construction crews began to drink the water. A maintenance worker, responsible for checking the safety of the tunneling equipment, thought the water tasted so good, he proposed that Japan Railways bottle the water instead of pumping it away. The idea was implemented, and the brand Oshimizu became so popular that Japan Railways installed vending machines for it on its nearly one thousand railway platforms. A subsidiary was formed that now offers home delivery of the water.

The objective of this chapter is to present easy-to-implement methods and tools for gathering product ideas from employees to spark innovation for new product development. We begin with an overview and brief history of employee idea suggestion programs. The chapter then presents several working examples of such programs and how they have evolved to enhance and sustain the new product development process. We conclude with step-by-step recommendations for designing and implementing an employee new product idea program for your firm.

OVERVIEW OF EMPLOYEE NEW PRODUCT IDEA SUGGESTION PROGRAMS

Background

The ultimate objective of all employee idea suggestion programs is to pull together the unique knowledge of the workforce in order to identify new product opportunities and process improvements. What varies among idea programs is how they are crafted and managed in order to meet this objective. Factors such as corporate structure and culture influence the need for and design of the new product idea program. Program characteristics such as scope and the idea evaluation process influence how the program is managed. Understanding the key characteristics of employee idea programs and how they interact is helpful in evaluating an existing program or creating your own new program.

The *scope* of an employee idea program defines the idea subject matter and the target audience. One program, for example, may focus on incremental process improvements, while another emphasizes new product ideas. *Idea ownership* defines the role the idea creator plays in implementing his or her idea. This can range from programs that allow the submitter to remain anonymous to programs that require the submitter to actually implement his or her idea. The *idea evaluation and approval process* defines how submitted ideas are reviewed and approved. Finally, the *framework* defines the tool used when recommending ideas, from traditional paper forms to sophisticated Web-based applications.

History of Employee Idea Suggestion Programs

Obtaining suggestions and new product ideas from employees is not new. While the majority of today's employee idea programs trace their roots to the total quality movement of the late 1980s, similar programs have existed for over a hundred years. Idea programs have changed in both form and substance over time, primarily due to advances in technology. New technology has made idea sharing and collaborating possible. The historical progression of employee idea programs and working examples are presented below.

The Employee Idea Suggestion Box

The first known idea suggestion box program was implemented in 1880 by Scottish shipbuilder William Denny and Brothers. Its mechanics were simple.

> **EXAMPLES**
>
> In 1895 John Patterson, founder of National Cash Register (NCR), initiated the first documented employee idea program in the United States. The goal of his "hundred-headed brain" was to capture the creativity of the workers and identify ways to improve efficiency. Initially, adopted ideas were awarded $1, and twice each year lavish ceremonies were held to reward employees with cash prizes up to $30. From 1899 to 1903 the size of the awards more than doubled, and in 1904 alone the program received more than seven thousand ideas, one-third of which were adopted. Patterson's strong support of the program, excellent awards, and practical results made the program a success. Patterson's program was revolutionary in design, the first sanctioned program to give a voice to the hourly worker (Robinson and Stern 1997).

A wooden box was placed in an area easily accessible to all workers and employees were asked to submit any ideas they thought might help the firm. Similar employee idea-gathering programs soon sprang up throughout Europe. The first U.S. employee idea suggestion programs were introduced in the 1890s by pioneering firms NCR and Kodak.

These first-generation employee idea suggestion programs were paper-based, process-intensive, and broadly focused. The employee suggestion box is traditionally a paper-based system in which submitters write their ideas on a standard form and drop it in a local box or mail it to a central address. Suggestion box forms request that authors identify themselves, but submissions can be anonymous. The questions on suggestion box forms tend to be open-ended and allow the writer to provide a range of ideas, from new products and/ or process improvements to changes in company policies. Idea submissions are periodically reviewed, recorded in a master idea log, and routed to others for further review and implementation. Feedback to the employee submitter, if any, is usually limited to an acknowledgment of the suggestion.

Employee suggestion box systems remain in use today, many still in their original format. Some companies have updated their paper forms with computer databases that simply replace the existing process with a format that is easier to use (see Table 9-1).

TABLE 9-1.
Suggestion Box: Strengths and Weaknesses

Strengths	Weaknesses
Easy to implement	Ideas easily lost or misplaced
Inexpensive to administer	Cumbersome and slow paper-based process
Allows for anonymous suggestions that might otherwise be withheld	Difficult to keep submitters abreast of action taken on their idea
Can target a narrow or broad audience by managing the distribution of the submission forms	Submitters have little input in implementing the idea; outcome may differ from what was envisioned

TABLE 9-2.
Kaizen Teian: Strengths and Weaknesses

Strengths	Weaknesses
Incremental ideas easiest to implement; promotes continuous improvement	Focus on incremental changes, not breakthrough ideas
Fits best for rapid action	Limits creativity
Management can be decentralized; closest to knowledge holder	Separation of small ideas that could otherwise be combined; program can be misused
Employee involvement promoted, program assessed by participation rate	Submitter may not be capable of implementing the idea

Kaizen Teian

The next generation of employee idea programs appeared after World War II in Japan. These programs, known as *kaizen teian,* were first introduced in Japan in 1945 by the U.S. Air Force as part of the Allied occupation's Training Within Industries program. *Kaizen teian* programs focus primarily on obtaining ideas related to continuous improvement. The *kaizen teian* program's fundamental goals are the promotion and reward of employee participation. Typically, monetary rewards are small. Motivation for this type of program is based on recognition and intrinsic awards, not monetary or extrinsic rewards.

In Japan these programs have become commonplace and are very successful in outperforming traditional suggestion box programs in terms of both the number of ideas received and employee participation rates. In 1996, for example, the average Japanese company received eighteen ideas from each employee, with nearly 90 percent of those ideas being implemented. U.S. firms, by comparison, averaged less than one idea for every five employees, with only one idea in three being adopted. The true success of *kaizen teian* programs is measured by participation rates rather than revenue contribution. A recent study revealed that nearly 75 percent of Japanese employees contribute to employee suggestion programs, compared to 10 percent for U.S. firms (Robinson and Stern 1997). (See Table 9-2.)

Employee-Driven Idea Systems (EDIS)

Employee-driven idea systems (EDIS) are a variation of the *kaizen teian* idea program. EDIS programs concentrate on quality and continuous improvement. The distinguishing trait of these programs is that once an idea is approved, the idea submitter becomes responsible for driving the idea from concept to implementation. The success of these programs is based on employee involvement. The book *Employee-Driven Quality: Releasing the Creative Spirit of Your Organization Through Suggestion Systems* quotes 60 percent participation rates in the first year versus 10 percent with traditional suggestion approaches (McDermott, Mikulak, and Beauregard 1993).

EXAMPLES

Honda of America Manufacturing of Marysville, Ohio, has one of the best-known EDIS programs in the country. This company, with nearly nine thousand employees, has over 50 percent of its hourly workers participating in the EDIS suggestion program. In recent years, over 75 percent of all ideas were implemented. A key characteristic of this program is that local management can approve the majority of ideas, and most approved ideas are implemented directly by the worker. Employees receive token monetary awards and points that accumulate over their career for special awards. Most importantly, idea submitters are publicly recognized for their suggestions (McDermott, Mikulak, and Beauregard 1993).

An EDIS program is driven by recognition rather than by monetary rewards. As with the suggestion box program, the employee prepares an idea form. The difference is that the idea form is reviewed locally with the manager for further detailing. Also, the submitter receives a nominal recognition reward from the manager for the idea. If the manager approves the idea and it is within the manager's budgeting capabilities, the idea is implemented (see Table 9-3).

WEB-BASED IDEA COLLABORATION PROGRAMS

With the advent of Web technology, employee idea programs are experiencing a dramatic rebirth. New technology has made collecting and managing ideas much easier and has radically improved the process, from linear data collection to communities of interactive collaboration between the submitters and those empowered to implement new products.

Web-enabled tools improve upon the past in three significant ways. First, Internet technology makes implementation, dissemination of idea information,

TABLE 9-3.
Employee-Driven Idea Systems: Strengths and Weaknesses

Strengths	Weaknesses
Idea owned by submitter, who manages its implementation	Works if idea submitted is within the scope of idea submitter's responsibilities
Promotion of total quality and continuous improvement.	Focus on incremental changes, not breakthrough ideas
Review and implementation often carried out in submitter's decentralized unit.	
Teamwork promoted	
Rapid idea evaluation turnaround	
Rewards based on recognition rather than monetary awards	

TABLE 9-4.
Web-Based Idea Collaboration Systems: Strengths and Weaknesses

Strengths	Weaknesses
Less administration needed since ideas flow directly to any area that might benefit from them	High volume of ideas can create constant flow of e-mails; potential to overwhelm users
Automatic feedback to the submitter upon receipt and any updates	All ideas viewable by anyone using the system, which might be anyone in the company
Anyone viewing the idea can comment or seek additional comments from the submitter	
Easily produces reports on last update, status, or disposition.	

and management of programs less time-consuming and more cost-effective. Second, employees find the interface easy to use and convenient, which makes them more likely to submit ideas. Third, because of the unique interactive capabilities of the Internet, idea programs now permit easy collaboration between the idea submitter, reviewers, and any other interested parties. Similar to participating in on-line message boards or virtual communities, any employee can build upon an idea by adding comments or asking questions. Internet technology provides the opportunity to integrate virtual brainstorming into the idea collection and review process (see Table 9-4).

Bank One's One Great Idea Program

Bank One Corporation, the nation's sixth largest bank holding company, with assets of more than $270 billion, offers a full range of financial services to large corporate and middle-market commercial customers and retail consumers. Bank One's Cash Management division develops, sells, and services sophisticated products that allow commercial customers to transact financial business. Given the rapid changes in banking services, this division places a premium on product innovation and constantly looks to its customers and employees for new product suggestions.

The Cash Management division recently implemented an employee new product idea program called One Great Idea. The goals of the program are:

♦ Promote the collaborative exchange of new product ideas among employees

♦ Collect new product ideas in a single repository for tracking and prioritizing product development efforts

♦ Generate better product ideas more quickly and easily

The One Great Idea program begins with a Web-based front-end application, the Idea Database, which is accessed through the company intranet. Because it is a critical form of communication at Bank One, all employees have access to the intranet, either on their desktop PC or at a shared computer in a public area such as a lunchroom. Employees with dial-up capabilities can even access the program from home or while on the road. There are no paper forms associated with the program.

The Idea Database is more than an on-line employee idea box. Unlike the idea programs described earlier, Internet technology provides an environment for a lively and creative exchange of employee ideas. The Idea Database is an interactive, centralized repository for collecting, sharing, and tracking ideas. Employees are encouraged to submit their original ideas or represent the voice of the customer by communicating product ideas customers may have suggested. The Idea Database is also used to collect ideas for improving the division, such as operational efficiencies, competitive product gaps, or training programs. The Idea Database is easy to use, with three basic functions: enrollment, entering or updating ideas, and viewing idea information.

Enrollment

The first time an employee enters the Idea Database, there is a prompt to enroll by creating a user profile and personal log-on. This quick step establishes basic data on the user, such as e-mail address, mail address, phone number, and department. There are no passwords for users.

A distinctive feature of the Idea Database is the subscription function. Users of the system can subscribe to receive automatic notification and updates of idea submissions by selecting from various product categories. Employees are automatically subscribed to receive updates on ideas they have submitted.

Entering or Updating an Idea

Employees enter ideas directly into the Idea Database using simple commands and drop-down menus. Submitters begin by classifying their idea to define the type of idea (e.g., new to the world or product extension), the source of the idea (e.g., customer request), and products that might be affected. Next, the employee describes the idea in as much detail as required for complete explanation. Finally, there are prompts to complete several additional fields to estimate market demand, target customers, and unique product properties such as technology requirements. As soon as the idea is submitted, an automatic e-mail is sent to all other users who have subscribed to that idea type.

Updates to existing ideas follow a similar flow. Any user can add more detail or related ideas to any existing new product idea. A message is then sent to the original employee submitter and any other subscribers alerting them that an idea has been updated. This dialogue continues, adding richness to the idea, until the idea is either implemented or killed (see Figure 9-1).

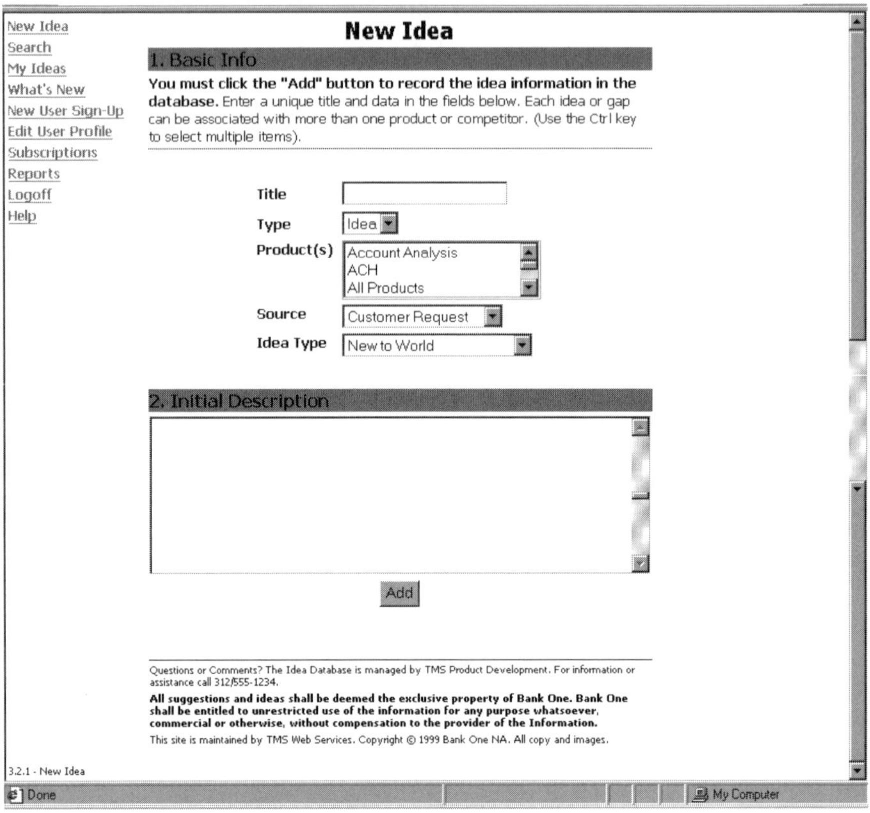

New Idea   **New Idea**
Search
My Ideas   **1. Basic Info**
What's New   **You must click the "Add" button to record the idea information in the**
New User Sign-Up **database.** Enter a unique title and data in the fields below. Each idea or gap
Edit User Profile can be associated with more than one product or competitor. (Use the Ctrl key
Subscriptions   to select multiple items).
Reports
Logoff   **Title**   [   ]
Help
                              Type   [Idea ▼]

                              Product(s)   [Account Analysis ▲]
                                                [ACH   ]
                                                [All Products   ▼]

                              Source   [Customer Request ▼]

                              Idea Type   [New to World   ▼]

                  2. Initial Description

                  [   ]
                  [   ]
                  [   ]
                  [   ]
                  [   ]
                  [   ]

                              [Add]

                  Questions or Comments? The Idea Database is managed by TMS Product Development. For information or
                  assistance call 312/555-1234.
                  All suggestions and ideas shall be deemed the exclusive property of Bank One. Bank One
                  shall be entitled to unrestricted use of the information for any purpose whatsoever,
                  commercial or otherwise, without compensation to the provider of the Information.
                  This site is maintained by TMS Web Services. Copyright © 1999 Bank One NA. All copy and images.

3.2.1 - New Idea
🖹 Done   🖳 My Computer

FIGURE 9-1. Bank One's New Idea entry screen.

Viewing and Tracking Ideas

Idea Database users have several options for viewing idea information. They can run a report for a single idea or see a list of the most recently submitted ideas. A search function permits users to look for specific topics or idea types. Summary reports help the system administrator produce usage and idea tracking reports for management (see Figure 9-2).

Fit with the Bank One New Product Development Process

The Idea Database is a vital component of the fuzzy front end engine that fuels the Bank One stage gate product development process. This employee new product idea program integrates seamlessly with new product development, portfolio management, and development tracking. Product managers, systems managers, and senior management can actively participate in the journey of an idea from fledgling suggestion to detailed concept. This sharing of data has rewritten the rules for idea evaluation, since in many cases the submitters and

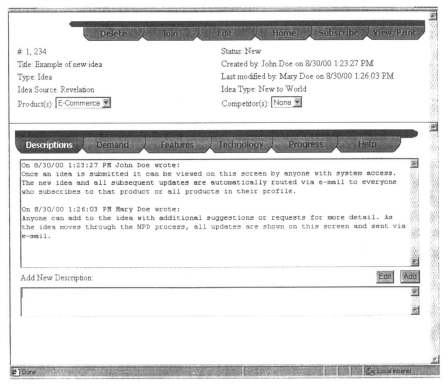

FIGURE 9-2. Bank One's Viewing and Idea screen.

eventual implementers collaborate to deliver the idea without the intervention of a formal review process. In some cases, the idea submitter ends up owning and implementing the idea.

Idea submitters receive token cash awards and recognition awards for their desk. Submitters whose ideas are adopted receive special public recognition.

BENEFITS OF EMPLOYEE IDEA SUGGESTION PROGRAMS

As seen in the examples above, idea suggestion programs can *promote creativity throughout the organization* and be very successful at getting all employees to participate in the new product development process. This can lead to more and better product ideas. Typically, the job of identifying innovative product concepts is assigned to marketing, engineering, or R&D professionals. Unfortunately, many firms overlook other employees as a source of creative ideas. Companies that attempt to centralize creativity and innovation could increase their probability of finding a new product winner by providing a way to decentralize creativity and seek new product ideas from all employees in the organization.

Employee idea programs can capture and integrate the internal and, indirectly, external knowledge of those closest to the products and services. In particular, Internet-based programs *distribute and record the knowledge base* of the organization. Ideas from employees or customers are retained until implemented or deleted, even if the original submitter leaves the company. Ideas that were not appropriate at the time they were proposed can be retrieved years later.

Having an idea suggestion program can *ensure that the right idea gets to the right person* quickly. An effective program has a simple and understood process for submitting new product ideas from anywhere in the company. Without such a program, employees with valid product ideas might not be encouraged to submit an idea, nor know how to submit an idea or whom to contact. Both knowledge and speed would be lost. This is especially problematic in big firms. The larger the company, the more likely it is there are creative ideas to be found, but the less likely it is that such ideas will ever reach the right person.

Finally, employee new product idea programs, particularly Web-based systems, can be used to *foster teamwork and enhance communication* throughout a company by tracking progress on ideas submitted. Effective idea suggestion systems provide feedback and metrics and track progress on ideas submitted.

DESIGNING AND IMPLEMENTING A SUCCESSFUL EMPLOYEE NEW PRODUCT IDEA PROGRAM

New product ideas are the trigger to beginning a product development effort. Often, collecting and evaluating ideas is downplayed because managers believe they have ample ideas. However, the idea collection and evaluation stage of the development process is so pivotal to the development of successful new products, a program for gathering ideas and criteria for prioritization must be developed. A new product development idea generation and capture program should be designed to provide a steady flow of high-quality ideas that is sustainable over time (see Figure 9-3).

We have provided the history, analytical framework, and examples of employee idea programs. These can be used as a benchmark for designing or enhancing your new product development program. In this section, we show you how to apply this information to the unique product innovation requirements of your company.

Whether you are improving an existing program, consolidating several programs, or designing a new program, there are many components to consider in order to ensure the creation of the best employee idea program for your organization. For each component, we have provided a series of questions to help guide you through designing your employee new product idea program. Additionally, we have summarized all the questions in a table format as Table 9-5, "Designing an Employee New Product Idea Program."

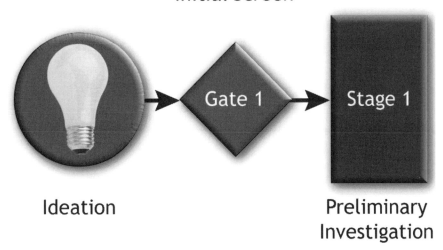

Initial Screen

Ideation · Gate 1 · Stage 1 · Preliminary Investigation

FIGURE 9-3. Initial idea screen gate.
From Product Development for the Service Sector *by Robert G. Cooper and Scott J. Edgett.*
Copyright © 1999 by Robert Cooper and Scott Edgett. Reprinted by permission of Perseus Book
Publishers, a member of Perseus Books, L.L.C.

Step 1: Review Current Idea Capture Methods

Every company has some method for gathering suggestions and ideas from employees. The method may be a formal, documented process or an unwritten, ad hoc approach. Perhaps there are several formal suggestion programs in your organization or some that have been tried but no longer exist. Performing a self-audit of the present employee new product idea collection process allows you to leverage what has already been used, learned, and developed. A series of interviews and meetings with the users and administrators of existing programs is a good way to gather the information. Modifying an existing idea program may be all that is necessary to start your program. Even if you do not have a formal new product idea collection process, it is useful to recognize how the idea collection process occurs at your company.

Questions to consider:

◆ How are new product ideas from employees captured in your organization today?

◆ Are you satisfied with the quality and quantity of new product ideas submitted?

◆ Does your company have a formal employee new product idea program?

◆ Does the existing formal program work for your needs?

◆ What were the last five winning new product ideas, and where did these originate?

TABLE 9-5.

Designing an Employee New Product Development Program

Questions	Response
Review Current Methods	
◆ How are new product ideas from employees being captured in your organization today?	
◆ Are you satisfied with the quantity and quality of new product ideas submitted?	
◆ Does your company have any formal employee new product idea programs?	
◆ Does the existing formal program work for your needs?	
◆ What were the last five winning new product ideas and where did they originate?	
Gain Management Support	
◆ What benefits could a structured employee new product idea program offer?	
◆ How does a new product idea program fit with your corporate strategy and culture?	
◆ What human, financial, capital, or other resources would be required?	
◆ What would management's role be in the process?	
Purpose	
◆ What are your employee idea program objectives? Will your program be a way for employees to brainstorm new product ideas, voice client ideas, or store ideas so they will not be lost?	
◆ Why should a new product idea program be implemented? For example, have ideas have been falling by the wayside? Do people leave and take ideas with them? Is your primary purpose to represent the voice of the customer?	
Scope	
◆ Who will your new product idea program reach?	
◆ What kind of new product ideas do you want to capture?	

TABLE 9-5. *(continued)*

Questions	Response
Idea Ownership/Intellectual Property	
◆ Do you want to know and track who is submitting ideas?	
◆ Who will implement the new product ideas submitted?	
◆ What are your company's policies on intellectual property?	
Framework and Concept Validation	
◆ What are the design functions required for the program to be successful?	
◆ What methods will be used for submitting ideas? Will submissions be paper-based, automated, collaborative?	
◆ What are the special technology requirements?	
◆ How will you pilot-test the concept? Who will be part of the test? How long will it last?	
Idea Evaluation Process	
◆ What form will be prepared when submitting a new product idea?	
◆ What are the program and idea submission guidelines?	
◆ What is required to process an idea? Who will screen ideas? What are the responsibilities of the administrator, idea evaluator, idea review committee, and senior management review board?	
◆ What is the time frame for evaluating and criteria for screening an idea?	
◆ What is the approval process once an idea has been evaluated?	
◆ What method will be used to communicate results of the idea evaluation and progress to the idea submitter?	
◆ What method will be used to track and monitor ideas?	
◆ How will the process integrate with the new product development process?	

TABLE 9-5. (*continued*)

Questions	Response
Measurement/Goal	
◆ What measurable goals would you like to achieve? Higher participation rates? Faster turnaround? Increased number of products launched?	
◆ How will you measure human factors, such as whether the program is easy to use and efficient?	
Awards	
◆ Do you want to reward employees for their new product ideas?	
◆ Who will be eligible to receive awards?	
◆ What type of awards and methods to distribute awards will be used?	
◆ When will awards be distributed—at the time of submission, or upon implementation?	
Implementation	
◆ What training will be provided to teach people how to begin thinking of ideas?	
◆ What training will be provided based on specific roles, such as submitter, evaluator, administrator, or manager?	
◆ Who will perform the training: the training department, human resources, product development, or other?	
◆ What identity and design logo will the program have?	
Administration	
◆ How much time is needed to manage the employee new product idea program?	
◆ What resources will be needed for managing the program?	
◆ What reporting information will be provided to management? With what frequency?	
◆ What will the ongoing motivational and promotional plan consist of?	

Step 2: Gain Management Support

From an understanding of existing programs, you can more clearly define your new product idea approach. But before you begin designing or enhancing your program, you need to gain management support. It is the role of management to establish a culture that values quality, innovation, and continuous improvement. Only management can create an environment in which employees are comfortable enough to suggest new product ideas. With management buy-in, employees are more likely to participate by investing time in the program. The program will have credibility, and resources can be provided with management's backing.

Questions to consider:

◆ What benefits could a structured employee new product idea program offer?

◆ How does a new product idea program fit with your corporate strategy and culture?

◆ What human, financial, capital, or other resources would be required?

◆ What would management's role be in the process?

Step 3: Define the New Product Idea Program Purpose

Like a mission statement, the purpose for a new product idea program gives a sense of direction. The program purpose can be defined in terms of why the program exists and the program's objectives and goals. A new product idea program may focus solely on capturing new product ideas. The purpose for establishing a new idea program will be different for every company. For example, a collaborative idea program goal might encourage employees to work together to build and refine ideas. The objective could be to identify five new product ideas to be implemented in a prescribed time frame. Another goal might be to provide a vehicle for all employees to voice their ideas. Objectives of the program should be stated in measurable terms.

Questions to consider:

◆ What are your employee idea program objectives? Will your program be a way for employees to brainstorm new product ideas, voice client ideas, or store ideas so that they will not be lost?

◆ Why should a new product idea program be implemented? For example, have ideas have been falling by the wayside due to shifting short-term priorities, or have people left and taken ideas with them? Is your primary purpose to represent the voice of the customer?

Step 4: Determine Scope

Scope defines the range of ideas you wish to capture and the audience you want your program to reach. New product ideas to capture include *customer product ideas* (ideas new to the company, ideas new to the world, enhancements, line extensions), *strategic product ideas* (to improve the company direction, policies, or practices), and *product improvement ideas* (to resolve specific process and product issues).

The target audience could be defined by functional area or specific topic, or it could be open to all employees for all topics. Programs designed to solicit ideas from all members in the corporation generally seek to gather new product ideas on many topics.

Questions to consider:

- ◆ Whom will your new product idea program reach?
- ◆ What type of product ideas do you want to capture?

Step 5: Establish Idea Ownership and Intellectual Property

Idea ownership refers to the role the idea submitter will play in implementing the idea. Most new product ideas are broad in nature and cross job boundaries. Ownership of the idea is usually passed to a product or project expert, who will carry it through implementation. When idea ownership is passed on, the submitter's participation can be active (working directly on the development effort) or passive (receiving progress updates). Whenever possible, it is recommended that the idea submitter be involved through implementation to keep employees interested in the program.

When an idea is submitted anonymously, there is no mechanism for recognition or for the submitter to participate in implementing the idea. Ownership of the idea must be passed to another individual for evaluation, often resulting in an outcome that is very different from what the submitter intended. For this reason, programs should avoid anonymous idea submissions.

Intellectual property relates to the legal ownership and right to use the idea. Usually companies include a statement of intellectual property ownership in the employee handbook. Company ownership of intellectual property, such as ideas, needs to be clearly stated in the employee idea program materials. Be sure to check with your legal department regarding the company's policy on intellectual property.

Questions to consider:

- ◆ Do you want to know and track who is submitting ideas?
- ◆ Who will implement the new product ideas submitted?
- ◆ What are your company's policies on intellectual property?

Step 6: Choose Form and Validate Concept

The framework chosen for your employee idea program details the way an idea is to be submitted and affects other employee idea program components.

Manual programs can be as simple as preparing a paper idea form and logging ideas on a master document for tracking. Administering a manual program is labor-intensive. Therefore, manual forms work best if the company and the expected number of ideas are small.

An on-line approach whose structure relies on powerful database technology for managing employee idea submission data is more effective, especially in larger organizations with multiple locations. With the technology available today, the ideal automated system should include data field search functionality, permit multiple idea entries for idea development, provide e-mail feedback notification of progress to the submitter, track the award program, provide evaluator information, integrate with standard word-processing software, and offer comprehensive award and idea reporting. These systems can be built or purchased from a number of vendors specializing in product development software. The cost of initial development of an Internet-based system can range from about $75,000 to $300,000, determined by the amount of sophistication and complexity of internal integration. Ongoing maintenance for hosting, assuming existing servers, could add another $10,000 to $20,000 a year. The cost for ongoing administration and management of the program will vary by the size of the audience.

The best framework for your organization may be a combination of the types listed. Some companies use more than one program framework.

In all scenarios, the idea evaluation process should be efficient, with fast turnaround time. Acknowledgment of receiving the idea should be immediate and include a description of the next steps that will be taken. The idea evaluation stage is typically thirty days. In addition, it is important that the form chosen make it easy to submit ideas. A universal approach where employees have only one place to enter their ideas will be more readily accepted. The more difficult it is to submit ideas, the fewer ideas you will receive.

To validate the idea suggestion program concept and the program components you have chosen, you may want to host a pilot or focus group with a smaller subset of the audience you plant to reach. Pilot and focus group members should represent all roles in the evaluation process, such as submitters, evaluators, administrators, implementers, and managers. Feedback from pilot participants will provide the opportunity to flush out problems and make improvements. Focus groups are a good way to find out what people in the organization think about an employee new product idea program. Results from the focus group can help in anticipating what can be expected when the program is fully launched.

Questions to consider:

♦ What are the design functions required for the program to be successful?
♦ What method will be used to submit ideas? Will submissions be paper-based, on-line, collaborative, voice-recognition-based, some of the above, or all of the above?
♦ What are the special technology requirements?

♦ How will you pilot-test the concept? Who will be part of the test? How long will it last?

Step 7: Implement an Idea Evaluation Process

Management of the implementation and evaluation process is needed to ensure the effectiveness of the idea program. The idea evaluation and approval process defines how ideas are reviewed and approved. Usually an initial screening of ideas is assigned to a focal person, such as a program administrator, or to a review board. This party usually reports to product development, strategy, or research and development. Some companies establish innovation centers or program offices. Typically this person or team is responsible for establishing the guidelines for submitting ideas, assigning ideas to specialists for concept development, facilitating review meetings, and reporting on the progress of all ideas in process. Establishing idea guidelines helps gauge the eligibility and type of ideas wanted. The ideal attributes for the individual performing the focal person role include excellent oral and written communication skills, experience in product development process and project management, technical and administrative skills, and a passion for innovation and creativity.

Response timeliness and feedback to the idea submitter are key to the idea evaluation process. In more structured programs, an idea is escalated to a subject matter expert for review. The idea is described in enough detail to obtain approval from a steering committee that manages the priorities and the product development project portfolio. The idea evaluation process can be directly tied to the product development methodology used to deliver products to market. An example of a systematic idea evaluation diagram, showing the process flow from receiving an idea and idea evaluation management through a focal person or team to integration with the product development process, is displayed in Figure 9-4.

Questions to consider:

♦ What form will be prepared to submit a new product idea?

♦ What are the program and idea submission guidelines?

♦ What is required to process an idea? Who will screen ideas? What are the responsibilities of the administrator, idea evaluator, idea review committee, and senior management review board?

♦ What is the time frame for evaluating and criteria for screening an idea?

♦ What is the approval process once an idea has been evaluated?

♦ What method will be used to communicate results of the idea evaluation and progress to the idea submitter?

♦ What method will be used to track and monitor ideas?

♦ How will the idea capture process integrate with the new product development process?

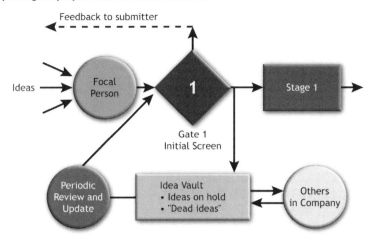

FIGURE 9-4. Example of a systematic idea capture and handling process.
From Product Development for the Service Sector by Robert G. Cooper and Scott J. Edgett.
Copyright © 1999 by Robert Cooper and Scott Edgett. Reprinted by permission of Perseus Book
Publishers, a member of Perseus Books, L.L.C.

Step 8: Establish Measurements and Goals

Expected outcomes should be identified for the employee idea program. Once
the program has been launched, actual results can be matched to the expected
outcomes as a measure of the return on ideas. Measurements act as a barometer
to confirm success or identify program improvements. Commonly, statistical
measurements are used, such as the percentage of employees participating, the
number of ideas (either per employee or per a prescribed time frame), the
number of new products implemented that originated from an idea submission,
and the number of new ideas launched. Human factors such as how useful
employees find the program should also be considered in measuring program
effectiveness.

Questions to consider:

◆ What measurable goals would you like to achieve: higher participation
rates, faster turnaround, or increased number of new products launched?

◆ How will you measure human factors, such as whether the program is
easy to use or efficient?

Step 9: Define the Awards Program

Choosing an awards program approach for an employee idea program is no
simple task. Approaches to reward ideas differ based on the program type and
purpose. The awards program approach you choose should include recognition
as well as other rewards.

EXAMPLES

BIC offers a $1 cafeteria coin for every idea submitted. The company also has awards for best idea of the week and month. BIC has an amazing participation rate of 90 percent in its employee idea program.

Rewards

Rewards can be intangible, material, or financial. Rewards programs can take many forms; however, cash is the most widely used, followed by merchandise. Idea program reward structures are typically based on a large cash award to one idea submitter. On the high end of the spectrum, Sony Pictures has offered awards up to a maximum of $250,000 for new revenue enhancement ideas. Suggested merchandise could include traveling trophies, humorous gifts, plaques, or free lunch.

While the concept of sharing the cost-saving benefits with an idea submitter seems logical, EDIS and *kaizen teian* program proponents believe a large cash award discourages continuous improvement and gives employees the sense that they should be paid for their ideas. EDIS and *kaizen teian* approaches use nominal cash rewards and place more emphasis on recognition and teamwork.

Recognition

Recognition plays a key part in the success of any employee idea program. Often the human side of implementing a program and intangible awards are overlooked. Key to keeping employees motivated to submit ideas is a response recognition program. All ideas should be recognized in a timely fashion with a thank-you and a response. There are many inexpensive public recognition items such as ceremonies, newsletters, or premium merchandise.

Questions to consider:

♦ Do you want to reward employees for their new product ideas?
♦ Who will be eligible to receive awards?
♦ What type of awards and methods to distribute awards will be used?
♦ Will awards be distributed at the time of submission or upon implementation?

EXAMPLES

ExxonMobil has developed an employee idea program using Lotus Notes. Recognition letters have been built into the program's design. The system automatically generates letters to:
♦ Thank an idea submitter
♦ Thank the evaluator for responding on time
♦ Thank the individual or team that was part of the implementation
♦ Thank the manager if he or she supported the effort
♦ Thank a submitter for an established number of ideas (e.g., five ideas in one month)

Step 10: Implement Program, Communicate, and Train

Once you have completed testing your approach and program materials, you are ready to introduce the program to the larger audience. If the program is companywide, it should be part of the employee handbook and communicated in new employee orientations.

Training

Initial and ongoing training will need to be defined and scheduled. Training classes should provide content specific to each target audience (administrators, evaluators, managers, submitters). For example, first-line managers should be aware of the critical role they play in promoting creativity among their staff. All employees need to understand:

- The kind of ideas you want
- How ideas are submitted and the process around submitting an idea
- What they will receive in return

Training to stimulate the creative process, using idea-generating techniques, would be valuable if the program is new for the organization. This will make people more comfortable submitting ideas and improve the likelihood of a successful program.

Questions to consider:

- What training will be provided to teach people how to begin thinking of ideas?
- What training will be provided based on specific roles, such as submitter, evaluator, administrator, or manager?
- Who will perform the training: the training department, human resources, product development, or other?
- What identity and design logo will the program have?

Step 11: Set Up Ongoing Administration/Program Maintenance

Administration refers to the ongoing management of the program, such as tracking ideas, performing an initial idea screening (qualified/nonqualified), preparing ideas for the screening review, and providing performance updates. This should be a designated job rather than a part-time effort.

Plans to keep the program fresh and ensure that objectives are being met can be achieved through the use of motivational techniques and ongoing training. Periodic audits are often needed to refine the program as business needs change.

Details about the employee idea program, such as information about the ideas received and future program plans, should be shared with all employees.

This can be done through special promotions, company newsletters, and management reports. Celebrating an idea successfully implemented will go a long way toward encouraging employees to take an active role in the program.

Questions to consider:

- How much time is needed to manage the employee new product idea program?
- What resources will be needed for managing the program?
- What reporting information will be provided to management and with what frequency?
- What will the ongoing motivational and promotional plan consist of?

All questions are summarized in Table 9-5, pages 230–232.

ENSURING IDEA PROGRAM SUCCESS

Once you have made the commitment to implement a new idea program, there are several key points to keep in mind to ensure program success:

- *Gain management support.* Management must take an active role in reviewing ideas and providing resources. Many grassroots programs have started and failed without the appropriate support.
- *Design a program for your company.* The program should be customized to your company's culture and work processes.
- *Make it easy to submit ideas.* Keep it simple—make it fun. Inspire, stimulate, and facilitate the flow of ideas from employees. Build a program culture to cultivate teamwork, promote learning, and increase commitment. Don't assume ideas can be shared easily across departments. Remember that technology cannot replace face-to-face meetings.
- *Measure success.* A program should be developed with the expectation that it will produce tangible results in the near term. It is important to the long-term survival of an idea capture program that it begin with a purpose connected to real business issues. Both outcome and process measures should be tracked.
- *Provide results.* Results produced should be communicated. Keep track of the number and types of ideas submitted by employees. This will help identify areas where program participation is low. Promotional activities can be targeted to low-participation areas to increase overall involvement.
- *Offer feedback and immediate acknowledgment.* Reinforce the program by recognizing all ideas submitted and keeping idea submitters informed. Response to the submitter should be fair and timely. Continually communicate idea evaluation status progress to submitters
- *Recognize and communicate success stories.*

♦ *Organize launch-related and ongoing promotional efforts.* Ongoing training and promotion are essential to retain employee interest. In some companies, the pilot test gets a lot of attention because of the newness of the program, but the real work begins after the pilot, in sustaining a flow of ideas.

SUMMARY

Employee idea suggestion programs, over one hundred years old, have evolved through various forms and are experiencing a rebirth due to changes in the economy. Such programs can and should be used for more than quality programs or incremental process improvements. Employee idea programs provide an efficient and effective way to source new product ideas from those closest to the production, distribution, and servicing of your company's products. These programs provide many benefits and can ultimately be used to develop more and better product ideas. We believe the trend toward collaborative idea suggestion programs will continue to expand internally and, eventually, externally.

To win in the new economy, organizations must tap into their most valuable resource—their employees and the valuable, creative ideas they hold. When a collection of brilliant minds, hearts, and talents come together, expect a masterpiece. Organizations that are committed to creative collaboration will ride the wave of the future; they will go beyond the realm of assumptive thinking and welcome the dawn of innovation. There is nothing more empowering than an idea whose time has come.

REFERENCES

Cooper, Robert G., and Scott J. Edgett. 1999. *Product Development for the Service Sector.* Cambridge, MA: Perseus Books.

McDermott, Robin E., Raymond J. Mikulak, and Michael R. Beauregard. 1993. *Employee Driven Quality: Releasing the Creative Spirit of Your Organization Through Suggestion Systems.* White Plains, NY: Quality Resource.

Robinson, Alan G., and Sam Stern. 1997. *Corporate Creativity: How Innovation and Improvement Actually Happen.* San Francisco: Berrett-Koehler Publishers.

ADDITIONAL RESOURCES

Bassford, Robert L., and Charles L. Martin. 1996. *Employee Suggestion Systems: Boosting Productivity and Profits.* Menlo Park, CA: Crisp Publications.

10 Lead User Research and Trend Mapping

Lee Meadows

How can we understand what a new market will require when the market doesn't yet exist? How could a company in the ophthalmology industry understand what market requirements would emerge from laser eye surgery before the procedure was approved by the Food and Drug Administration (FDA) and launched into the market? How could 3M determine where the next major opportunities would arise in its diverse business units? The rapid explosion of new technologies and new types of products makes this a pressing problem for those responsible for innovation. Because new technologies and new markets are also responsible for the largest new business opportunities, new product developers are particularly interested in how to anticipate their impact and new product requirements. This chapter deals with two synergistic research tools, lead user research and trend mapping, that address these needs. Lead user research helps new product developers find emerging user needs and solutions to those needs. Trend mapping is an analytical tool for finding and projecting business trends.

Lead user methodology is a research process for finding innovative users on the leading edge of market change. Eric von Hippel at the Massachusetts Institute of Technology's Sloan School of Management was the first to identify lead users and their important role in new product invention (von Hippel 1988). Von Hippel found that lead users are more than early adopters; they also create new solutions to address the issues and problems that they encounter. Thus lead user research can be used to define the needs that future users will have as a result of these changes, as well as find initial solutions to those future needs. Not only can lead users articulate needs that the market doesn't comprehend yet, but they also develop prototypes of new products and services to meet those needs.

Lead users invent their own solutions because their needs are not yet common enough to represent a viable market for development of commercial products. Because lead users are technically sophisticated and have significant incentives to create prototype solutions, they often are the first to invent such new products. Eventually the number of potential users grows, and manufacturers or service providers note the new market opportunity. Often lead user inventions are observed and incorporated into a commercial new product. Von

Hippel has shown that a majority of seminal new products were created by lead users in industries as diverse as computers, chemicals, and scientific instruments (von Hippel 1988).

An example of a lead user that was observed during a project at Nabisco was the company's own design engineering group. This department did initial manufacturing equipment design for bakery cooking and packaging equipment that incorporated innovations not yet offered by baking equipment vendors. When the design offered unique plant efficiency potential they would contract for prototype construction of their design from an external equipment manufacturer that refined and built their invention for use at Nabisco baking plants. Eventually equipment manufacturers adopted the Nabisco innovations into their general product line. Nabisco then purchased these commercial versions of the machines they created since they were less expensive than the custom-built prototypes. In another example, von Hippel shows how a similar bread-baking plant was considered a problem client by its bakery equipment manufacturer because of frequent service calls for a packaging machine that the bakery had modified. The machine was modified to run three times faster than its original design specified. Eventually the manufacturer noted the changes and adopted them to offer improved speed on its packaging equipment (von Hippel 1988: 62).

The Arm & Hammer Division of the Church & Dwight Company offers an example of how lead users can become an important source of new product development ideas. Virtually all Arm & Hammer's new products are based upon tracking consumer and commercial applications of its basic baking soda (bicarbonate of soda) product. The company receives a constant stream of user letters and market research describing innovative ways that consumers use baking soda. The company then formulates and markets enhanced versions of their consumers' home remedies. The company's consumer division grew by twenty times between 1972 and 1992 by consistently following this lead user business-building scenario. Improving on lead user practices generated consumer products such as Arm & Hammer detergent, toothpaste, deodorant, and cat litter. The industrial division of the same company has entered large new markets for airplane cleaning and printed circuit board cleaning by following customer-initiated innovations as well. For each of these new industrial products customers came to the company and showed how they adapted baking soda to meet their needs. Then the company created new products consisting of baking soda and other ingredients that provided better solutions than baking soda alone offered.

The Arm & Hammer and Nabisco examples of lead users show the power of lead users without a definite methodology for finding and using their innovations. Eric von Hippel evolved methods for finding and using lead users to accelerate the development of seminal new products. In von Hippel's work with the 3M company a four-step process was created to find and use lead users knowledge. This process is captured in *Breakthrough Products and Services with Lead User Research: Methods for Uncovering the Ideas and Prototypes of Leading Edge Users*, a handbook for 3M applications (von Hippel, Churchill, and Sonnack 1998). The four steps include:

◆ *Stage 1: Preparing to launch the lead user project.* In this stage 3M forms a small (4–6 people) cross-functional team to implement the lead user project. The team sets boundaries, secures resources, creates an informal sponsor group, agrees on a business area to focus on, and outlines its development plan.

◆ *Stage 2: Identifying trends and key customer needs.* In this exploration stage the lead user team searches for market and technology trends that are changing the targeted business area. By the end of the stage the team has focused on one or two trends that are changing their target business category and the related customer needs that will be the focus of the project.

◆ *Stage 3: Explore lead user needs and solutions.* In this phase of the project the team follows knowledge and communication networks to those leading world experts who understand the market need and potential solutions best. In interviewing individuals along these knowledge networks the project team develops deep understanding of emerging customer requirements and potential solutions. By the conclusion of this stage the team has identified innovators that have already invented new prototype solutions for their own use.

◆ *Stage 4: Improve concepts with lead users and experts.* In the final stage of the 3M lead user research process two to three day workshops are hosted at 3M andattended by lead users, external technical experts and 3M product developers. The workshop teams create improved new product solutions and preliminary new business cases to support development proposals within 3M. By the conclusion of this stage new product concepts are finalized for broader customer testing, and business plans completed for senior management review and approval.

A recent study of the impact of lead user research at 3M confirmed its ability to find major new product opportunities. According to the research's lead author, Gary Lilien, from Penn State's Smeal College of Business Administration:

> Annual sales of lead user product ideas generated for the average lead user project at 3M are conservatively projected to be $146 million after 5 years—more than eight times higher than sales for the average contemporaneously conducted "traditional" project. Each funded lead user project created a major product line for a 3M division. As a direct result, divisions funding lead user project ideas experienced their highest rate of major product line generation in the past 50 years. (Lilien et al. 2001: 1)

The remainder of this chapter will examine these lead user project steps in greater detail. The steps are slightly modified from the 3M process, reflecting experience in a more diverse range of industries. The modifications address several issues. The first is 3M's willingness to commit 25 to 50 percent of four to six managers' time to lead user projects. Many companies lack the internal staffing resources needed to make such innovation commitments. The second issue that leads to process modification is the adoption of additional trend-mapping tools to identify where the lead user search should be focused.

PHASE I: FRAMING THE LEAD USER PROJECT

The success of any new product project is determined in large part by the quality of its participants and resources. The initial phase of a lead user research project is to staff a cross-functional project team, secure a budget, and establish management participation levels that optimize the chances of success.

Forming the Lead User Project Team

Lead user projects require a relatively small, experienced cross-functional project team that is comfortable outside traditional industry boundaries. The team needs to be relatively small because it will require a significant amount of time and commitment from its participants. Most teams at 3M consist of four or five members. At other companies the team size has ranged from three to six. The participants should represent the management functions that drive the business. These always include a representative of at least the marketing, technology, and manufacturing functions. The titles change with each industry and type of product or service delivered. Cross-functional teams bring together these management functions to bypass slower sequential departmental development. Successful lead user teams typically bring together senior managers and directors. Such managers should have at least three years' experience with the company and ten or more years in their functional area. This ensures that the team understands how their firm and functional areas work at both formal and informal levels.

Lead user projects require that project team members reach outside their normal knowledge and practice universe, because they will deal with new customers and technologies. Members who are uncomfortable leaving their work environment and its comfort zone limit the ability of the lead user team to find emerging new markets and opportunities. Therefore, it is productive to make team membership voluntary and limit it to those who have manifest curiosity about processes and markets outside their current business arena. In all the teams that we have worked with (and in all of the 3M lead user projects) team members have been previously trained in cross-functional team practices as well.

The team members must commit a significant portion of their time to the lead user project. At 3M 25 to 50 percent of team members' time is committed to the project during its three-to-five-month life (von Hippel, Churchill, and Sonnack 1998: 3:9). Some of this time can be transferred to external consultants where budgets permit. However, at least 10 to 20 percent (two to four work-days during each month) of the team members' time needs to be committed to the lead user project even when extensive use of consultants is planned.

Securing a Budget

The team needs to capture a working budget for the lead user project in its initial phase. The typical out-of-pocket budget at 3M is $30,000 to $40,000, covering travel, workshop honoraria, and other miscellaneous charges (von Hippel, Churchill, and Sonnack 1998: 3:10). In addition, there is an internal facilitation/trainer charge at 3M that is similar to what other companies pay external consultants.

In our project experience an honorarium of $500 to $1,000 is needed to recruit each lead user for participation in a two- to three-day workshop. Each workshop contains five to seven lead users. It is advisable to plan two lead user workshops as part of the project, with at least one week between workshops. Travel expense budgets for workshop participants vary with travel distance and the number of days in the workshops. For international lead users, the work-shop can be held at an international conference that is likely to draw many of the target participants. For example, sessions have been held at the beginning and end of major world medical conferences when surgeons were the targets. Most workshops, however, are hosted at the company's conference facilities. This is efficient from a budget viewpoint, and it permits broader company personnel participation.

Setting Project Boundaries, Focus, and Objectives

Most contemporary project management texts and training programs stress the critical need to establish a common understanding and agreement within the team on its objectives and mandate. Because lead user projects seek new prod-uct breakthroughs, their boundaries are broader and the objectives more aggressive. From our observations seminal new products come from either find-ing technologies that permit delivery of new benefit levels to existing customers or identifying unmet needs among current or new customers. Lead user project boundaries must allow the project to focus on these gray areas around the current business. The project objective should be framed to find large new market opportunities and new-to-the-world products that can address these opportunities better than current offerings.

An example of such gray areas and objectives from a recent financial serv-ices project was to create a new type of financial service that addresses unmet needs of elderly adults in assisted-living residences. This project's objective and boundaries focus on a new market that the developer new little about. It rec-ognizes that the unmet or undelivered financial service needs of this emerging market were not known at the project outset. Importantly, the team's objective statement did not limit the type of solution that could be created to types of products that the company already offered.

Research by the Product Development and Management Association (PDMA) and others show that such innovative products take longer to develop. The PDMA best practices research shows that new-to-the-world products take twice as long to develop when compared to major product revisions (Griffin 1997: 15). Based upon this work, the time horizon in lead user research needs to be longer than is the case in more incremental innovation efforts. However, our experience is that lead user projects accelerate development of new-to-the-world products because lead users can more easily define product requirements and often bring prototypes of solutions to developers.

Developing a Project Plan

The team needs to plan its research and project phases at the onset of the program. Because business trend identification and lead user tracking require group efforts, it is desirable to set mutual schedules to undertake these activities. Approximate time requirements for the project team are:

   ◆ For the first phase, project planning and staffing, five to seven person-days are usually required from each team member. This will normally require two to five weeks to complete
   ◆ For the second stage, trend mapping and prioritization, we recommend team members commit six to seven person-days over a one-month period. These should be common days when the entire team is working simultaneously on trend mapping.
   ◆ For the third phase, lead user tracking, each team member should again commit six to seven person-days. This can take three to five weeks depending on the availability of team members.
   ◆ For the fourth and final phase, a lead user workshop to develop product concepts, we recommend the team commit five to six common person-days. This will occur over two to three weeks around the workshop dates.

It is strongly suggested that the lead user team schedule workshop dates and key team meetings for all phases of the project during the initial planning phase of the project. These project milestones ensure that the project moves forward in a timely fashion.

Creating an Informal Senior Management Sponsor Group

Stage gate systems create review points for senior management as each new product evolves through its development process. 3M has found it useful to

Core Tasks	Outputs	Duration
Team selection/recruiting	Recruit three to six managers from core functions	One to three weeks
Set objectives/boundaries and focus	Written paragraph	One to three weeks
Create project plan	One-page milestone and date targets for each phase's key task	One to three weeks
Secure budget	Project full project costs and manpower needs and gain approval	One to three weeks
Secure sponsor group	Identify senior managers to review phase outputs	One to three weeks
Phase I completion	All of the above	Two to five weeks

FIGURE 10-1. Phase I: Framing the lead user project summary.

create an informal senior management review process as lead user projects pass through each of the four phases of the process. To do this the project team identifies key executives who control resources that will be critical to any new product development program resulting from the lead user research. These are executives who are often one or two levels above the lead user team's immediate supervisors. The team will seek to review the project with these executives at the end of each lead user phase, as well as gain any inputs that the sponsors may offer. This informal lead user stage gate process ensures that senior executive concerns are identified and addressed as the project unfolds. It also removes the discomfort that often occurs when a radical new project is initially proposed. Identifying and meeting with this informal group of sponsors starts at the end of the first (framing) phase of the lead user project and then recurs with each successive phase. It will be easier for the team to do these briefings individually with each senior executive, rather than trying to find a common meeting time. As informal gates, the reviews should not stop the initiation of a new phase, but rather should be completed as soon as feasible.

PHASE II: TREND MAPPING

Lead users are customers at the leading edge of a trend that will soon impact a broader general market. The second phase of the lead user project identifies and validates the trends that the project will focus on. The objective of this phase is to identify user or technology changes that already are altering the target market. The problem is that there are many more perceived trends than real trends. The challenge is to validate which trends are real, and among these which are actually changing the marketplace. There are two tools that can be used to find and prioritize trends that will become the focus of the project. The first is every consultant's favorite learning tool: interviews to capture company and industry knowledge (see Chapter 3 for details on this tool). The second is a business trend research tool created at Motorola in the 1980s that Motorola called Product/ Technology Roadmapping. Both tools are used to find long-term changes (trends) in users, user needs, or the technologies that are used to satisfy users.

Step 1: Capture Existing Company and Industry Trend Knowledge

We suggest that every trend-mapping effort start by capturing current industry trend knowledge and expectations. Executives and managers who have spent their careers focused on a single market have gathered and sorted large amounts of data to address the question of how their market will change in the future. To capture this knowledge the lead user team typically interviews fifteen to twenty internal managers and executives, as well as an additional fifteen to twenty outside the company. Internally, it is necessary to interview several vertical levels of management as well as different functional areas within the organization. Senior management receives different types of industry information than do individuals at the director and manager levels. Similarly, the marketing, R&D, and manufacturing functions all see the future of the industry through different lenses. Externally, suppliers, customers, industry-focused academics, and industry financial analysts capture different information sets, too. To access their knowledge we ask several simple questions: "What trends do you believe will change the target business area over the next five to seven years? What data can you direct us to that support this belief? Which trends that were identified are likely to change the business the most?" Ideally, these interviews

---

The snack product development team is undertaking a category scan of likely future changes in the snack food business. We would like to capture the changes that you anticipate in this category over the next five to seven years, and the drivers of those changes. Please insert your responses and return this questionnaire by e-mail.

1. What changes do you anticipate in the snack food category in the next five to seven years? What drives the changes that you anticipate? Are there any data you can direct us to that highlight the change drivers that you identified?

2. What will be the most important change in consumer usage or behavior in the snack food category during this period? Can you direct us to any data that brought these changes to your attention?

3. What will be the most important change in technology or manufacturing for the snack food category in the next five to seven years? How will this impact the business?

FIGURE 10-2. Sample trend questionnaire.

should be done in person. In cases where this is difficult, phone and Internet interviews are feasible.

Upon completion of the trend interviews the team should aggregate the data to see how the answers vary by function, level, and location. This can be accomplished by creating a spreadsheet that captures each anticipated change driver and the function and level of each interviewee who mentions it. Often this aggregation shows that R&D managers are much more focused on an emerging technology's likely impact than are their counterparts in marketing or manufacturing. Conversely, marketing and manufacturing are likely to project changes in customers or production that have gone unnoticed by their counterparts. Visualizing these differences is often the first step in the teams' recognition of the scope of potential change in their future business.

The output of the interviews is a list of change driver candidates that must be quantified, validated as real trends, and correlated with market changes to see if they are actually altering the targeted market. Trends do not start tomorrow. They are already under way. The important ones are growing exponentially and are already changing the target business. Often trends are hidden by artificial business boundaries that ignore purchase of products from several categories or distribution channels for the same user need. An example was Ralston Purina's belief that dog foods purchased from veterinarians did not compete with those sold in grocery stores. Only when sales through veterinarians passed $1 billion did Ralston acknowledge that their products competed with products in this alternative distribution channel. Differing channels of distribution or product category definitions can obscure significant business trends and opportunities.

The team's core task is to take each change driver candidate and demonstrate that it is already impacting the target business. This requires identification of how the trend can be quantified over time from secondary research or primary data that the company has periodically collected.

Often the results provide startling new insights that can pay off immediately. In a project in the early 1990s a leading worldwide oral care company learned for the first time that market segments in developing countries were growing in opposite directions from those in developed markets. In North America, Western Europe, and Japan premium-priced products were gaining market share. However, in developing nations value-priced products were growing faster than the mid- and premium-priced alternatives. This had been ignored because the trends were visible only when the data cycle was extended to ten years from the company's usual three-year frame. The company also discovered through vendor data collected during the trend-mapping process a dramatic market trend that its own selling practices masked. The former discovery resulted in new R&D strategy that created new products for the developing-country market rather than waiting for developed-country innovations to trickle down. The latter discovery changed the sales practices of the company in all developed markets in a way that increased sales significantly.

But the most important trend came from analysis of forty years of dental office treatment data collected by the American Dental Association. This showed that topical cavities normally suffered by the young were disappearing. This was offset by the rapid rise in gingival cavities that result from the age-correlated receding of the gum line. Changes in both professional and consumer products were needed to address this demographically driven change.

In the dental care application thirty managers within the multinational company were interviewed. The internal company interviews included the chief operating officer, domestic and offshore marketing managers, regional directors, regulatory affairs executives, and technology developers from the bench to the chief technology officer. All of the trends cited above were noted in these interviews. But a dozen other possible trends were also cited. The key was to find quantitative data that showed which trends were real and if they were already changing the oral care market.

The second step in the industry trend survey is to find quantitative data that show how fast the trend is growing through time and whether it is correlated with actual market change. For each trend identified in the survey, secondary research sources are searched for historical change markers. In the oral care case the change in the type of dental cavities suffered in the population came from epidemiological studies and from forty years of American Dental Association survey data concerning what dentists were treating patients for. Forecasts of change utilized population age shifts to demonstrate the explosive increase in gingival cavities that the team projected over the next several decades. It is essential to find data that clearly demonstrate that the trend is real, will grow significantly in the future, and is changing the business. Only this type of data can convince the entire organization to invest in new products to address the trend.

EXERCISE: TREND MAPPING

To further understand trend mapping try the following exercise.

Background: You have been asked by a leading automotive manufacturer to identify trends that will change the automotive market in the next seven to ten years. Complete the following questionnaire as a part of this assignment. You should find at least one answer for each probe:

1. List any important changes that will occur to automobile buyers during the next ten years.
   Demographic changes_____
   Economic changes_____
   Physiological changes_____
   Other buyer changes_____

2. List any trends in automobile attributes over the past decade that you are aware of.
   Changes in auto size and physical characteristics_____
   Changes in auto power_____
   Changes in auto costs/pricing/financing_____
   Changes in auto convenience_____
   Changes in auto regulations_____

Changes in alternative transportation_____

Changes in other auto attributes_____

3. What are the differentiating attributes of successful new types of autos launched in the recent past?

Compact SUV (e.g., Jimmy, Explorer)_____

Full-size SUV (e.g., Expedition, Grand Cherokee, Yukon)_____

Nostalgia cars (e.g., VW Beetle, Prowler)_____

Other successful new types_____

4. Where would you find quantitative data to validate that three of the trends above are real over the past decade?

	Trend Name	Data Source
1.		
2.		
3.		

Step 2: Roadmapping

Roadmapping is simply an analytic process for finding and validating the trends that are changing markets. The earliest description of roadmapping appears in a 1987 *Research Management Journal* article, "Technology Roadmapping," by two Motorola managers (Willyard and McClees 1987). The rapid development and dissemination of roadmapping occurred in the electronic and semiconductor-based industries. These industries use roadmapping as a strategic analysis tool that brings marketing, technology, and manufacturing functions into agreement on how and when market demand will differ from current demand. Kim Clark and Steven Wheelwright at Harvard built upon roadmapping in Chapter 3 of their 1992 book *Revolutionizing New Product Development*. Their vacuum manufacturer case application shows how roadmapping can be applied and expanded to more prosaic industries.

Rich Albright, former head of technology strategy at Lucent, created the roadmapping approach described below. The case example comes from a presentation made at a PDMA conference by Albright in 1998. The first step in this roadmapping process is to capture historical category data in several charts, including studies of customer/consumer product ratings and priorities, competitive strategies, product life cycle data, and product share and unit sales data. Figure 10-3 shows an example of these charts for the cellular handset market. (Note that the data in each of these charts have been altered to protect confidential company information.)

The next step is to create product experience curves. One of these data maps captures how product price change has impacted unit volume (Figure 10-4). This is augmented with product attributes maps over time (Figure 10-5).

COMPETITIVE STRATEGY AND DIFFERENTIATION

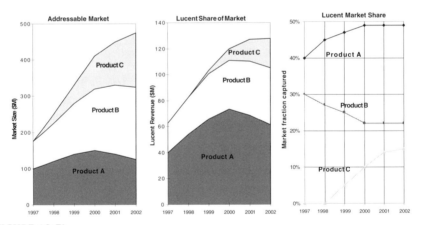

FIGURE 10-3a.
Used with permission of Richard E. Albright.

MARKET SHARE AND GROWTH

FIGURE 10-3b.
Used with permission of Richard E. Albright.

The electronics industry focuses on discovering how fast product costs are falling in relation to demand increases, as illustrated in the cell phone example (Figure 10-4). The 70 percent slope reflects that as the average cost of handsets falls by 30 percent the demand for handsets increases by 70 percent. The example below uses log values to create linear chart data.

The analysis then determines which product attribute change is most correlated with the changes in demand and unit cost. In the examples below it is

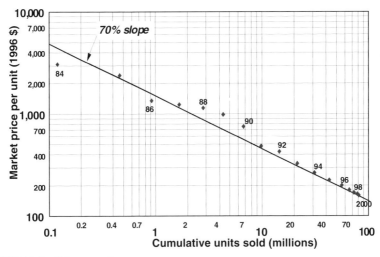

FIGURE 10-4. Price experience curve.

evident that only the falls in handset weight and size are closely correlated to the demand slope. The tight fit of the historical data indicates that it is feasible to project continued reduction in size and weight of the cell phone at its current pace. This also provides validation of which cell phone attribute changes are critical to both cost and demand changes. Exponential smoothing provides a projection of when additional market changes are anticipated in the critical size attribute. For the company this results in timing targets for when it must offer smaller units, and how small they must be to lead the market. The term *road-mapping* comes from the product improvement timing plan that is developed based on these projections.

For lead user research, roadmapping often identifies and provides strong evidence of the change drivers that must be focused on in the Phase III search for lead users. In the cellular phone example, the trend toward smaller handsets would become the focus of the lead user search in the next phase.

Step 3: Tracing Historical New Product Success Drivers

There is one additional trend test that the lead user team may find useful. It parallels the roadmapping process in several important ways. To initiate this analysis the team should identify all new product introductions over a significant historical period (ten to thirty years). The team then isolates what benefit each new product offered. Finally, it captures the maximum share achieved by each new product. By aggregating the new products by core benefits, it is possible to determine if any single benefit dominates new product success. Figure 10-6 shows that a single benefit in the oral care category dominated new product success. This new product trend assessment often provides new insights to

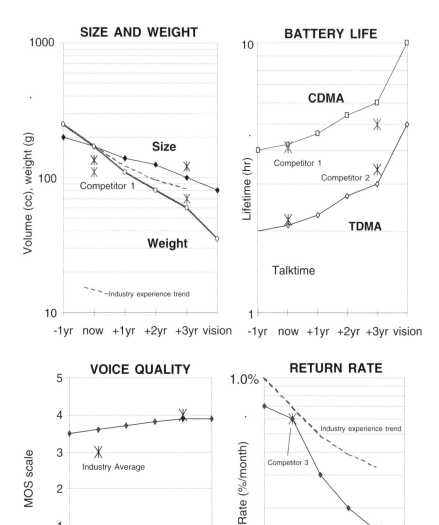

FIGURE 10-5. Product attribute experience curves.

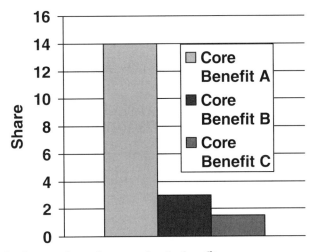

FIGURE 10-6. Average share of new product by benefit.

the lead user team. The example shown in Figure 10-6 is taken from one of the oral care product categories.

PHASE III: FINDING LEAD USERS AND TREND KNOWLEDGE

Phase III has two related objectives. The first is to find lead users who are already at the front edge of the trend and are creating new solutions for trend-related needs. The second objective is to learn about the trend and its impact in much greater depth. To accomplish this, von Hippel identified knowledge and communication networks as the most suitable research vehicles.

To understand why this novel research process is necessary, we need to examine lead users in more detail. Von Hippel has noted that three types of

Tasks	Outputs	Duration
Capture existing change expectations	List of change drivers from internal and external managers and experts with quantified data to support impact	Four to eight weeks
Roadmap product characteristics	Map historical market/product/ manufacturing changes and project	Four to eight weeks
Trace new product success drivers	Analyze what benefits drove historical new product successes	Two to four weeks
Review and select trend or change to focus on	Project team and sponsors review above and agree on need or opportunity to focus lead user search	One week
Phase II completion	All of the above	Five to ten weeks

FIGURE 10-7. Phase II: Summary chart.

Hello, my name is [name] and I am a [give functional area] with [name of company]. I am part of team at [company] that is investigating the nutritional needs of athletes and how they might be met with improved offerings.

I was given your name by [source] as an expert on nutrition and athletes. Would you have a moment now to answer several questions, or would there be a better time to call?

1. What are the key nutritional requirements for elite athletes, and how do they differ from an average person's needs? [Probe several times.] How would a recreational athlete's needs compare?

2. What are the chief challenges in meeting the nutritional needs that you outlined for an elite athlete? [Probe until topic exhausted.] What are the challenges for dedicated recreational athletes in meeting their nutritional needs? [Probe again until all problems and issues are shared.]

3. What solutions have elite athletes found to overcome the challenges that you mentioned in meeting their nutritional needs? What types of athletes have created the most innovative solutions? [Probe for examples and names.]

4. Of the foods that you just mentioned [cite examples], which would offer the most benefits to the dedicated recreational athlete? Why do you say that? [Probe.]

5. Whom else do you recommend that I talk to about the nutritional needs of athletes and how best to meet them? Are there any innovators in this area that I should make a special effort to talk with?

Thank you very much for your time. You have certainly helped me understand this area.

FIGURE 10-8. Sample phone interview script: Snack food for recreational athletes.

lead users are potentially useful in our search for seminal new product proto-types and related knowledge. These three types include lead users in the target market, lead users in similar markets that have advanced further on the trend, and lead user experts on some key attribute of the trend (von Hippel, Churchill, and Sonnack 1998: 6:4).

Lead users in the target market are the classic lead users whose needs fore-shadow those of the general market and who are motivated to create prototypes to solve their personal problems. In a Nabisco project that targeted foods for recreational athletes, a useful example of this type of lead user was a body-building couple who had jointly authored several cookbooks for athletes. They offered novel snacks that they prepared for their own use.

Lead users in similar markets that have advanced further on the trend can often be found by looking outside the target market to markets that address similar needs with advanced solutions. For the Nabisco project, an example of this type of lead user was the head of nutrition for the Navy SEALS. The lead user had developed several interesting nutritional products that the SEALS could use in the field to sustain their performance.

Lead user experts on some key attribute of the trend can often be found by pinpointing the key attributes of a successful solution for the emerging market. Through this the team can identify technical experts who can bring new insights to the lead user invention workshop. For Nabisco, such an expert was a physiologist who worked with NASA. This expert could specify the sources of protein and carbohydrates that could be used most effectively by those under stress.

Examples of each type of lead user are offered in the 3M application of the process for the surgical drape market (von Hippel, Thomke, and Sonnack 1999: 8). Surgical drapes are materials used in operating rooms to create a barrier to infection around the surgical wound. The 3M team identified the growth of drug-resistant microbes and the growing market focus on more economical health care as their trend focus for the project. An example of lead users in the target market that were particularly useful is surgeons in Asian clinics that served the poor. Their needs foreshadowed those in advanced markets because they were already dealing with minimum budgets. A valuable innovator from a parallel or similar market was a leading veterinarian surgeon who held patents on low-cost surgical materials and devices for animals. Finally, an expert on a particular attribute of the ideal surgical drape was a Hollywood makeup artist who had extensive knowledge of materials that can easily conform to the body's diverse contours, as the issue of how to keep surgical dressings in place was critical to the 3M team's ideal solution.

Each of these three types of lead users can make contributions to the development of new types of products that address emerging market needs. But the number of lead users in the universe is quite small. Traditional sampling procedures that are used in market research are impractical in searching for lead users. You are looking for a needle in a haystack. Interviewing thousands of people will not find a lead user.

Von Hippel has recommended that we follow communication/knowledge networks to their apex to find lead users. This can be accomplished through literature searches followed by sequential telephone interviews that trace the knowledge network. Project team members are searching for those who know the most about the trend and the types of need that it creates. Industry journals and research journals provide quick starting points. Authoritative industry experts on each trend can be found in these journals using electronic literature searches. Phone interviews start with these public experts, who can provide both knowledge about the trend and suggestions of other individuals who have the greatest knowledge about the trend and its related problems. Sequential calls to ten or twenty experts quickly bring the project team member to leading world experts on the project's target trend. Such experts can direct the interviewer to lead users who have created interesting prototype solutions for problems associated with the trend. In following this trail, team members also become more knowledgeable about the trend. By sharing their learning with each other, the team codifies the learning.

An example of this search process occurred in applying lead user research to refractive laser eye surgery prior to approval of the eye lasers by the FDA. The medical literature search and telephone interviews quickly narrowed to

Tasks	Outputs	Duration
Find lead users in target markets	Identify at least five lead users and capture incremental knowledge from interviews	Two to four weeks
Find lead users in similar markets	Identify at lead five lead users and capture relevant knowledge from interviews	Two to four weeks
Find lead user experts on key product attribute	Identify at least five lead user experts and capture relevant knowledge from interviews	Two to four weeks
Phase III completion	All of the above	Four to six weeks

FIGURE 10-9. Phase III: Finding lead users and trend knowledge.

several dozen corneal surgeons who were involved in the early clinical tests of laser machines. These were customers who were the first to use the new machines and who understood the new problems and issues that they created. Further, they were already modifying equipment and applying drugs approved for other applications to address these new issues. Despite the seven-figure income of these surgeons, they still answered the researchers' phone calls and agreed to attend the lead user workshops. They cooperated because they were very interested in access to better products.

The second type of lead user identified in the refractive surgery project came from allied medical applications. When the team asked the corneal surgeons about users who were concerned with similar issues, the responses included neurosurgeons, plastic surgeons, and practitioners in several other medical specialties who had already applied lasers in their field. Again, from surveying the medical literature it became evident which clinicians had the most experience in working with new lasers. Telephone interviews quickly led to the apex of laser surgery knowledge in these parallel medical fields. Lead users were those clinicians who had created new surgical techniques, instruments, and drugs to address the new issues they encountered in these advanced laser surgery applications.

The third type of lead user sought for refractive laser surgery was technical experts dealing with some important attribute of the ideal solution. Experts on eye tracking measurement and wound healing proved useful in the development of new pharmaceutical products to aid laser surgery. These experts were again identified through a detailed review of technical journals followed by phone interviews to determine who could provide the best insights on product development issues.

In implementing the lead user search, the first step should be a literature search to gain both a better understanding of the issue and a list of potential contacts. Team members then construct a simple open-ended questionnaire to use in the telephone survey. Each interview starts with an explanation of who the caller is and the trend he or she is researching. This is followed by probes of what the interviewee sees as the key issues resulting from the trend. Finally, each interview ends with the question of who might know even more about this topic. The team meets regularly during the research to review their findings and discuss issues or problems in their interviewing. Market research professionals are useful in helping the team to develop and refine their telephone questionnaire.

At the conclusion of the third phase the lead user project team should be able to construct preliminary product concepts, definitions of product requirements that come from their research, and initial estimates of target market size and possible rates of consumption. These are useful in starting to build the business case. Most importantly, the team should have identified a dozen or more lead users who they believe could help them construct even better new products.

EXERCISE: IDENTIFY POTENTIAL LEAD USERS

Background: Ford and GM have recently announced long-term efforts to improve SUV gas mileage. As part of a lead user team, you are asked where useful lead users might be found for this issue. Complete the following questionnaire as a part of this assignment.

1. Lead users in the target market who are technically sophisticated (e.g., race car builders)
   a. _____
   b. _____
   c. _____

2. Lead users in similar markets that have advanced further on the trend (e.g., airplane manufacturers)
   a. _____
   b. _____
   c. _____

3. Lead user experts on some key attribute of the trend
   a. Weight reduction (e.g., builders of yachts for the America's Cup)_____
   b. Energy efficiency (e.g., printer manufacturers)_____
   c. Air and surface friction reduction (e.g., turbine makers)_____

PHASE IV: LEAD USER WORKSHOPS AND JOINT INVENTION SESSIONS

This phase starts with team agreement on the lead users and company developers it will invite to participate in the lead user workshops. The workshops usually bring six to eight lead users together with six to ten company developers. A facilitator should also be recruited who has experience with group dynamics and moderating creative work sessions.

The team issues telephone invitations followed by faxed or Internet-based descriptions of the workshop, its participants, and the basis for participation.

Tasks	Outputs	Duration
Select and invite five to seven lead users to each workshop	At least two workshops scheduled with best lead user candidates	One to two weeks, with one month additional time until workshops can occur
Conduct co-invention workshops	Definition of several new types of products and supporting technologies	Two to three days each
Validate lead user innovations	Confirmation of market acceptance	Varies with industry
Phase IV completion	All of above	Two to six months

FIGURE 10-10. Phase IV: Summary chart.

External participants should be offered a meaningful honorarium for their time. Such payments typically fall between $500 and $1,000. In addition, the company should offer to pay their travel expenses. The lead users are faxed a confidentiality agreement and the terms of participation. The agreement is one to two pages in length. Under this agreement any inventions created at the workshop become the property of the host company in return for the honoraria and travel expenses. Additionally, the lead users agree to keep the workshop and information shared during the session confidential. If a participant is not willing to agree to these conditions, move on to your next choice. In our experience 70 to 80 percent of those invited agree to participate.

Often the company is surprised that lead users are willing to take the time to participate in the workshop. Von Hippel, Churchill, and Sonnack (1998) have noted three motivations for lead users to attend the workshop and help the sponsor:

- They expect that they will learn from the experience and enjoy playing a role in the company's product development process.
- They know they are unlikely to develop the particular innovations that may be created in the workshop without the stimulus of workshop participants.
- They have a very high interest in seeing new products developed and thus are very eager to contribute innovative ideas they may have developed.

Lead user workshop participants often know each other personally or know about each other's work. The invitee wants to meet and work with the other participants because of their expertise. As an example, dental clinicians who both worked in Veterans Administration hospitals and were on the faculties of leading universities participated in one of the oral care workshops focused on dental trends among older consumers. The stature associated with those invited, plus the chance to discover new solutions to their patients' problems, led all of those invited to attend the workshop.

Workshops start with a dinner the night before and then continue for one or two days. The dinner conversation breaks the ice and develops rapport. It should include all company and lead user participants. At the dinner each participant introduces him- or herself to the other workshop participants and provides an overview of his or her work focus.

The workshop is organized around first capturing the lead users' knowledge about the trend issues and then jointly inventing solutions to the new issues that they have encountered. Lead users often show or describe their prototype solutions at the workshop. The lead users' prototypes offer a starting point for joint invention of improved new products.

The workshop facilitator should start by establishing rules for working together and help the group agree on workshop objectives and agenda. The facilitator is then responsible for securing full input from all group members, keeping the process moving on schedule, and making the process enjoyable.

The outputs from the lead user workshops are:

1. *Several new product concepts that address the needs of those who will be impacted by the selected market trend.* In our applications of lead user research at least three new types of product concepts have been identified at the workshop. These concepts are captured in written statements that are often accompanied by drawings done at the workshops.

2. *Definition of solution requirements that the general market can't articulate because they have not encountered them yet.* By capturing both the new problems created by the trend and characteristics of the ideal solution in written form, the workshop provides a basis for further refinement and new product invention.

3. *Identification of new technology that is not currently used in the industry to enhance solutions.* The technical expertise of lead users has led us to describe the workshops as qualitative R&D. These inventive product users and product attribute experts are more likely to look outside the market for solutions than their corporate counterparts are. As an example, in a study focused on recreational performance foods a lead user responsible for the diet of the Navy SEALS identified new soy varieties with superior nutritional benefits that the host company researchers were not familiar with.

The central challenge after the lead user workshops is to validate that the new products identified in the workshop warrant development and launch. This is a significant issue because trends develop through time, and the new products may be launched before there are enough potential users. This leads to a two-step new product validation process. The first step is to test the new product concepts and prototypes with those who already have the need. The second step is to use the roadmapping and trend modeling from Phase II of the lead user project to forecast how fast the available pool of potential users will increase. This allows the company to match the product development and investment plan to the projected growth in the new market. Simulated market tests have also proven to be effective in estimating the growth of market for new-to-the-world products, such as Nextel's radio cell phone, PRK/Lasik refractive eye surgery, and new types of computer products, before they are introduced.

OTHER TYPES OF LEAD USER APPLICATIONS

A majority of the early applications of lead user research have been to find new product opportunities in existing or emerging markets. The general lead user process outlined above addresses this type of objective. However, lead users have addressed other types of new product challenges as well. These applications include forecasting how fast new technologies will develop and finding the best markets in which to apply new technologies.

Forecasting Technology Development

Historically, technology development has been forecasted using projections of historical rates of development, as in roadmapping, or by interviewing panels of technology experts. We have found that lead users are superior to traditional technology experts in forecasting when technologies will be able to deliver new levels of product performance. Technologists tend to be overly optimistic in projecting when current levels of performance will be enhanced and negatives eliminated. Lead users, who are early customers working with the technology, lack the ownership bias that technology creators suffer from. They also understand the limitations and disadvantages of the current technology.

In the laser eye surgery case discussed earlier, we found that panels of lead user eye surgeons were 40 percent more accurate than laser manufacturer personnel in estimating when new technology would eliminate side effects of the procedure. They were 60 percent more accurate over a six-year period in estimating when FDA approval would be given to launch new or improved medical devices. The technologists and companies who manufacturer the lasers had greater mastery of the laser technology, but they did not understand the physiology issues as well as the clinicians did. The clinicians also lacked bias about which technology would solve the problem. Those making the lasers could not balance their personal need for early success with the need to forecast accurately.

To use lead users for technology forecasting, the identification processes follow the Phase III steps outlined above. In the workshop setting lead users are asked to estimate if and when manufacturers will be able to reach specific performance levels over the next seven years. We find that seven years has been about the limit of forecasting that lead users are comfortable addressing. Then each lead user is asked to explain why he or she has predicted a specific time horizon. After group discussion a consensus forecast is sought, reflecting the new learning that often comes from the discussion.

SHOULD YOUR FIRM USE TREND MAPPING AND LEAD USER RESEARCH?

Both trend mapping and lead user research consume time and budget resources that are currently committed to other activities at most companies. The processes are neither easy nor painless for the lead user participants. However, the rewards can be significant. The benefits of application are evenly distributed between improving business focus (strategy), identifying new technology, and capturing new market opportunities. As a result, it is critical that senior management, R&D, and marketing commit to participating in these processes either as sponsors or by providing appropriate team members.

Keys to Lead User Project Success

- ◆ Staff the project with inquisitive, enthusiastic managers.
- ◆ Create an informal senior manager sponsor group that reviews the output of each lead user project phase.
- ◆ Actively participate in each step rather than passing off research to outside consultants.
- ◆ Quantify trends to validate that they are real and already impacting the business.
- ◆ Find and interview world experts on the trend before starting to identify lead users.
- ◆ Approach lead user solutions with the question of how can they be adapted for use by a broader target group.
- ◆ Blend all three types of lead users into each workshop.
- ◆ Build a business case for each lead-user-suggested new product before approaching management.

BIBLIOGRAPHY

Albright, R. 1998. "Roadmaps and Roadmapping for Commercial Applications." Presentation at the PDMA/T2S conference "Linking New Technologies to Market Windows," Atlanta, March 6.

Griffin, A. 1997. *Drivers of NPD Success: The 1997 PDMA Report.* Moorestown, NJ: Product Development and Management Association.

Lilien, G., P. D. Morrison, E. von Hippel, K. Searls, and M. Sonnack. 2001. Penn State news release (review of research results on 3M lead user process), http://www.newswise.com/articles/2001/3/LEAD.SCB.html.

von Hippel, E. 1998. *The Sources of Innovation.* New York: Oxford University Press.

von Hippel, E., J. Churchill, and M. Sonnack. 1998. *Breakthrough Products and Services with Lead User Research: Methods for Uncovering the Ideas and Prototypes of Leading Edge Users.* Minneapolis: LUCI Press.

von Hippel, E., S. Thomke, and M. Sonnack. 1999. "Creating Breakthroughs at 3M." *Harvard Business Review,* September-October, 47–57.

Wheelwright, S., and K. Clark. 1992. *Revolutionizing Product Development.* New York: The Free Press.

Willyard, C., and C. McClees. 1987. "Motorola's Technology Roadmap Process." *Research Management* 30, 5: 13–19

11

Technology Stage-Gate™: A Structured Process for Managing High-Risk New Technology Projects

Greg M. Ajamian and Peter A. Koen

Traditional Stage-Gate™ (SG) (Cooper 1993) or PACE® processes (McGrath and Akiyama 1996) assume that there is little uncertainty associated with the technologies to be utilized. However, the inability to manage high-risk technologies as part of product development is frequently the cause of canceled or significantly delayed new product development projects. Unlike product development, the ultimate outcomes of technology development efforts are unpredictable. Prematurely introducing a technology into the product development process when there is high uncertainty that the technology will ever meet the desired specifications often leads to project delays, project uncertainty, and project cancellation.

We have found that a different management process, the technology Stage-Gate™ (TechSG) process, is needed to manage technology development efforts when there is high uncertainty and risk. TechSG, initially described by Eldred and Shapiro (1996) and Eldred and McGrath (1997), brings a structured methodology for managing new technology development without thwarting the creativity needed in this early stage of product development.

The overall objective of this chapter is to provide the reader with more insight into the process and to describe the tools and methodologies that make up the TechSG process. The process is specifically intended to manage high-risk technology development projects when there is uncertainty and risk that the technology discovery may never occur and therefore the ultimate desired product characteristics might never be achieved.

To understand the difference between TechSG and traditional SG processes and the need for the TechSG process, consider the following extreme example. Imagine that due to unanticipated changes in regulations a business has decided that it will need a new product in its line that would best be described as "water without hydrogen." If the business immediately began a traditional SG process, it would very likely get bogged down in the product development stage, since no one had ever produced such a molecule. The product development team would be trying to complete the product design while their scientists were still trying to discover how to make the molecule in the laboratory. By using a TechSG process to focus on the technology development issues (with a long-term view toward business strategy, plans, and needs), the business can manage the technology development effort separately. The effort would continue until it is feasible to start product development or it becomes evident that the risks are too high or the rewards too low to pursue the technology. In this example, the TechSG process would be used to manage the technology development effort until the risks have been substantially reduced by demonstrating the ability to at least produce such a molecule in the laboratory.

The overall innovation process may be divided into three areas: the fuzzy front end (FFE), new product development (NPD), and commercialization, as indicated in Figure 11-1. Many companies utilize traditional SG systems to manage the NPD process. For most companies, even for a totally new offering, product development is essentially a known, predictable, and repeatable process. From 70 to 85 percent of the process is the same from product to product, e.g., planning, laboratory trials, preliminary design, internal and external testing, manufacturing scale-up, support training, market launch planning, and commercialization (Cooper 1993). Creativity is less of an issue in the actual development and commercialization of the product. Project leaders need good project management skills and decision makers need to concentrate on the overall costs and benefits to the business. Traditional SG processes have brought dramatic reductions in product development cycle time in many companies by bringing structure and an overall business process to the NPD process. However, a frequent complaint about and limitation of the traditional SG process is that product development begins when the ultimate outcome of the technology development efforts still contains significant uncertainty and risk. This has often been a source of frustration, causing significant delays and wasted

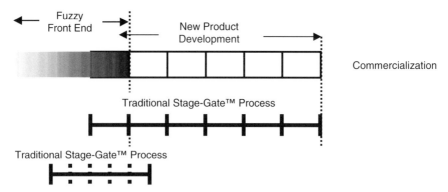

FIGURE 11-1. Overall innovation process may be divided into three parts: the fuzzy front end, new product development, and commercialization.
Traditional SG processes are utilized to manage projects in new product development. TechSG is used to manage high-risk projects within and at the transition between the fuzzy front end and new product development.

effort as the advantages and limitations of the new technology are determined. The technical uncertainty and continuing missed project deadlines associated with introducing a yet-to-be-completed technology discovery often results in canceled projects.

TechSG lies within and between FFE and the traditional SG process. The FFE represents the initial part of product development, from idea generation to development of a concept that includes primary features and customer benefits combined with a broad understanding of the technology needed. The traditional SG process typically begins with a well-defined concept, although many companies include the FFE in the initial part of the traditional SG process. Due to high uncertainty—especially in high-risk projects—we prefer *not* to include the initial part of the FFE in the traditional SG process. Further, we recommend that high-risk projects *do not* enter the traditional SG process if the project contains technologies where the ultimate outcome of the technology development effort has elements of both high uncertainty and high risk. Conceptually, the traditional SG portion of the product development process will not start until the TechSG process is completed and the technology development risks have been substantially reduced. In actuality, many projects start the traditional SG process in the later stages of the TechSG process, when significant risk has been reduced, in order to get to the marketplace earlier.

New technology development is by definition new, different, and unpredictable. It is difficult to capture and leverage past experience for future efforts, making cycle times difficult to estimate. One cannot "schedule" technology discovery. The range of possible experiments and their outcomes is almost limitless. Detailed overall project planning is therefore impractical. Too much structure or repetition of past work can severely inhibit creativity. It is often difficult to determine when the new technology is ready to transition to product development. This can be a very subjective decision, arrived at through informed discussions. Much more than in traditional SG, during new technology development, project leaders need the ability to manage uncertainty and do "good science" while focusing on project goals. (Examples of project goals for a next-generation copier [see Table 11-3] could be higher productivity, environment-friendliness, and lower operating cost.) Decision makers need to be able to balance risk and probability versus business needs and potential rewards.

The overall purpose of the TechSG process is to bring both scientific and business rigor into the technology discovery process, to better select and allocate resources to high-risk projects, and to reduce technology development times. The hallmark of a well-functioning TechSG is a project that is being executed using sound scientific principles and is properly resourced. Traditional SG projects also need to be properly resourced, but the importance of doing good science is less of an issue since the technology is already known. Initial reluctance to implement a TechSG process is related to the presumption that the added structure will inhibit creativity. In fact, we have found just the opposite. Technology development teams embrace the TechSG since it brings a sci-

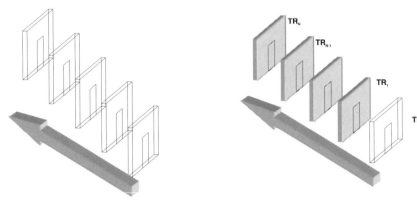

Traditional Stage-Gate™   Technology Stage-Gate™

FIGURE 11-2 In traditional SG, shown on the left, the gates are transparent. The product development team can "see" all the deliverables at the gates. In contrast, in TechSG, shown on the right, the gates are opaque. The technology development team can only "see" to the next gate and understands that the deliverable may change as the technology is developed.
From Eldred and Shapiro 1996. Reprinted with permission.

entific rigor often missing in many of their projects due to inadequate manpower allocation and financial resources, lack of the correct scientific expertise, and lack of peer review.

The differences between the traditional SG process and TechSG may be conceptualized in Figure 11-2. In the traditional SG process the gates—represented by walls—are transparent, since most of product development is predictable. The gates can be identified, clearly defined, and planned for, and their outcomes are known right from the beginning of the project. The product development team can "see" through the walls to the end of the project and can clearly envision the end results. Like the traditional SG process, the TechSG process consists of a series of gates or reviews. However, the details of the development plan are known only to the next stage. Although there is always a rudimentary overall plan for the entire project, the number of gates is only an estimate at the start. In new technology development, you know only what you are going to do next, not how it is going to turn out. So you cannot plan subsequent actions in great detail. Thus the gates in the TechSG process are opaque for all but the next stage. In contrast to the traditional SG process, in the TechSG process the gates are different in nature and the number of gates is highly variable. There is overall agreement about the desired outcome—the last gate—but there is uncertainty about how to get there. The TechSG process recognizes that the deliverables at each of the subsequent gates may change due to the need to take a different approach, unexpected discoveries, or scientific barriers or limitations uncovered during the technology development process.

The primary benefits of using a formal business process are to ensure that critical, limited resources are allocated to the projects of most importance to the business and to reduce overall the time to market of new technologies and the products that they enable. The technology planning and review process highlights business, process, and investment issues very early in the project and allows them to be more effectively addressed. Technology goals are well understood by all involved, are aligned to the product/platform strategy, and are revalidated from stage to stage. The TechSG process creates a collaborative environment between the technical and business communities, whether the entities are within one company or a joint venture between companies.

The remainder of this chapter is broken up into two main parts. The first part explains each of the six elements of the TechSG process. The second part discusses how this process may be implemented.

DESCRIPTION OF THE PROCESS

The TechSG process consists of six elements, as represented in Figure 11-3: the project charter, the technology review committee (TRC), the technology review process, structured planning, the technology development team, and the process owner. While similar in name, in most cases these elements are different from those used in the traditional SG process for managing NPD. A comparison between each of the elements is shown in Table 11-1.

All TechSG projects start off with a project charter to define the initial scope and objectives of TechSG effort. Although the TRC has business representatives, it is primarily a technology-focused team. Representatives from both the marketing and R&D divisions along with internal and external scientific peers are part of this decision-making group, which is responsible for making a go/no-go decision at each of the gates. The technology review process mainly

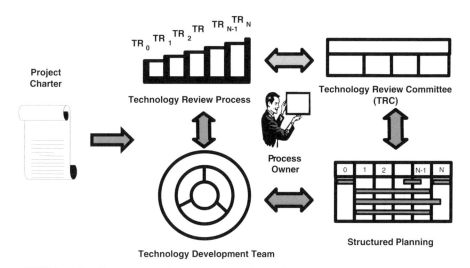

Project Charter

Technology Review Process

Technology Review Committee (TRC)

Process Owner

Technology Development Team

Structured Planning

FIGURE 11-3. Six elements of the technology Stage-Gate™ (TechSG) process.

TABLE 11-1.
Comparison of the Traditional SG Process with the TechSG Process

	TechSG	Traditional SG
Project Charter	Formal pre-agreement with upper management on the technologies to be investigated and the high-level approach.	Formal pre-agreement with upper management on the entire project, including the timing expected for project completion.
Review Process	Emphasizes the technology portion of the project and focuses almost entirely on the technology work and timing to the next gate. The gates are relatively opaque, with the deliverables known only to the next gate. The number of gates is unknown and varies significantly between projects.	Discusses the entire project, deliverables, and timing expected for the entire project. The gates are relatively transparent, with well-accepted deliverables known for all gates from the start of the project. Number of gates is usually the same for all projects.
Review Committee	Representation focused on technology. The chairman typically is the technology leader, with business representatives and scientific peers.	Broad representation of key functions of the business: R&D, marketing, regulatory, and manufacturing. The chairman is usually the division business leader.
Structured Planning	Detailed project plan exists only to the next gate and is very specific to each project.	Detailed plan throughout all of the gates. Same general plan for all projects; fairly detailed and known from the start of every project.
Development Team	Primarily consists of R&D or R&E.	Multifunctional. Typically made up of representatives of R&D, marketing, regulatory, and manufacturing.
Process Owner	Responsible for making sure that the TechSG process is adhered to.	Responsible for making sure that the traditional SG process is adhered to.

focuses on the next stage of work rather than on the deliverables for all of the gates, since the range of potential outcomes is vast. In addition, the gates lack the predictability that occurs in the traditional SG system. Structured planning utilizes technology development and performance tables combined with anchored scales so that all of the stakeholders understand the technologies being worked on and the project risks. The fifth element is the technology development team, which typically consists of scientists and engineers doing the actual development work. The final element is the process owner, who shepherds the methodology, assists process users, and ensures the continued use and improvement of the process over time. Each of these elements is discussed in more detail in the sections to follow.

EXAMPLE

For an example, we will use a modified version of an actual technology development project that was designed to enable XYZ Chemical Company to produce a molecule for sale that other companies were already producing. (Certain aspects of the project have been modified to maintain confidentiality.) Market analysis indicated that demand would increase significantly in future years and that the other companies would have difficulty meeting the increased demand. Furthermore, applying their expertise in certain processes and new technological advances might enable lower-cost production than traditional methods. Although other companies were already using the three different possible routes (chemical reaction, thermal reaction, and a patented process), XYZ Chemical had never produced this molecule, with its unique properties, hazards, and risks. It would therefore have to "discover" the specific details of how to make a production process work effectively and to investigate the new technologies that no one had yet used successfully in this application. Therefore, it was appropriate to start a project using the TechSG process before entering into a traditional SG process. This example will be used throughout the chapter to better explain the characteristics of the TechSG process.

Project Charter

All projects begin the TechSG process with a charter that is typically developed by senior executives, members of the TRC, and key members of the technology development team. This is similar to the product innovation charter discussed by Crawford and DiBenedetto (2000) but focuses on areas of technology improvement. The purpose of the charter is to provide a clear set of expectations for the project for both management and the development team, the resources available to begin work, and the expected timing for the first review (TR_0). Specifically, the charter should (1) define the scope and overall objective of the project and confirm its fit with the overall business strategy, (2) describe the specific technologies to be investigated, (3) indicate the key technical, business, regulatory, and marketing assumptions and risks associated with the technical and anticipated product development efforts, and (4) state the proposed resources that will be needed to develop the T_0 plan (usually a lead scientist/engineer, with some assistance from a planning/marketing/strategy representative and possibly from technical services/application support). The chartering process provides clarity around both the project and business goals *before* the work begins and saves considerable time and wasted effort by the project team. An outline of a project charter is in Figure 11-4.

The impetus to draft a charter can be driven by the business strategy, portfolio plan, or long-term technology plan, or it can be precipitated by some technical breakthrough or business development. The charter is usually written by the lead scientist or engineer based upon a request from technical management and presented (in person) at a meeting of the senior business leadership. The purpose is to get alignment on the technology to be investigated and its fit with the business needs prior to utilization of limited resources.

Although the focus of the charter is clearly on the technology development

Chartering Teams

To be successful, you must define who writes the charter, whom it is delivered or presented to, and who decides to accept, modify, or reject the proposal. Whoever is chosen to make the decision must have the authority to free up and reassign the resources (people, time, dollars, equipment, etc.) to make things happen. The goal is one sheet of paper (front and back). The charter must identify the key assumptions, expectations, deliverables, roles and responsibilities, and timing, plus provide a framework, a process, and a forum for discussion and understanding.

Charter Outline

1. Initial scope and objectives of work (very briefly comment on each)
   - Fit with the business strategy
     - What business strategy (competitive, growth, other) does this project support?
     - What markets are addressed?
     - How will the business win in the marketplace?
   - Fit with the product/platform strategy
     - What product/process platform is improved or created?
     - What is the market impact (productivity, performance, cost position)?
     - What is the impact on manufacturing? (manufacturability, contract, supply chain)
2. Primary areas of technology improvement
   - Brief description of the technologies to be investigated / pursued / developed
3. Key assumptions and risks
   - What are the key assumptions behind this development?
   - What are the key risks associated with this project?
4. Recommended process: normal (full) process, scaled-down process, other (specify)
5. Proposed resources to do the first-stage work
   - Team leader (usually the lead scientist/engineer at 100 percent)
   - Other technical resources (often a process engineer at 10–25 percent)
   - Business resource (often a marketing manager at 5–10 percent)
6. Proposed timing for first stage review

FIGURE 11-4. Technology development charter outline.

work being proposed, the business reasons why such work should be undertaken will not have been addressed anywhere else. Since the technology development project charter precedes any work on either the technology or a product, senior management will have seen no other documents describing the fit with the overall business strategy, business plans, technology plans, or product platform plans. The charter is specifically intended to justify the technology development work. However, business, product, and market assumptions *must* be included to get the leadership to consider assigning the resources to begin this work.

Unlike in the traditional SG process, the output of a technology development project is not a specific product or system. It is often a demonstration of an enabling technology or a proof of concept for a manufacturing process. Although the project may begin with a product (or more likely a product family

> **EXAMPLE**
>
> In the example, a product line manager at XYZ Chemical Company discussed the need for the molecule with the research director, who then asked a senior researcher to draft a charter. That senior researcher presented the charter to the business committee a few weeks later and answered questions with the limited information available at the time. The team leader was given the resources to complete the stage T_0 work in the TechSG process.
>
> The business committee is typically made up of the senior executives of the business, who determine if this is a valuable enough concept to charter a project and allocate resources. The business committee is not the same as the technology review committee. While the latter contains some members from the business committee, the technology review committee's focus is on technology and its membership comes more from the technology community.

or platform) in mind, at the end of the technology development project the business may still have nothing that resembles the product that it eventually wants to take to market.

Technology Review Committee

Just as in the traditional SG process, in the TechSG process there is a decision-making body with clear roles and responsibilities for the TechSG process that oversees development projects. The decision-making body in traditional SG systems typically consists of the senior executives of the business unit: general manager or CEO, VP of marketing, VP of research and development, VP of finance, and VP of operations (McGrath 1996). The leader of the decision-making body is typically the general manager or CEO. In many companies the members remain unchanged from project to project even though other senior executives may attend the meeting for one stage or one project.

In contrast, the TRC has considerably more variability. The authors' experiences in both chemical and consumer products companies have been that some, if not all, of the members of the TRC are different for every technology development project undertaken by the business. Since each project typically works on new or different technologies, and since the review committee's primary function is technical in nature, the experts needed to make decisions and recommendations will be different for each project. The makeup of the TRC may change many times as the project evolves, with new technologies required and others no longer needed.

Ostensibly the variability of the TRC within and between projects could make continuity and consistency a challenge. In reality, the most important issues in the authors' experience have been associated with the initial TechSG implementation, when the company is unsure of the process. After the first several projects the TRC and the TechSG process begin to work more efficiently as the company gains experience. Later projects, which may use a different TRC, typically do not have the same start-up difficulties, since the company has learned how to achieve quality technology plans and conduct effective Stage-Gate™ reviews. In addition, the chairperson as well as a few of the senior technology people remain on many of the technology review committees.

Since the focus of the TechSG process is on technology development, the chairperson of the TRC is often the technology leader for the business or business unit, not the business leader, as in traditional SG systems. There is a representative of the business who usually comes from the marketing organization to ensure a link to the downstream business and to provide input on customer needs and wants.

In the case of technology development, there are also additional members to provide advice about the key technologies involved. These advisors may be senior research fellows or advisors from inside and/or outside of the company. These advisors, scientists who are acknowledged to be leaders in their field, subject the technology development plan to a scientific peer review. The presence of scientific peers on the TRC elevates the scientific aspirations of the project and helps ensure that the science involved meets the necessary standards of excellence and rigor. Many technology projects in companies are not accomplished with the correct scientific rigor as a result of inadequate resource allocation. Peer review forces the TRC and project team to address the hard scientific issues, and this will typically result in sounder scientific plans. Scientific peer review represents a fundamental characteristic of the TechSG process that is essential for ensuring technical rigor. While scientific peers may exist within the company, the authors recommend that companies utilize external scientific peers at the gate reviews. External peers are both more likely to provide a fresh view of the project and typically are more forthright in their evaluation of the technical risks associated with the project. The external peers invited to participate typically are required to sign a confidentiality agreement, which includes a non-compete clause and assigns any invention that occurs as a result of this engagement to the company. The scientific representation on the TRC will also change depending on the expertise required for the technology being developed.

In traditional SG the decision-making committee members are truly the gatekeepers; they focus on the proper completion of all deliverables and milestones, adherence to the process, and the cycle time commitments. In contrast, the TRC is more of a review board. This committee does make specific decisions about allocation of resources and approval of plans for the next stage of work—the link to the ultimate needs of the business and the customers. But its primary focus is advising the team about their technical approach, proposing alternatives, and suggesting other factors to consider and other experts to contact.

The logistics of the technology review process are schematized in Figure 11-5. In order to ensure timely decisions, the TRC must make a decision to either fund, not fund, or redirect the project at the end of the technology review. The actual stage reviews are often divided into two parts. The first part is the technology review, and the second focuses on the business issues. During the first part, the scientific merit of the project is reviewed with a focus on the

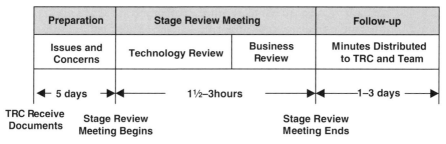

Preparation	Stage Review Meeting		Follow-up
Issues and Concerns	Technology Review	Business Review	Minutes Distributed to TRC and Team

← 5 days → ← 1½–3hours → ← 1–3 days →

TRC Receive Documents   Stage Review Meeting Begins   Stage Review Meeting Ends

FIGURE 11-5. Logistics for a typical technology review process.

overall quality of the science. Often the business managers do not participate in this initial session. The second part of the review focuses on business and resource issues.

At the conclusion of the meeting the process owner, as indicated in Figure 11-5, has the responsibility for producing a memorandum that summarizes the meeting and the resulting action items within three days of the meeting (many companies require completion of the memorandum within twenty-four hours).

Technology Review Process

The review process consists of a series of reviews (TR_0 to TR_N) where the details of the development plan and the technical approach are known only through the next stage. Periodic reviews help the team to think about the entire project and its impact upon and value to the business. And, like other processes, preparing for the review and planning the next stage's work helps the project team to regroup, rethink, and refocus on common goals. In contrast to traditional SG, neither the project team nor the decision makers can say how many reviews there will be for a given project except that the number of reviews will probably be different from project to project.

Technology development teams are asked to provide an overall plan and an estimate of how many gates there will be and how long the work will take,

---

EXAMPLE

The technology review committee at XYZ Chemical Company was chaired by the director of R&D for the business that would eventually use the molecule. The remaining members of the TRC consisted of two senior research fellows from the central research laboratory, a senior researcher from within the business, and one scientific peer from another business who was recognized as an expert in the synthesis path for the molecule being developed. The business perspective in this TRC would be the responsibility of the chairman, who would continue to update the divisions' senior executives who initially approved the project for the TechSG process and provided the funding for this effort. Other TRCs have more representation from people on the business side—though nontechnical senior executives sometimes lose patience when technology issues are critically discussed.

but it is necessarily *only* an estimate since, as many researchers are quick to point out, technology discovery cannot be scheduled. The endpoint for most stages often represents a significant technical milestone, such as the demonstration of a critical piece of the technology or process. In the extreme example given earlier, it could be that a single molecule of "water without hydrogen" was actually produced in the laboratory, proving that it could be accomplished. This would not be enough to begin construction of the new production facility, but it would be promising enough to fund the next block of possibly costly experiments.

Milestones, which might trigger a gate review, could be successful in achieving an intermediate goal. For example, if the current process was only capable of producing something with 55 percent purity, and the ultimate goal was 99 percent, the technology review committee might ask for a review when the team could demonstrate 75 percent purity or six months from the last review, whichever came first. For the former condition, they might want to discuss the next steps or techniques to investigate; in the latter condition, they might want to decide if the project is worth further effort or investment.

The failure of a key assumption would also trigger a review to immediately rethink any future efforts. In the earlier example, everyone assumed that when the TechSG team got to a certain point in the experiments, the use of a certain chemical process would yield at least 60 percent purity. But when they finally performed the experiment, best efforts could yield only 20 percent purity. That would require a rethinking of the project and agreement on a new path forward, if any, and would therefore trigger a gate review.

Of course, a major change in any critical parameter would also trigger a review. Examples might include (but are not limited to) a change in anticipated market need, a change in estimated development cost or time, a shift in the market window of opportunity, a new patent, or an unforeseen disclosure or action by a competitor.

On occasion a gate may be a specific actionable event, such as signing a letter of intent or filing a patent application. Ideally the gate should occur at a significant and measurable milestone so that clear deliverables and kill points may be established. A kill point might be at the end of six months; if there is no progress toward resolving a critical issue, the project might be canceled and other high-risk projects begun. Alternatively, a kill point might occur when all conceivable routes to resolving a critical need (agreed to at a previous review) have been exhausted without any signs of potential future success. A third kill point might be a significant drop in the probability of success as more details are learned from experiments or the realization that a critical safety or economic factor could not be overcome.

However, the measurable milestone may be too long a time for the TRC to comfortably wait. Thus in more complex projects the endpoint may simply be an agreed-upon "time in grade" or the information necessary to commit to the next level of effort. Most technology development teams plan their work for a block of time, usually two to six months, depending upon the complexity of the work, the nature of the industry (for example, pharmaceuticals development and Internet businesses have different time scales), and the risk tolerance of the TRC.

Using options theory to assess the potential of the project at each of the gates may also help the TRC committee in assessing projects. Three recent articles discuss applying "real options" assessment, with examples in the chemical and pharmaceutical industries (Boer 2000; Angelis 2000; McGrath and MacMillan 2000). McGrath and MacMillan (2000) advocate a scoring methodology that may be used to assess the option value of the project. This scoring model could be used at each gate in the TechSG process to determine if the option is increasing in value, as a result of what is learned in the preceding stage, or decreasing in value, as might occur when the development costs and technology risks are beginning to exceed expectations.

In the traditional SG process, the activities and deliverables for all stages are common to all projects and are expected from the moment the project is begun. In TechSG, only the first stage (T_0) and the last two stages $(T_{N-1}$ and $T_N)$ have enough predictability from project to project to be clearly defined in advance. The first stage (T_0) focuses on planning future work, clarifying goals and boundary conditions, and aligning expectations. The initial gate (TR_0) defines the overall technology development plan, including the overall technology strategy and approach, program team structure, resources, a detailed plan (including deliverables) for proceeding to the next gate (TR_1), and an overall risk analysis.

The next-to-last stage (T_{N-1}) is typically focused on the planning for the technology development work to achieve all of the agreed-upon feasibility points and to begin the transition to the NPD process. In stage T_{N-1}, it is recommended that the future NPD team leader be added to the technology development team to begin the transition. The final stage (T_N) is focused on transferring the knowledge about the technology to the product development team. Technology transfer often begins before the final gate (TR_N) in order to get to the marketplace as early as possible, although technology development may continue after the transition in order to approach the ultimate performance goals.

Approval of the T_0 plan represents the launch of the technology development program. Subsequent technology reviews $(TR_1$ to $TR_{N-1})$ focus on the experimental outcomes of the stage just completed and the revised plans for going forward with agreed-to kill points. If and when a kill point is reached, the project team has a review with the TRC to decide if the project should be continued despite the condition, redirected, redefined, changed in scope, or if all work on the project should be stopped and resources reassigned.

The use of agreed-to kill points creates an environment of "fast failures." The problem with many high-risk projects are that they are allowed to linger in the hope that the technology barriers will be overcome. An environment of fast failures allows the company to investigate more technologies. There is always the concern that a technology may be killed too soon. While this certainly may happen, it is the authors' experience that most companies have too many projects that should have been stopped earlier. The second value of having agreed-to kill points is that it creates a learning environment. It is important that the project be stopped without "killing" the people involved in the project. Having the team define the kill points helps change the paradigm for stopping technology projects from what may appear to be an arbitrary decision by management to a situation where the development team, along with advice and consent from the TRC, defines the characteristics.

Early involvement of key members from the traditional SG product development team during the TechSG process helps ensure a smooth transition. Transitions are even more seamless when the members of the NPD team have been involved in preparing for the TR_0 review and when the key technology members (who were part of the TechSG process) transition to the NPD team before the first traditional SG. In addition, the TRC should always contain executives from the division(s) that eventually will utilize the technology. In these cases, the leaders responsible for future product development will be aware that certain key technology discoveries have yet to occur and that the NPD effort may represent wasted effort if the remaining technology hurdles cannot be surmounted.

Structured Planning

Once the business committee has approved the charter, the designated resources begin work to prepare for the first stage review (TR_0). The purpose of this first stage is to flesh out the plan for the project, to provide the next level of detail about the proposed future technology work, and to get agreement quickly on specific work to be done and the resources required. The structured process provides a framework for communicating the project plan and executing it. In preparation for the TR_0 meeting, a technology development plan is developed that will be updated for all subsequent reviews. The team leader needs to expand upon the information contained in the charter to clarify the details of what technologies will be developed and why, and which approaches will be investigated. The details of the technology development plan need to be defined by the company so that they are congruent with its culture and existing planning systems. However, there are a number of differences between a standard business plan and the technology development plan. Specifically, it includes much more about the technology and its development, but much less detail about business plans, projections, and financial analysis. The technology development plan document usually includes the sections shown in Table 11-2. While the focus of the TechSG process is on technology development, the business strategy, organization, marketing, financial, and regulatory assumptions and risks will also need to be addressed, since the TR_0 review is typically the first time a business plan is developed for the project.

Unlike a traditional SG process, the TechSG process emphasis is not on a specific product and its financial contribution to the future of the business. There is a great deal less detail about the expected financial performance of any specific product in the TechSG process. There is a lot more emphasis on the

TABLE 11-2.
Outline of a Typical Technology Development Plan

Section	Description	# of Pages
I	Executive summary	1
II	Introduction, including fit with strategy, market need, and why this project will win	1–3
III	Competitor analysis	2
IV	Project objectives, including high-level financial analysis	1
V	Technical approach(es)	2–12
VI	Summary of previous stage plan activities	1–5
VII	Technology performance and potential development tables	1–3
VIII	Future stage objectives and timing	1–2
IX	Updated stage plan	1
X	Project organization	1
XI	Conclusions, risk assessment, and decisions required	1–2
	Total	13–33
Appendix A	Original project charter	
Appendix B	Glossary of terms	
Appendix C	Status of action items from previous stage review	

technology needed for future business plans and the work required to demonstrate that the technology is feasible, rather than on the production of a prototype of a specific product design. Although the plan for the next stage must be sufficient to justify the dedication of the necessary resources for the next stage of work, just like in the traditional SG process, only a small portion of the deliverables is devoted to financial justification. There is much more emphasis on technology issues in the TechSG process.

A technology performance table like the one shown in Table 11-3 is prepared to help explain the specific performance capabilities to be developed and to help manage the expectations of everyone associated with the project. It is important to note that this is *not* a final product specification. Rather, it is a guide to help manage the technology development effort. It has also been found to be the document that is the most difficult to properly develop, the one that typically gets the most attention, and the one that provokes the most discussion.

The example shown in Table 11-3 identifies specific performance criteria needed to enable the key product capabilities for a new product (in this case, a new office copier). Note that the table is *not* a complete set of design specifications, but simply a definition of the technology feasibility points that must be met in order for the project to proceed into product development. For example, it does not specify how to move the paper within the device, but rather the technologies required to control the static and dry the ink in order to meet the market need for more copies per minute. In order to avoid assumptions and

TABLE 11-3.
Example of a Technology Performance Table for a Next-Generation Office Copier Capable of Reproduction Speeds of 25 Copies per Minute
For this example, assume that current copiers can only make ten copies per minute. Ultimately, the goal would be to develop a copier capable of producing two hundred copies per minute. The table specifies the overall performance criteria, technology feasibility points, and confidence level for each of the desired characteristics.

Market Need	Desirable Performance	Technology Performance Criterion	Technology Feasibility Point	Confidence Level	Ultimate Performance Criterion	Confidence Level
Higher productivity	Significantly more copies per minute	High evaporation rate for faster ink drying	Drying time of 30 microseconds	50%	Drying time of 2 microseconds	<30%
	Significantly more copies per minute	Infrared-sensitive ink for faster drying	Infrared absorbency by test #41X > 25%	<50%	Infrared absorbency by test #41X > 75%	<30%
		High-speed paper transport	25 feet per second	70%	195 feet per second	<30%
	No paper jams	Less static electricity to reduce jamming	<50 volts at 50% relative humidity by test #21	50%	<2.1 volts at 50% relative humidity by test #21	<30%
	No waiting	No warm-up time	1 second preheat	50%	0.1 second preheat	<30%
Environmentally friendly	No dangerous fumes	No environmentally restricted solvents	0.0 ppm hydrocarbons	30%	0.0 ppm hydrocarbons	30%
Lower operating costs	Longer-wearing brushes	Maintains stiffness	<20% stiffness loss after 1 year	90%	<10% stiffness loss after 1 year	50%

TABLE 11-4.
Anchored Scales for the Technology Confidence Levels Used in Table 11-3

Level	Overall Expression	Influencing Variables	Information Sources
<30%	Uncertain	Totally uncontrollable; many unknown variables and unpredictable experimental results	Instinct and intuition, belief of the technology team, and few if any experiments
30–50%	Possible	More uncontrollable than controllable; some unknown variables and low predictability of experimental results	Experience in a few analogous areas, some preliminary experiments
50–70%	Probable	Somewhat uncontrollable; few unknowns and moderate predictability of experimental results	Extensive experience, theoretical and experimental foundation combined with broad internal input
70–90%	Highly Probable	Controllable; most variables are known and understood, experimental results are predictable	Preliminary database, independent confirmation with broad multifunctional internal input
>90%	Certainty	Totally controllable; variables are known and understood, experimental results have been reproduced	Large database and familiarity, multiple independent confirmations with broad multifunctional external input

misinterpretations, a set of anchored scales, as indicated in Table 11-4, is used to communicate the technology development team's confidence level around each of the performance goals and feasibility points. The TechSG project should be managed through the gates so that the overall confidence level is increased through succeeding gates. The technology development table also contains the ultimate performance criteria and confidence levels for each of the technical criteria that may be required for future products.

In order to increase the confidence level at succeeding gates, resources and focus should be directed at the performance criteria with the lowest confidence level. In most cases, as the performance of one of the parameters approaches the feasibility point, the tendency is to try to complete the work on that parameter. The level of understanding is typically higher and the difficulties fewer, and there is a desire for a sense of achievement and a certain degree of closure. However, if the feasibility level of some other key parameter is never achieved, there is no hope of success for the overall project. Therefore it is important to restrict or even stop work on the parameters that are close to feasibility and to concentrate on raising confidence around the parameter with the lowest confidence level. Resources during most stages should be focused on the technology feasibility characteristic with the lowest confidence level, in contrast to spreading the resources across all the technology initiatives. During later stages, efforts may be required to increase the confidence level in multiple technologies. In

EXAMPLE

Figure 11-6 shows a typical stage plan for the XYZ Company as it might appear for the first gate. The plan for the first stage was to better define the problem, focus the technology development work, extensively study the available literature, and to begin doing serious experimental work. A portion of the Technology Performance Table developed by the team is shown below. The team proposed three known ways to produce the molecule as part of the first stage. The goal in the second stage was expected to be to identify the least attractive technical route and to recommend at the TR_1 review that future work would concentrate on only two of the three original routes. After the TR_1 review, the plan would be to experiment on the remaining two routes with the intent of selecting the best remaining option. The last stage would focus on the completion of the laboratory experimental programs, the completion of the early scale up work, and the technology transfer to the Product Development Team and the traditional SG process.

At the beginning of the Technology Development Project, the team did not know which route would eventually be chosen. Whether or not a new test facility would need to be built was not initially known since the synthesis route was not identified. Because a large capital outlay might be needed for the test facility and no specific plans for the commercial process could be made,

Technology Performance Table for XYZ Company

Market Need	Desirable Performance	Technology Performance Criteria
More of Molecule X available	Mfg. process yield	Chemical process yield
		Thermal process yield
Environmentally friendly	No Mfg. waste stream	No environmentally restricted by-products
Safe Process	No chance of explosion	No combustible by-products

some instances the confidence level may even decrease at the next gate as additional knowledge is obtained.

Projects at TR_0 are also required to evaluate both potential and actual development along various dimensions of the project, including team organization, marketing, financial, regulatory, and technology issues, as indicated in the potential/development plot shown in Figure 11-8. Anchored scales for each of the factors are indicated in Table 11-5. These potential/development plots help evaluate project uncertainties beyond just the technology factors reviewed in the technology performance table. Project potential characterizes the ability of the project to ultimately be successful in the marketplace. A project with a very high potential will typically be characterized by some or all of the following criteria: It can be protected by patents; it can be launched in a large, fast-growing market with few competitors; and it has few regulatory requirements

it was impossible to calculate a single NPV estimate. But, the high level plan that was presented allowed the business leadership to understand its strategic value and what was needed in order to plan the specific work for each subsequent stage. The TechSG process facilitated the appropriate focus and discussion at each point in the project and helped to manage the expectations of all parties.

As shown in Figure 11-7, the last stage plan for the project shows that it required building a test facility and modifying it to evaluate both of the last two possible routes before a decision could be made between them. The product development team leader was identified early enough to be involved in the scale-up testing of the selected route even though the product development Stage-Gate™ process did not begin until the end of stage T_4 because all of the feasibility points had not yet been demonstrated. Note that even though the Product Development project had begun work on the business plan and product definition, the technology project still had one more stage of work. That last stage was to complete certain planned experiments and to transition knowledge about the technology to the product development team.

Technology Feasibility Point	Confidence Level	Ultimate Performance Criteria	Confidence Level
75% conversion of raw materials in lab	50%	95% conversion of raw materials at plant scale	<30%
50% conversion in lab	70%	98% conversion of raw materials at plant scale	<50%
5 ppm molecule Y	50%	0.0 ppm molecule Y	30%
<10 ppm molecule Z	50%	<1 ppm molecule Z	<30%

that will prevent its introduction. Project development characterizes the uncertainty along these same dimensions but evaluates the internal corporate risk factors that may jeopardize the project. Although the major focus of the TechSG process is on technology development, there may be additional activities needed to reduce market, regulatory, and financial uncertainty in parallel with the technology development effort before the traditional SG process begins.

Subsequent plans (TR_1 to TR_{N-1}) focus on project planning and execution to the next stage. The marketing, customer, and competitor portions of the business analysis are typically not redone unless there has been a significant change that will affect the direction and course of the program. Both the technology performance table and the project technology/development plot are presented at each gate with the expectation that overall confidence levels will all be increasing.

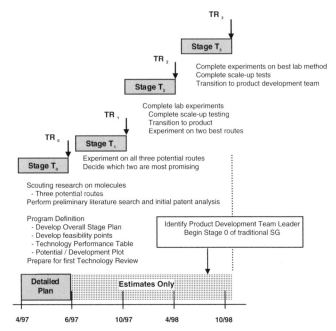

FIGURE 11-6. Example of initial stage plan for XYZ Company.

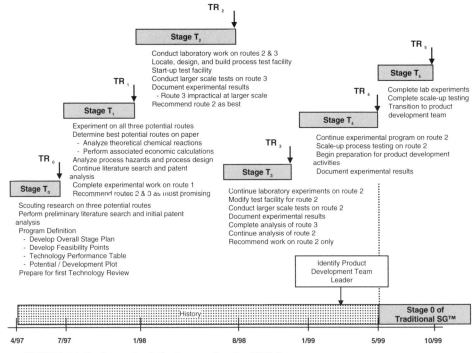

FIGURE 11-7. Example of final stage plan for XYZ Company.

286

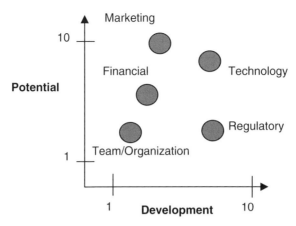

FIGURE 11-8. Example of a potential/development plot.

TABLE 11-5.
Anchored Scale Used to Determine the Project Potential/Development as Shown in Figure 11-6

Team/Organization		
Criteria	Team Potential	Organizational Development
1–2	No people/facilities; must hire.	No project champion exists.
3–5	Acknowledged shortage in key areas.	There is a project champion, with business people acting as advisers.
6–8	Resources are available but in demand; must prioritize. No team member is committed to project 100% of time.	Project has an executive champion at the department head level.
>9	People/facilities are immediately available. Key team members are committed to project 100% of time.	This project is a high priority to the CTO and company CEO, with at least one acting as a champion for the project.

Marketing		
Criteria	Market Potential	Market Development
1–2	No apparent need; market is in decline.	Market is entirely new to the company; extensive studies are required. A considerable challenge for the company.
3–5	Need must be highlighted for customers; mature/embryonic market.	Market study is required. A challenge but doable.
6–8	Clear relationship between product and need; moderate growth.	Market study is required but is straightforward.
>9	Product immediately responsive to customer needs in rapid growth market.	Market is well known to the company; no additional studies are needed.

TABLE 11-5. *(continued)*

Financial

Criteria	Financial Potential	Financial Development
1–2	Declining product demand and large competitive pressure. Lower-than-acceptable gross margin. Product being developed as a defensive response.	Unknown start-up costs. More information is needed in order to estimate unit and capital cost requirements.
3–5	Slow product demand and competitive response expected, which may force company to accept low gross margin.	Unit and capital costs are guesses. Extrapolation of costs from products currently in production may not be reliable.
6–8	Modest product demand and competitive pressures. Acceptable gross margin.	Unit and capital costs may be estimated with reasonable degree of confidence.
>9	High product demand, low competitive pressures. Higher-than-normal gross margin.	Knowledge of unit costs and capital costs are well known. Similar to other products in production.

Technology

Criteria	Technology Potential	Technology Development
1–2	Easily copied.	Technology new to the company (almost no skills).
3–5	Protected but not different.	Some R&D experience, but probably insufficient.
6–8	Solidly protected, with trade secrets, patents.	Selectively practiced in company.
>9	Solidly protected (upstream and downstream) through a combination of patents, trade secrets, raw material access, etc.	Widely practiced in company.

Regulatory

Criteria	Regulatory Potential	Regulatory Development
1–2	No regulatory claims are needed to sell this product.	Controversy around the correct clinical and regulatory tests needed to support claims.
3–5	Regulatory claims provide no product impact but are necessary to sell product.	Clinical/regulatory tests have been designed but have yet to be started.
6–8	Regulatory claims will provide a high impact to sales but may be easily copied by competitor.	Enough clinical/regulatory tests have been completed so that desired claims are expected to be achieved. However, there is no legal approval.
>9	Regulatory claims will provide a high impact to sales and a barrier to competition.	Clinical/ regulatory tests have been completed, with legal approving claim.

Technology Development Team

The technology development team is a small technical team empowered by the TRC to develop and evaluate a technology or group of technologies against a set of performance criteria. In the beginning the technology development team is charged by the technology and/or the business leadership to develop a team charter. The charter is then presented to the senior business management for their advice and approval. Once the charter is agreed to, the technology development team then begins planning their work for presentation at the first TR_0 review.

A technology development team will typically consist of a team leader and a small group of scientists and laboratory staff as an inner circle. Where corporate-level research or external partners are significantly involved, individuals from those groups are often also included on the inner circle of the team. In some cases, the technology development team may be composed of leaders of subteams representing groups working on various aspects of a project. The T_0 team often also includes marketing and product development members who assist the team in formulating the business rationale of the project. These members typically are not part of the inner circle in stages T_1 through $T_{N-1}.$ In most cases, there is also an outer circle of downstream functional advisors who are assigned to provide a broader perspective to the inner circle members. They may be technical advisors, technical experts who will be consulted on specific topics, and/or other subject-matter experts in areas such as legal, finance, regulatory issues, manufacturing, and marketing. These outer circle members play more of a supportive role to the project, in contrast to the inner-circle members, who spend major portions of their time focused on work for their one specific program.

Team leaders in the traditional SG process need to excel in project planning and management and to focus on overall business issues. Team leaders in TechSG need a mastery of the scientific method and the ability to manage uncertainty. As with any team in any process, by clearly identifying roles and responsibilities, this team structure facilitates effective communication, coordination, and issue resolution. The team leader is often the lead scientist for the technology development work. As such, he or she will design, plan, and direct experiments that will lead to understanding the capabilities and limitations of the technology. The team leader schedules and leads peer reviews when appropriate to support the project work. He or she also manages the project work by providing overall project leadership—drafting overall project objectives and plans within the agreed-to budget and other constraints. The team leader is also responsible for scheduling stage reviews with the TRC, ensuring high-quality experimental procedures, and the completion of the review.

EXAMPLE

The technology development team at XYZ Chemical Company consisted of a team leader from central research who had excellent interpersonal skills and was skilled in the synthesis techniques involved. Also on the team were two senior scientists. For the initial T_0 stage, representatives from marketing and technical service were also assigned to help the team develop the initial T_0 plan. The marketing and technical service people were no longer active team participants after the TR_0 review. The team then consisted of a team leader and two senior scientists through stage T_1. During stage T_2 two engineers were added to the team who had responsibility for developing and running the test facility (see Figure 11-7 for more details). During stage T_4 an additional engineer was added as scale-up of the process continued. People from marketing, R&D, and operations from the business side, who were also destined to be part of the traditional SG process, joined the team as they prepared for their TR_4 review.

Process Owner

Someone must "own" the TechSG process. That person must understand all of the details of the process and be ready, willing, and able to train and advise the technology development team and the TRC at any time. The process owner must be able to focus on the process, especially during development team meetings and technology reviews, without becoming overly involved in the content of the meeting or the details of the technology. This person's primary focus must remain on the process; otherwise he or she becomes another member of the development team, leaving no one to monitor the process itself. Senior management must actively champion the process owner, since the process will not be followed without real tangible support.

The ideal process owner should have superb interpersonal skills and be well networked throughout the company so that he or she knows where to find and/or send people for information. This implies that the process owner should be someone with both experience and credibility whose opinions and insights are useful and valuable. Ostensibly one might have different process owners for the traditional SG and TechSG processes, with the former having considerable knowledge about product development and the latter having considerable experience in technology development. However, the typical $200 million–$300 million division can usually support only one full-time process owner. The skills of the process owner are dictated by the type of projects he or she is most likely to facilitate. For example, a process owner in a research division would typically have more technology development experience as compared to a process owner in a more traditional product development division. However, the process owner's most important skills are interpersonal abilities, which enable him or her to facilitate and motivate the product team, and help the team solve problems through his or her people network and experience.

The process owner also has the responsibility for continuous improvement. He or she must be observant to identify what is working well and

EXAMPLE

Responsibility for the TechSG process was assigned to the process owner for the traditional SG process that already existed at XYZ Chemical Company. This was believed to be an excellent choice, since the current process owner possessed the key requirements for a successful TechSG process owner: excellent interpersonal skills, a good reputation among well-respected scientists in the laboratory and engineers in the division, a strong technical background, and excellent contacts and linkages throughout the corporation.

what is not. Where appropriate, the process owner needs to document procedures, techniques, and tools that might be of use to future teams. The process owner also helps teams that are struggling with the process and seeks to make the instructions clearer or to propose changes to the process in order to be more responsive to the teams' needs. As in product development, the process owner also needs to be a visible, outspoken champion of the process, spreading the word and reinforcing its value and benefits wherever possible. He or she must continuously update the process, provide recurrent training, and continue to facilitate the projects. In a benchmarking study Shapiro and Gilmore (1996) indicated that companies with annual sales between $300 million and $500 million were devoting ten person-years of effort every year to maintain their traditional SG process. Experience in traditional SG processes has shown that without a process owner to keep the process alive, it will quickly wither and die despite any benefits the business or individuals may have seen in the past. Similar to Cooper's (1999) and Shapiro and Gilmore's (1996) experience in the traditional SG process, the authors have never seen the successful implementation, necessary rigor, and continuation of the TechSG process without a process owner.

IMPLEMENTATION METHODOLOGY

We have found that successful implementation follows three chronological steps: (1) understanding the need for change, (2) getting started, and (3) full implementation. The following sections will discuss these steps in more detail.

Understanding the Need for Change

Like any new business process, the work of implementing the TechSG process requires a significant amount of change management. Successful implementation needs to be preceded by both senior and functional management understanding that high-risk technology development projects are being poorly managed. Typically these high-risk technologies are often prematurely introduced into the traditional SG process. As a result, there is considerable delay and frustration for the product development team, as the actual performance characteristics of the technology are found to be different from the initial assumptions, resulting in the need to significantly redo many of the product development activities. Some companies realize that this problem is occurring and are proactive in implementing the TechSG process. Unfortunately, implementation often occurs following a crisis when a major project was significantly delayed or canceled because the development was poorly managed, the technology never met the desired characteristics, and/or the inherent risks and assumptions concerning the technology were poorly understood and communicated. The TechSG process should be implemented whenever the technology is new to the company, even if it is not new to the world. The project should start in the TechSG process rather than a traditional SG process as long as the technology represents a significant risk or uncertainty due to the need for discovery or invention by the company.

Getting Started

TechSG is best implemented in a division or company that already has a well-functioning traditional SG process in place. Based upon the authors' experience, implementation typically takes approximately six to twelve months, depending on the size of the business, the type of projects, the business environment, and the organizational readiness to change. Implementation usually begins with a one-time process design team led by the process owner and consists of four to six people, including one marketing or business representative and one or more researchers. It is beneficial to have the team leader for the first project also included as part of the design team. The design team has the responsibility of developing the overall planning documents, procedures, and processes that the technology development teams will follow.

Implementation should begin with one or two projects that are typical of the kind of project undertaken by that business. They must be real projects that are meaningful and significant to the business. They should not be projects that are created just to test the process, and they should not be projects that have already been worked on for a long period of time without any formal management process. The purpose of the first project(s) is to test the TechSG process design, evaluate the effectiveness of the process, identify improvements, build confidence in the process, and ensure that it fits with the culture and environment of the organization.

The TechSG process will involve senior researchers who may be accustomed to working alone or with minimal management oversight. They often take great pride in their work and their ability to eventually make the breakthrough discovery that helps the business, but they may have little patience for meetings, reports, and customer-focused plans. They may get great satisfaction from solving a difficult technical problem and little from preparing for a review.

They may feel threatened or even insulted to have other people question their choice of experiments, procedures, or approaches. On the other hand, many technology development team leaders and members often recognize the benefit of having an impartial review and getting suggestions for alternative routes and additional contacts.

The process owner and those responsible for implementing the process will need to help the early users see the value of using the process and immediately address concerns and problems. Many teams later admit that getting ready for reviews helped them organize their thoughts, facilitated valuable discussions, and identified potential pitfalls and solutions much earlier than in the past without such a process. But until some success can be demonstrated, the process will need visible support from senior management for its value and usefulness.

Full Implementation

Only after the first one or two projects have each had at least two stage reviews, and any minor adjustments to the process design have been made and tested, should the process be rolled out to the rest of the organization. Projects that are nearing completion without any formal process or with whatever older process may have been in place should be allowed to finish without forcing them to use the new process. Of course, the tools and techniques of the TechSG process can provide valuable upgrades to their work, and their use should be encouraged.

It is best if eventually *all* technology development projects use the TechSG process. Small projects might be allowed to informally use the process, while those with significant risk, cost, or need for resources would use the process in its entirety and with the most rigor. Each business must decide how much flexibility it will allow based upon its risk tolerance and the past behavior of the organization. Like any other business process, it has been demonstrated over and over that allowing too many shortcuts too early in the implementation will keep the process from ever being used to its full extent and its full benefit from ever being enjoyed by the business. Until the new methodology becomes well known and well used, it is strongly recommended that shortcuts *not* be allowed.

The first few projects are often focused upon technology development for a specific end product. Documents in the first few projects tend to be lengthy and verbose. In addition, the project teams initially struggle with the technology performance table and the potential/development plots. As the process becomes more widely utilized, the business begins to see the benefits of reduced development time, better-focused projects, aligned expectations, and a smoother-running TechSG process. At that point, many businesses begin to focus on long-term technology planning. It is then that the TechSG process begins to be used for the development of enabling technologies and core competencies needed for the future health of the business, with less focus on a single end product.

CONCLUSIONS

The primary benefits of using the technology Stage-Gate™ (TechSG) process are to ensure that critical, limited resources are allocated to the projects that are of highest importance to the business and to reduce overall the time to market of new technologies and the products that they enable. The technology planning and review process highlights business, process, and investment issues very early in the project and allows them to be more efficiently addressed. Technology goals are well understood by all involved, are aligned to the product/platform strategy, and are revalidated from stage to stage. The TechSG process creates a collaborative alignment between the technical and business communities.

In addition to significantly reducing overall cycle time for high-risk projects, the tools used in the process have brought additional benefits. The increased clarity around both project and business goals—before work or even planning begins—that results from formally chartering project teams saves time and wasted effort by the project team. The use of the technology performance table and the potential/development plots as well as other parts of the structured methodology enables and enhances communications, eliminates confusion and misunderstanding, and manages expectations of both the project team and the business leadership.

Periodic reviews, as in the traditional SG process, help the team to evaluate the entire project and its impact upon and value to the business. And, as in other processes, preparing for the review and planning the next stage's work helps the project team to regroup, rethink, and refocus on common goals. The advisors and experts at the reviews provide a fresh view of the work and can provide possible alternatives and additional contacts and references. The clarity of the information developed and the efficiency of the review process significantly increase the probability that the company will allocate resources to higher-risk and potentially more profitable projects. By using the structured process, possibilities discovered in the course of one project have often led to new, even more exciting projects.

Without such a process, many companies either shy away from doing any high-risk projects, begin product development before technology discovery has occurred, or poorly manage the scientific resources of the company. All of these situations are unsatisfactory. In the first case, the company does only incremental-type projects, which typically have a low profit margin. In the second case, the lack of knowledge causes frequent project delays, frustration, and redesign when the characteristics of the required technology unfortunately do not become clear until the middle of product development. In the third case, the scientists often have inadequate resources, a lack of adequate scientific peer review, and a tendency to continue projects without any measurable kill points. Based upon the authors' experience, the TechSG process changes the paradigm and creates an environment of "good science," which encourages fast failures. With effort focused on demonstrating a basic understanding of the technology first, subsequent product development does not get stalled waiting to make the technology perform as planned. In addition, the use of scientific peers helps elevate the scientific stature of the project to further ensure that good, rigorous scientific approaches are being followed. Furthermore, TechSG holds the scientists accountable for demonstrating scientific goals while clearly recognizing that one cannot schedule technology discovery.

BIBLIOGRAPHY

Angelis, D. 2000. "Capturing the Option Value of R&D." *Research Technology Management* 43, 4: 31–35.

Boer, P. F. 2000. "Valuation of Technology Using Real Options." *Research Technology Management* 43, 4: 26–30.

Cooper, R. G. 1993. *Winning at New Products*, 2nd edition. Reading, MA: Addison-Wesley.

Cooper, R. G. 1999. "The Invisible Success Factors in Product Innovation." *Journal of Product Innovation Management* 16: 115–33.

Crawford, C., and A. DiBenedetto. 2000. *New Products Management*. Boston: Irwin/McGraw-Hill.

Eldred, E. W., and M. E. McGrath. 1997. "Commercializing New Technology—I." *Research Technology Management*, 40, 1: 41–47.

Eldred, E. W., and A. R. Shapiro. 1996. "Technology Management." In M. E. McGrath, ed., *Setting the PACE in Product Development*. Boston: Butterworth and Heinemann.

McGrath, M. E. 1996. "The Phase Review Process and Effective Decision Making." In M. E. McGrath, ed., *Setting the PACE in Product Development*. Boston: Butterworth and Heinemann.

McGrath, M. E., and C. L. Akiyama. 1996. "PACE: An Integrated Process for Product and Cycle Time Excellence." In M. E. McGrath, ed., *Setting the PACE in Product Development*. Boston: Butterworth and Heinemann.

McGrath, R. G., and I. C. MacMillan. 2000. "Assessing Technology Projects Using Real Options." *Research Technology Management* 43, 4: 36–49.

Shapiro, A., and D. P. Gilmore. 1996. "Implementing PACE: How to Make It Real and Make It Lasting" In M. E. McGrath, ed., *Setting the PACE in Product Development*. Boston: Butterworth and Heinemann.

12 Universal Design: Principles for Driving Growth into New Markets

James L. Mueller and Molly Follette Story

DEFINITION

Universal design *is the design of all products and environments to be usable by people of all ages and abilities, to the greatest extent possible.*

In the 1980s Sony released a very small (for that time) 8 mm video camera. With the camera came an instruction manual. Even without the French and Spanish translation, the instructions were eighty pages long. The manual weighed exactly one-third what the camera weighed (including battery), and the camera was smaller. Owners who studied this manual carefully learned that the viewfinder had sixteen different indicators to keep track of while also supposedly viewing the subject. By comparison, the head-up display (HUD) of an F-16 jet fighter has only thirteen indicators.

Certainly, videotape recording was not meant to be as demanding a task as flying a jet fighter. The typical video camera user, far less skilled and trained than a fighter pilot, was probably more confused than aided by all these indicators.

WHAT IS UNIVERSAL DESIGN?

Anyone who has struggled to learn how to operate a new product has probably wondered, "Just who is this designed for anyway?" Mass-produced products are supposed to be designed for the largest potential user market, yet "normal" users often find new products physically or mentally challenging.

These challenges are magnified for persons with mental, sensory, or physical limitations due to age or disability. As the population ages and health care improves, product developers are finding that elders and people with disabilities are growing segments of their potential customer market. But until recently,

All figures in chapter 12 copyright Center for Universal Design, reprinted with permission.

297

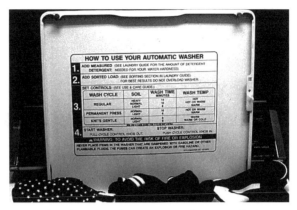

FIGURE 12-1. Instructions in large, high-contrast print on the inside of whirlpool clothes washer lid.

there has been very little guidance about how to include these new market segments in product planning and development. Consequently, product developers tended to shy away from this new market potential.

For some companies, their own customers helped them to realize the market potential. A steady stream of letters and hotline calls from elder customers caused Whirlpool Corporation to establish their Appliance Information Service in 1981. This service met the needs of customers with arthritis, decreased vision, and other limitations by offering product adaptations as well as helpful hints for making kitchen and laundry jobs easier. Whirlpool's Appliance Information Service became so popular that it began to earn the company regular coverage in widely circulated publications such as *Better Homes and Gardens*. The company estimated that this coverage alone was worth hundreds of thousands of dollars every year.

A common complaint among Whirlpool customers was difficulty in keeping track of the washer's instruction booklet and reading it with failing eyesight. The Appliance Information Service responded by asking Whirlpool's designers to print the most important instructions right on the inside of the washer lid in large black lettering on a white background. This was just what customers with limited vision needed, but Whirlpool didn't receive any complaints from other customers with dimly lit basements, where washers tend to be (see Figure 12-1).

Whirlpool and other companies that began to pay attention to the design needs of customers with limitations learned that these needs were not so different from those of their "normal" customers. They began to realize that, rather than creating special products for unique user groups, they could integrate the accommodations into their regular product lines and improve usability for everyone. This approach to design for people of all ages and abilities has become known as universal design and increasingly is being applied by product developers around the world.

The Ford Focus sedan, developed by Ford of Europe to be a true "world car," incorporates design features for drivers of all ages (although the car's advertising is clearly targeted to the younger driver). These features are hardly accidental. The development program for the Ford Focus coincides with heightened awareness of the needs of older drivers among designers and engineers at Ford. "The numbers show that mature and elderly drivers are becoming an increasingly large percentage of the motoring public," explains Richard Perry-Jones, Ford's vice president for product development.

"When you're young and fit enough to leap out of a car without effort, it's hard to appreciate why an older person may need to lever themselves out of the driver's seat by pushing on the seatback and the door frame," says Mike Bradley, ergonomics specialist in Ford's design center in Dunton, England.

Jeffrey Pike, a design analysis engineer at Ford, describes how Ford is beginning to design for the growing population of senior drivers by sensitizing staff to the ergonomics of aging: "As we grow older, our vision changes. We're more susceptible to glare, and we don't adapt as quickly to changing conditions. It's harder for elderly drivers to use their peripheral vision."

The Ford Focus has been cited by a major consumer magazine as "very easy to get in and out of; the cabin has a spacious, airy feel, and the driving position is high, which makes for good visibility." Also noted was the spacious trunk, rear seat room, and "climate-control switches that are easy to use and radio buttons that aren't a stretch to reach." The same magazine rated it above the Nissan Sentra, Dodge Neon, and Saturn S.

Ford's television ad could just as easily have demonstrated the ease of entry and exit for a cane user, with the Focus's extrawide doors and a number of other features especially useful for drivers with limitations due to age or disability. But Ford chose instead to follow a long-held automotive marketing approach: "You can sell a young man's car to an old man, but you can't sell an old man's car to a young man."

Who Benefits from Universal Design?

The relative ease of use of a product depends as much on the situation and the environment as on the user's abilities. Universal design benefits users with and without disabilities alike because it addresses situations everyone faces in using products, whether due to the environment or the individual's own limitations.

For example, product managers at Leviton Manufacturing Company were surprised to learn in 1993 that their Decora rocker wall switches were considered a model of universal design. Architect Ron Mace, an expert in architectural accessibility for people with disabilities and the recognized father of universal design, was specifying Decora switches in his designs for accessible homes. Leviton learned that these switches, designed to be elegant and easy to use for persons carrying packages, were also easy for persons with arthritis or other physical limitations. This led Leviton to a series of universal design products. Their universal design approach eventually became the subject of a marketing teaching case distributed by Harvard Business School (Mueller and Ingols 1997).

The Value of Achieving Universal Design

The examples offered above briefly outline how some companies have benefited from including the needs of elders and people with disabilities in the planning and development of new products. In some cases, these very populations have themselves created new business opportunities.

OXO International

In 1960, Sam Farber founded the successful kitchenware maker Copco, Inc. Before this, he had worked for eleven years for his father, Louis, who owned Sheffield Silver. Farber's uncle Simon had founded Farberware in 1900. After thirty-nine years in the kitchenware business himself, Sam Farber retired in 1988 at age sixty-six. Despite so many years of experience in the kitchenware business, Farber's ideas of inclusion of elderly and disabled customers did not come into focus until his personal experiences brought the importance of the idea home.

Shortly after retirement, Sam and his wife, Betsey, rented a home in Provence, France, for two months. Betsey had developed arthritis, and the kitchenware available at their rented home was difficult and painful for her to use, especially as more friends came to visit. "You don't know how many friends you have until you rent a place in Provence," Sam said. The more cooking they did together, the more inadequate the utensils seemed. Betsey's knitting hobby only added to her discomfort.

For years, kitchen tools such as vegetable peelers were designed to be manufactured in the easiest, least expensive way. They were better than paring knives, but only just. Betsey's discomfort forced Sam to wonder, "Why can't there be wonderfully comfortable tools that are easy to use?" In 1989 Sam Farber decided to come out of retirement and establish OXO International to produce kitchenware with older and disabled users in mind. Farber chose the name because it was universally readable, whether viewed horizontally, vertically, or upside down. The Farbers' son, John, took a leave of absence from his position as a vice president at Prudential Bache to help set up the business's finances.

Well-known transgenerational designer Patricia Moore was consulted for advice, along with Smart Design, Inc., with which Farber had worked before. In exchange for a small advance and a 3 percent royalty, Smart Design waived the usual fees to design the product line, OXO Good Grips, which generated immediate demand at its debut at a San Francisco trade show in April 1990.

The design incorporated plump, resilient handles for twist and push-pull tools such as knives and peelers, while squeeze tools such as can openers had hard handles. All handles were oval in cross section, to better distribute forces on the hand and enhance grip, even for wet hands. The measuring cups and spoons featured large, high-contrast markings for visibility.

In 1994 another line, Good Grips Sierra Club Garden Tools, began reaching the market. Next came a line of barbecue tools. Ideas for new products came from looking at common, everyday products and finding ways to make them better. By 1999, the OXO product line had grown to 350 items, with the Swivel Peeler, one of the original fifteen products, leading sales. Four lines were being produced for specialty stores such as Bed Bath and Beyond, Bloomingdale's, Crate & Barrel, Linens N' Things, and Lechters, department stores such as Kmart (the Touchables line), Target (Soft Works), and Wal-Mart (Sensables), as well as for mail-order catalog sales.

Since its debut, the Good Grips line has won worldwide acclaim, including awards from the Arthritis Foundation, Design Zentrum in Germany, Good Housekeeping Institute, *Metropolitan Home,* and the Industrial Designers Society of America. Good Grips have been selected for permanent collections at the Chicago Atheneum, the Cooper-Hewitt National Museum of Design, and the Museum of Modern Art (MOMA). The products enjoy so much media attention that an advertising budget has been all but unnecessary. Other kitchenware producers have followed OXO's success with similar large-handled utensil designs. To counter competition in the lower price ranges, OXO has established its own lower-priced line known as Good Grip Basics.

In 1992 the Farbers sold OXO International to General Housewares Corporation and retired again in the fall of 1995. In 1999, World Kitchen (formerly Corning Consumer Products, makers of Corelle, Corningware, and Pyrex) purchased General Housewares, along with EKCO Group, and adopted the World Kitchen name early in 2000. By 2000, OXO International enjoyed an annual growth rate of 37 percent, with about $60 million in annual sales.

Tools in This Chapter

Another industry example at the conclusion of this chapter will describe Tupperware Corporation's unique approach to universal design. Also featured in this chapter will be perspectives on the diversity of human abilities that affect product usability, tools to accommodate this diversity in product development, and an exercise in the use of these tools.

UNDERSTANDING THE ABILITIES OF REAL PEOPLE IN THE REAL WORLD

Because of the natural human tendency to adapt, most people succeed in making products work for them. Because of differences in abilities, some do this quickly and easily. Some take longer and expend more effort. Others may become frustrated or even injured in the process. Still others will find it impossible.

Each person is unique in age, size, abilities, talents, and preferences. Human abilities can be grouped into the following categories: cognition, vision, hearing and speech, body function, arm function, hand function, and mobility. The following descriptions show how variations in each of these abilities, as well as environments and circumstances, may affect design usability.

Cognition

Cognition in Daily Life

Cognition is the task of receiving, comprehending, interpreting, remembering, and acting on information. Imagine trying to assemble and use an unfamiliar new product, reading only the foreign-language instructions. If you've ever used a cellular phone while driving, you may have experienced distraction (likened by some to driving under the influence of alcohol). Have you ever taken medication that made it difficult to think clearly? Have you ever had difficulty recalling the name of a close friend or relative? (See Figure 12-2.)

Causes of Varying Abilities in Cognition

Cognition can vary widely due to the environment, temporary situations, or chronic conditions.

Environmental conditions
- Limited literacy
- Unfamiliarity with the local language or culture

Temporary situations
- Fatigue or distraction
- Childhood (limited vocabulary, grammar, and reasoning skills)

Chronic conditions
- Diminished memory and reasoning skills associated with aging
- Retardation
- Down's syndrome

 DOSAGE: Follow dosage below or use as directed by a doctor. Dosage cup provided. Do not exceed 6 doses in a 24-hour period.

 **ADULT DOSE
(and children 12 yrs and over):**
2 teaspoonfuls every 4 hrs.

 **CHILD DOSE
6 yrs to under 12 yrs:**
1 teaspoonful every 4 hrs.
2 yrs to under 6 yrs:
½ teaspoonful every 4 hrs.
under 2 yrs: Consult your doctor.

FIGURE 12-2. Dosage instructions made easier to understand with illustrations.

♦ Learning disabilities
♦ Head injuries
♦ Stroke
♦ Alzheimer's disease

When cognition is affected by one or more of these conditions, the following functions may be limited:

♦ Self-starting (initiating tasks without prompting)
♦ Reacting to stimuli (response time)
♦ Paying attention (concentration)
♦ Comprehending visual information
♦ Comprehending auditory information
♦ Understanding or expressing language
♦ Sequencing (doing things in proper order)
♦ Keeping things organized
♦ Remembering things, either short- or long-term
♦ Problem-solving and decision-making
♦ Creative thinking (doing things in a new way)
♦ Learning new things

Considering Cognition in Design

Whatever the cause, customers with varying abilities in cognition can be accommodated using product development tools such as the Principles of Universal Design (Center for Universal Design 1997) described later in this chapter. Especially relevant to accommodating these customers are Principle 1 (Equitable Use), Principle 2 (Flexibility in Use), Principle 3 (Simple and Intuitive Use), Principle 4 (Perceptible Information), and Principle 5 (Tolerance for Error).

Vision

Vision in Daily Life

Try reading a book at the beach without sunglasses, finding your way after walking out of a movie theater into bright daylight, or driving toward the sun. You will experience the limiting effects of glare. Try getting a key into your front door in the dark, reading a detailed road map in your car at night, or finding the light switch in a dark room. You will appreciate the limitations caused by inadequate light (see Figure 12-3).

When you're lost and struggling to find a specific road sign, all signs may seem small and hard to locate. How much more difficult would this be if your glasses or windshield were badly smudged? How difficult would it be to find the sign if you could not move your neck, used only one eye, or viewed the

FIGURE 12-3. Large telephone keypad is easier for everyone to see.

world through a cardboard tube? These situations may cause anyone to make mistakes, slow down, seek help, or avoid even simple tasks because the demand on visual capabilities is too great, whether temporarily or permanently.

Causes of Varying Abilities in Vision

Vision can vary widely due to the environment, temporary situations, or chronic conditions.

Environmental conditions
- A busy visual environment
- Colored lighting or very high or very low lighting conditions
- Adverse weather conditions

Temporary conditions
- Fatigue from excessive visual tasks
- Eye surgery

Chronic conditions
- Blindness
- Hereditary loss of vision
- Cataracts
- Glaucoma
- Retinitis
- Presbyopia (farsightedness after middle age)
- Macular degeneration
- Eye injuries

When vision is affected by one or more of these conditions, the following functions may be limited:

- Perceiving visual detail clearly
- Focusing on objects up close and far away

◆ Separating objects from a background
◆ Perceiving objects in the center as well as at the edges of the field of vision
◆ Perceiving contrasts in color and brightness
◆ Adapting to high and low lighting levels
◆ Tracking moving objects
◆ Judging distances

Considering Vision in Design

Whatever the cause, customers with varying abilities in vision can be accommodated using product development tools such as the Principles of Universal Design described later in this chapter. Especially relevant to accommodating these customers are Principle 1 (Equitable Use), Principle 2 (Flexibility in Use), Principle 3 (Simple and Intuitive Use), Principle 4 (Perceptible Information), and Principle 5 (Tolerance for Error).

Hearing and Speech

Hearing and Speech in Daily Life

Have you ever struggled to determine where a siren was coming from while driving with the radio on? Has the congestion from a head cold, especially if you did any airline traveling, ever left you temporarily impaired in hearing, speech, or even balance? Try giving directions to your home to someone across a busy street. Try following verbal instructions while listening to music through headphones. Much of the message may get lost or confused in the ambient sound (see Figure 12-4).

If you have ever used a pay phone at a shopping mall, airport, or train

FIGURE 12-4. Television captions for use in noisy areas and for users with hearing limitations.

station, you have had the experience of trying to hold a conversation amid background noise and other distractions. In addition, the variable quality of transmission often causes lapses in communication or even interference from other conversations. These situations can cause anyone to miss important information, repeat messages, rely on other sensory input, or just give up because the demand on auditory capabilities is too great, whether temporarily or permanently.

Causes of Varying Abilities in Hearing and Speech

Hearing and speech can vary widely due to the environment, temporary situations, or chronic conditions.

Environmental conditions
  ◆ Very noisy environments
  ◆ Use of audio or noise-suppression headphones

Temporary conditions
  ◆ Presence of several auditory sources
  ◆ Blockages in the route to the inner ear

Chronic conditions
  ◆ Deafness
  ◆ Hereditary loss of hearing
  ◆ Damage from prolonged exposure to excessive noise
  ◆ Diseases
  ◆ Presbycusis (reduction of hearing in older age)
  ◆ Head injuries or stroke

When hearing is affected by one or more of these conditions, the following functions may be limited:

  ◆ Localizing the source of sound
  ◆ Separating auditory information from background sound
  ◆ Perceiving both high- and low-pitched sounds
  ◆ Carrying on a conversation

Considering Hearing and Speech in Design

Whatever the cause, customers with varying abilities in hearing and speech can be accommodated using product development tools such as the Principles of Universal Design described later in this chapter. Especially relevant to accommodating these customers are Principle 1 (Equitable Use), Principle 2 (Flexibility in Use), Principle 3 (Simple and Intuitive Use), Principle 4 (Perceptible Information), and Principle 5 (Tolerance for Error).

Body Functions

Body Functions in Daily Life

Imagine working in a chair with one missing caster. With every change in posture, you might lose your balance. This would affect your concentration and productivity and might cause you to avoid changing body position. Try doing your job from a straight-back chair with your spine firmly against the seat back and your feet on the floor. Retain that position without twisting or bending as you try to retrieve materials from your desk, use the telephone, and perform other simple everyday tasks. Limitations to your reach, field of vision, and mobility make simple tasks more difficult and eventually cause fatigue and pain from the lack of range of motion (see Figure 12-5).

Perhaps you have carried a bulky object up or down a flight of stairs. The added weight made finding your balance more difficult, and the object may have prevented you from using the railings for support or even seeing the steps in front of you. Remember the last time you had the flu. Even the simplest tasks were exhausting, and it was difficult to concentrate on anything for very long. Getting up from the bed or a chair required a few extra seconds for you to clear your head and keep your balance. Consider the difficulty of strenuous exercise on a very hot summer day. In each of these situations, the demands of the tasks may exceed human capabilities to some extent. For some people these situations are minor or temporary inconveniences. For others they may be frustrating, exhausting, dangerous, or even impossible.

Causes of Varying Abilities in Body Functions

Body functions can vary widely due to the environment, temporary situations, or chronic conditions.

FIGURE 12-5. Reach, field of vision, and mobility change throughout the life span.

Environmental conditions
- ◆ Bad weather
- ◆ Extremes of temperature
- ◆ Poor air supply
- ◆ Unstable footing

Temporary conditions
- ◆ Childhood (limited physical development)
- ◆ Later stages of pregnancy, during which balance is affected by the weight of the baby
- ◆ Pain or limited range of motion due to temporary or minor injuries or illness
- ◆ Fatigue or illness

Chronic conditions
- ◆ Diminished stamina, balance, or other body functions associated with aging
- ◆ Extreme body size or weight
- ◆ Epilepsy or other seizure disorders
- ◆ Allergies
- ◆ Multiple chemical sensitivities
- ◆ Asthma
- ◆ Diabetes
- ◆ Arthritis
- ◆ Musculoskeletal injuries or illness
- ◆ Hernia
- ◆ Stroke

When body functions are affected by one or more of these conditions, the following functions may be limited:

- ◆ Physical stamina
- ◆ Achieving, maintaining, and changing posture
- ◆ Maintaining equilibrium
- ◆ Breathing

Considering Body Functions in Design

Whatever the cause, customers with varying abilities in body functions can be accommodated using product development tools such as the Principles of Universal Design described later in this chapter. Especially relevant to accommodating these customers are Principle 1 (Equitable Use), Principle 2 (Flexibility in Use), Principle 5 (Tolerance for Error), Principle 6 (Low Physical Effort), and Principle 7 (Size and Space for Approach and Use).

FIGURE 12-6. Automatic garage door openers reduce need for arm strength and range of motion.

Arm Function

Arm Function in Daily Life

Think of objects you regularly reach for, lift, and carry. Some ordinary household products weigh more than you might guess. A six-pack of 12-ounce cans and a ream of paper each weigh over 5 pounds. One-gallon containers of milk or juice weigh about eight pounds each, and cartons of detergent can weigh up to twenty pounds each. Could you move these products using only one arm? How would you reach them if you could not straighten your arms to reach forward, up, or down?

What about other ordinary tasks such as driving, cooking, or opening a window? Think about the last time you experienced pain in a shoulder or elbow. How did it affect the way you performed these everyday tasks? How would your strength and movements be limited if you wore a three-pound weight on each wrist? In each of these situations, the demands of the tasks may exceed human capabilities to some extent, making the task inconvenient, frustrating, exhausting, dangerous, or impossible.

Causes of Varying Abilities in Arm Function

Arm function can vary widely due to the environment, temporary situations, or chronic conditions.

Environmental conditions
- ◆ Only one free arm due to carrying things or performing another task
- ◆ Thick clothing (e.g., in extreme temperatures)

Temporary conditions
- ◆ Pain or limited range of motion due to temporary or minor injuries or illness

- Fatigue
- Childhood (limited physical development)

Chronic conditions
- Diminished joint range of motion or strength associated with aging
- Congenital loss or deformation of an arm
- Amputations
- Spinal cord injuries
- Cerebral palsy
- Post-poliomyelitis
- Muscular dystrophy
- Multiple sclerosis
- Lou Gehrig's disease (amyotrophic lateral sclerosis, or ALS)
- Parkinson's disease
- Arthritis, bursitis, tendonitis
- Stroke

When arm function is affected by one or more of these conditions, the following functions may be limited:

- Reaching up, down, forward, or behind
- Pushing
- Pulling
- Lifting
- Lowering
- Carrying

Considering Arm Function in Design

Whatever the cause, customers with varying abilities in arm function can be accommodated using product development tools such as the Principles of Universal Design described later in this chapter. Especially relevant to accommodating these customers are Principle 1 (Equitable Use), Principle 2 (Flexibility in Use), Principle 5 (Tolerance for Error), Principle 6 (Low Physical Effort), and Principle 7 (Size and Space for Approach and Use).

Hand Function

Hand Function in Daily Life

Consider how many tasks require use of both hands. Using only one hand, try hammering a nail, tying a shoe, or placing a telephone call. Try twisting a doorknob with oily or wet hands. Try using only your nondominant hand for precision tasks such as using scissors, cutting food, or shaving. Try doing these tasks while wearing mittens (see Figure 12-7).

Perhaps you have experienced a cut, sprain, fracture, or burn that tempo-

FIGURE 12-7. Food preparation can be demanding of hand strength and range of motion.

rarily limited your ability to open a jar, squeeze a tube of toothpaste, operate a faucet, or hold a cup of coffee. In each of these situations, the demands of the tasks may exceed human capabilities to some extent, making the task inconvenient, frustrating, exhausting, dangerous, or impossible.

Causes of Varying Abilities in Hand Function

Hand function can vary widely due to the environment, temporary situations, or chronic conditions.

Environmental conditions
- ◆ Use of gloves
- ◆ Water or grease on hands
- ◆ Only one free hand due to simultaneously performing another task

Temporary conditions
- ◆ Childhood (small hands and weak fingers)
- ◆ Pain or limited range of motion due to temporary or minor injuries or illness
- ◆ Hand fatigue from repetitive tasks

Chronic conditions
- ◆ Diminished joint range of motion or strength associated with aging
- ◆ Congenital loss or deformation of a hand
- ◆ Amputations
- ◆ Cerebral palsy
- ◆ Spinal cord injuries
- ◆ Post-poliomyelitis
- ◆ Muscular dystrophy

- Multiple sclerosis
- Lou Gehrig's disease (amyotrophic lateral sclerosis, or ALS)
- Parkinson's disease
- Carpal tunnel syndrome
- Arthritis
- Stroke

When hand function is affected by one or more of these conditions, the following functions may be limited:

- Grasping
- Squeezing
- Rotating
- Twisting
- Pinching
- Pulling
- Pushing

Considering Hand Function in Design

Whatever the cause, customers with varying abilities in arm function can be accommodated using product development tools such as the Principles of Universal Design described later in this chapter. Especially relevant to accommodating these customers are Principle 1 (Equitable Use), Principle 2 (Flexibility in Use), Principle 5 (Tolerance for Error), Principle 6 (Low Physical Effort), and Principle 7 (Size and Space for Approach and Use).

Mobility

Mobility in Daily Life

Consider driving your car without using your legs. Without walking, how would you get to work? Could you do your job without leaving a seated position? What if there are stairs along the way? Consider the difficulty of maintaining your balance while walking or standing in an airplane, subway car, or bus. Imagine having this difficulty even on stable ground (see Figure 12-8).

Remember the last time you walked a long distance or ascended a long flight of stairs and how the fatigue affected your stability. Did you tend to use the railings more toward the end? Consider how carefully you use stairs that are slippery with water or ice, and how dangerous it is when you lose your balance on stairs.

Notice the different ways people walk on different surfaces. Grass, sidewalks, loose gravel, carpeting, and tile floors each require a different gait to maintain balance and avoid tripping or slipping. When surfaces change unexpectedly, falls can result.

FIGURE 12-8. Mobility varies widely with age and ability.

If you've ever injured a leg and used crutches, you realize the additional time and effort required to cover distances, especially if stairs, revolving doors, or slippery floors are in your way. You may have also learned the importance of space to elevate or straighten your leg or maneuver a wheelchair. As you recovered, you learned the value of grab bars and sturdy surfaces to lean on. In each of these situations, the demands of the tasks may exceed human capabilities to some extent, making the task inconvenient, frustrating, exhausting, dangerous, or impossible.

Causes of Varying Abilities in Mobility

Mobility can vary widely due to the environment, temporary situations, or chronic conditions:

Environmental conditions
- Bad weather
- Uneven or unstable terrain

Temporary conditions
- Childhood (limited physical development)
- Pain or limited range of motion due to temporary or minor injuries or illness
- Fatigue

Chronic conditions
- Diminished strength, stamina, balance, range of motion in spine and lower extremities, or proprioception (sensing the positions of body parts and the motions of the muscles and joints) associated with aging
- Extreme body size or weight
- Congenital loss or deformity of a leg
- Amputations
- Spinal cord injury

- Cerebral palsy
- Post-poliomyelitis
- Muscular dystrophy
- Multiple sclerosis
- Cerebral vascular disease
- Stroke
- Diabetes
- Lou Gehrig's disease (amyotrophic lateral sclerosis, or ALS)
- Parkinson's disease
- Arthritis
- Asthma, emphysema, or other respiratory complications

When mobility is affected by one or more of these conditions, the following functions may be limited:

- Rising from a seated position
- Standing upright
- Walking
- Running
- Jumping
- Climbing
- Kneeling
- Balancing on one foot
- Operating foot controls

Considering Mobility in Design

Whatever the cause, customers with varying abilities in mobility can be accommodated using product development tools such as the Principles of Universal Design described later in this chapter. Especially relevant to accommodating these customers are Principle 1 (Equitable Use), Principle 2 (Flexibility in Use), Principle 5 (Tolerance for Error), Principle 6 (Low Physical Effort), and Principle 7 (Size and Space for Approach and Use).

PRINCIPLES OF UNIVERSAL DESIGN

Everyone, at some time in their lives, experiences a limitation in one or more of these abilities. Designing for such a wide diversity of abilities and situations can seem daunting, but universal design does not mean a single design must suit all users. Companies such as Whirlpool, Ford, and Leviton have proven that there are many creative ways to develop products that are usable for people of all ages and abilities. In these successful applications, universal design is unnoticeable, except to those whose limitations or situations require it.

In 1997 a group of architects, design researchers, engineers, and product designers collaborated to develop universal design principles to guide the design process, allow the systematic evaluation of existing designs, and assist in educating both designers and consumers about the characteristics of more usable products and environments.

Following are these seven Principles of Universal Design and their twenty-nine associated guidelines as well as examples of how each principle can be implemented. These principles, guidelines, and examples were adapted from *The Universal Design File: Designing for People of All Ages and Abilities* (Story, Mueller, and Mace 1998). The examples shown are not necessarily universal in every respect, but each is a good example of a specific guideline and helps to illustrate its intent.

Principle 1: Equitable Use

The design is useful and marketable to people with diverse abilities.

Guidelines

1a. Provide the same means of use for all users: identical whenever possible, equivalent when not. For example, a powered door with proximity sensors is convenient for shoppers in wheelchairs as well as those pushing carts (see Figure 12-9).

1b. Avoid segregating or stigmatizing any users. For example, AT&T public phones with volume controls accommodate users with hearing limitations as well as those in noisy locations.

1c. Make provisions for privacy, security, and safety equally available to all users. For example, ATMs with adjustable tilt screens enable cus-

FIGURE 12-9. Powered door with sensors is convenient for all shoppers, especially if hands are full.

tomers of varying heights and postures to conduct transactions with equal privacy.

1d. Make the design appealing to all users. For example, OXO's Good Grips kitchen utensils are popular worldwide among people with and without disabilities.

Principle 2: Flexibility in Use

The design accommodates a wide range of individual preferences and abilities.

Guidelines

2a. Provide choice in methods of use. For example, Herman Miller's height-adjustable work surface allows users to choose standing or seated work positions.

2b. Accommodate right- or left-handed access and use. For example, Fiskars Softouch scissors accommodate use with either hand and allow alternation between the two in highly repetitive tasks (see Figure 12-10).

2c. Facilitate the user's accuracy and precision. For example, AT&T's Big-Button phone accommodates users who don't see keys accurately, hurry through the process, or lack dexterity.

2d. Provide adaptability to the user's pace. For example, tape speed control on a dictation machine enables a transcriptionist to work at his or her best pace and also facilitates use by "talking book" users.

FIGURE 12-10. Large-grip scissors accommodate use with either hand and allow alternation between the two in highly repetitive tasks.

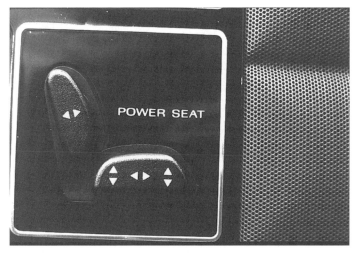

FIGURE 12-11. Automobile power seat control switch mimics the shape of the seat, enabling driver or passenger to make adjustments intuitively.

Principle 3: Simple and Intuitive Use

Use of the design is easy to understand, regardless of the user's experience, knowledge, language skills, or current concentration level.

Guidelines

3a. Eliminate unnecessary complexity. For example, operation of single-lever faucets is easily understood without instruction or previous experience.

3b. Be consistent with user expectations and intuition. For example, the design of an automotive power seat control mimics the shape of the seat, enabling the driver or passenger to make adjustments intuitively (see Figure 12-11).

3c. Accommodate a wide range of literacy and language skills. For example, Ikea's furniture assembly instructions eliminate translation problems of worldwide distribution by providing clear illustrations that don't require text.

3d. Arrange information consistent with its importance. For example, an illustrated and color-coded dosage label emphasizes precautions in taking cough medicine

3e. Provide effective prompting and feedback during and after task completion. For example, on-screen VCR programming takes the user through a step-by-step menu for setup and operations.

FIGURE 12-12. Modified round wall thermostat incorporates enlarged visual information, tactile lettering, edge texture, and audible click stops at two-degree temperature intervals.

Principle 4: Perceptible Information

The design communicates necessary information effectively to the user, regardless of ambient conditions or the user's sensory abilities.

Guidelines

4a. Use different modes (visual, auditory, tactile) for redundant presentation of essential information. For example, Honeywell's round wall thermostat incorporates enlarged visual information, tactile lettering, edge texture, and audible click stops at two-degree intervals (see Figure 12-12).

4b. Maximize visual, auditory, and tactile legibility of essential information. For example, a dark background on overhead signage contrasts with the lighted ceiling at Dulles International Airport in Virginia.

4c. Differentiate elements in ways that can be described (i.e., make it easy to give instructions or directions). For example, audio plugs and jacks differentiated by color make it easier to connect equipment, especially when using phone or on-line technical troubleshooting.

4d. Provide compatibility with a variety of techniques or devices used by people with sensory limitations. For example, Nokia's "loopset" makes the company's cellular phones compatible with hearing aids.

Principle 5: Tolerance for Error

The design minimizes hazards and the adverse consequences of accidental or unintended actions (see Figure 12-13).

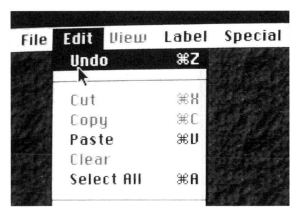

FIGURE 12-13. Undo option allows computer users to correct mistakes without penalty.

Guidelines

5a. Arrange elements to minimize hazards and errors. Place the most-used elements in the most accessible place. Hazardous elements should be eliminated, isolated, or shielded. For example, Bagel Biter shields hands from the blade while holding the bagel securely.

5b. Provide warnings of hazards and errors. For example, a red spout on a bottle of contact lens cleaner warns the user not to confuse it with a bottle of wetting solution that is of identical shape and size.

5c. Provide fail-safe features. For example, a clothing iron shuts off automatically after five minutes of nonuse.

5d. Discourage unconscious action in tasks that require vigilance. For example, a "dead-man" handle on a lawnmower requires the user to squeeze a handle and lever together to keep the engine running.

Principle 6: Low Physical Effort

The design can be used efficiently and comfortably and with a minimum of fatigue.

Guidelines

6a. Allow users to maintain a neutral body position. For example, a split, angled keyboard allows a computer operator to maintain a neutral position from elbow to fingers.

6b. Use reasonable operating forces. For example, Snap-Ware kitchen food containers by Flotool International require only a gentle movement to open or close them.

FIGURE 12-14. Free-rolling casters greatly reduce the effort of traveling with carry-on luggage.

    6c. Minimize repetitive actions. For example, quarter-turn Tylenol pain reliever bottles minimize repeated twisting.

    6d. Minimize sustained physical effort. For example, free-rolling casters reduce the physical effort of traveling with carry-on luggage (see Figure 12-14).

Principle 7: Size and Space for Approach and Use

Appropriate size and space is provided for approach, reach, manipulation, and use regardless of the user's body size, posture, or mobility.

Guidelines

    7a. Provide a clear line of sight to important elements for any seated or standing user. For example, Herman Miller's Milcare hospital furniture system incorporates a lowered counter section at the nurses' station to accommodate patients of varying heights and postures.

    7b. Ensure a comfortable reach to all components for any seated or standing user. For example, Whirlpool side-by-side refrigerator/freezers with full-length handles accommodate users of all heights and postures.

    7c. Accommodate variations in hand and grip size. For example, the loop handle on Copco's chopping knife accommodates hands of all sizes.

    7d. Provide adequate space for the use of assistive devices or personal assistance. For example, a wide vehicle door opening provides for a close approach to the seat with a walker or wheelchair (see Figure 12-15).

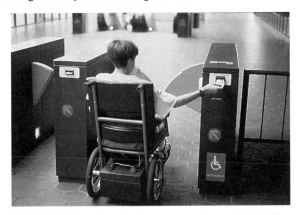

FIGURE 12-15. Wide gate at subway station accommodates wheelchair users as well as commuters with packages or luggage.

Summary: How Principles of Universal Design Accommodate Differences in Abilities						
Principle	Cognition	Vision	Hearing and Speech	Body Functions	Arm Function	Hand Function	Mobility
1 Equitable Use	X	X	X	X	X	X	X
2 Flexibility in Use	X	X	X	X	X	X	X
3 Simple and Intuitive Use	X	X	X				
4 Perceptible Information	X	X	X				
5 Tolerance for Error	X	X	X	X	X	X	X
6 Low Physical Effort				X	X	X	X
7 Size and Space for Approach and Use				X	X	X	X

NOW LET'S TRY IT

We've discussed the concept of universal design and how human abilities can be affected by the environment, temporary situations, age, and disability. We've studied the Principles of Universal Design and some examples of how some companies have successfully applied universal design guidelines. Now let's try it.

Select a product, preferably one unfamiliar to you and one that offers the experience of opening the package, using the product, storing it, and eventually disposing of it. Disposable kitchen storage containers and single-serving cereals are some simple, inexpensive choices.

Test how usable this product might be for persons of different ages and abilities, considering the Principles of Universal Design and their guidelines, as well as the variety of abilities presented above. For example, is the product useful, usable, and safe if you:

Use it without help or instructions?

Use it carelessly?

Use it in a very dark room or with your eyes closed?

Use it in a very noisy place or with your ears plugged?

Use it when very tired?

Use it with three-pound weights on your wrists?

Use it with only your nondominant hand?

Use it wearing mittens?

Use it only while sitting down?

How would you change this product to improve its usability for people of all ages and abilities? Do any of the universal design examples presented earlier give you some ideas?

How to Get Started: Universal Design Performance Measures for Products

To help product developers get started toward the goal of universal design, the Center for Universal Design has developed Universal Design Performance Measures for Products (Center for Universal Design 2000). This tool provides a procedure for evaluating how well products satisfy the seven Principles of Universal Design and their twenty-nine guidelines, described previously. The Principles of Universal Design are widely used in media coverage of examples of good design. Since they are based on these seven principles, the performance measures can be useful in identifying the universal design features of products for design competitions and award programs and in promoting the universal design features of products to potential customers.

Single printed copies of the Universal Design Performance Measures for Products are available free from the Center for Universal Design. Contact information is provided at the end of this chapter.

These performance measures were developed with the input of product designers, marketing professionals, and persons with disabilities. The performance measures were then tested in sixty-one households diverse in socioeconomic status and geographic region, and having members diverse in age, abilities, and attitudes about universal design. The test involved evaluating a group of everyday products using the Universal Design Performance Measures for Products.

Though designed as a simple checklist to uncover a product's universal design strengths and weaknesses, the Universal Design Performance Measures for Products are not intended as a substitute for real-world testing by individuals with personal experience in aging or disability. No evaluation tool can substitute for this insight. But product developers with some knowledge of the issues discussed in this chapter will find the performance measures useful in developing product testing and focus group methodologies for use with individuals of diverse ages and abilities.

No Substitute for Personal Experience with Disability

This chapter is designed to point out ways that age and disability affect product usability and to give product managers the tools to integrate the needs of elders and people with disabilities into product development. It should be apparent that meeting the needs of elders and people with disabilities is not special design but just good design—for everyone.

In addition to these tools, an important component of this process is direct input from elders and people with disabilities themselves. For product developers without experience in gerontology or disability studies, it may be daunting to seek out this input.

For many years, the Center for Universal Design at North Carolina State University and other research centers have conducted focus groups for this purpose regarding products as diverse as kitchen appliances, door hardware, and telephones. These focus groups are composed of people diverse in age and ability, as well as clinicians with expertise in gerontology and disability.

Two prevailing lessons of this research are that all these groups welcome the opportunity to participate and that no amount of well-meaning empathy or secondhand knowledge can substitute for personal experience with limitations due to age or disability. Product development managers are urged to seek out this experience within their community early in the product development process—before barriers are designed in that can be expensive or impossible to eliminate later.

Now is the time to get started, as many companies have already done.

Tupperware: Re-creating an American Icon

Tupperware has been a household word for generations. The brand was established in the 1950s by Earl Tupper, a self-educated engineer working for a DuPont chemical plant. Now headquartered in Orlando, Florida, Tupperware is one of the world's leading manufacturers and sellers of plastic food serving, storage, and preparation products. One of the most well recognized brand names in the world, its products are found in over 90 percent of American households, and the company is widely recognized for designing top-quality, innovative products.

Growing Up with the Baby Boomers

Unlike so many consumer products, Tupperware containers remain useful for decades after purchase. The same container that kept the baby's food fresh was still used years later to save dinner leftovers for that same child when she came home late from high school soccer practice.

In the ensuing years, young homemakers who purchased their first Tupperware in the 1940s reached middle age, while their children and their elderly parents used their Tupperware products as well. Though life changed considerably for baby boomers and their families through the next three decades, Tupperware design remained essentially the same.

Sealing Out Some Users

For many children, elders, and people with disabilities, the same airtight seal that had been Tupperware's trademark was also a barrier, because the narrow lip was difficult to open. At the same time, many who had been young homemakers in 1945 and were among Tupperware's most faithful customers had begun to experience arthritis and other natural effects of aging that made use of that classic seal difficult for them as well.

One of those users was the mother of Morison Cousins, vice president of design for Tupperware Worldwide. Like many of her contemporaries, she had found that the narrow lip around the edge of the seal had become difficult to use.

Usability Meets Durability

In 1990 Cousins faced a formidable challenge. Tupperware had decided that it needed to update its products to reach a new generation of homemakers. This would mean changing a design that had remained essentially unchanged since the 1950s while increasing in sales for three decades.

Cousins remembered the 1950s fondly, and Tupperware had been among the more popular and exciting home products during these years. Born in Brooklyn in 1934, Cousins had studied industrial design at Pratt Institute and had later opened his own design office, also in New York, before joining Tupperware in 1990.

That same year, Cousins undertook the redesign of Tupperware products. In developing the new One Touch Seal and the redesign of the classic Wonderlier bowls, Cousins had in mind users such as his eighty-seven-year-old mother. He replaced the narrow lip seals with larger seal tabs and double-arc handles that were easier to grasp (see Figure 12-16).

Strong color contrast between the lids and bowls increased usability for

FIGURE 12-16. Tupperware bowl lids have large, high-contrast tabs.

people with limited vision. The very features appreciated by museum curators also had a straightforward usability, even for people limited by age or disability.

Cousins's approach earned Tupperware products a place in design museums around the world. With Cousins's redesign of the classic seal, Tupperware products became capable not only of enduring through the user's life span but of remaining useful throughout that life span as well.

By 1999, company sales totaled over $1 billion through a sales force of about one million spread throughout more than a hundred countries worldwide.

MORE INFORMATION ABOUT UNIVERSAL DESIGN

Universal design is important to everyone who lives long enough to experience the limitations that come with age or through accidents or illness. The growing size of the potential markets among seniors and those with disabilities make the Principles of Universal Design an important addition to the literature on product development.

The Center for Universal Design has numerous publications on this subject, including case studies about some of the companies mentioned in this chapter. For more information, contact:

The Center for Universal Design
College of Design
Campus Box 8613
North Carolina State University
Raleigh, NC 27695-8613
1-800-647-6777

Or visit the Web site: http://www.design.ncsu.edu/cud

BIBLIOGRAPHY

Center for Universal Design. 2000. *Evaluating the Universal Design Performance of Products*. Raleigh: North Carolina State University, Center for Universal Design.

Center for Universal Design. 1997. The Principles of Universal Design (version 2.0). Raleigh: North Carolina State University, Center for Universal Design.

Investext Group. 1999. *Company Report, October 6, 1999—Tupperware*. New York: Wall Street Transcript Corporation.

McNeil, J. M. 1997. *Americans with Disabilities: 1994–95*. U.S. Bureau of the Census. Current Population Reports, P70–61. Washington, DC: U.S. Government Printing Office.

Mueller, J. M. 1998. *Case Studies on Universal Design*. Raleigh: North Carolina State University, Center for Universal Design.

Mueller, J. M., and C. A. Ingols. 1997. *Leviton Manufacturing Company, Inc.: Universal Design Marketing Strategy*. Case #996–063. Boston: Harvard Business School Publishing.

Pirkl, James J. 1994. *Transgenerational Design*. New York: Van Nostrand Reinhold.

Schneider, M. 1999. "Tupperware Moves Its Parties to the Mall." *Los Angeles Times*, May 7, Orange County edition, p. 7.

Story, M. F., J. L. Mueller, and R. L. Mace. 1998. *The Universal Design File: Designing for People of All Ages and Abilities*. Raleigh: North Carolina State University, Center for Universal Design.

Part 4

Portfolio Tools

Program managers are responsible for managing a portfolio of product development projects to support an entire organization or large business unit. Part 4 contains tools to help program managers characterize and manage across their portfolio of projects. The chapters in this part are organized from tools broader in nature and application to those narrower in nature.

Chapter 13, "Portfolio Management: Fundamental to New Product Success," addresses the broad issue of how the corporation should most effectively invest its R&D and new product resources—how it should manage its portfolio of projects over time. Portfolio management, defined as doing the right set of projects over time, is attained through achieving four goals: maximizing the value of the portfolio of projects for a given level of investment; achieving the right mix, or "balance," of projects; aligning the set of projects to the firm's strategies; and working on the right number of projects, given the limited resources available at the firm. Specific tools for achieving each of these four goals are presented in detail in this chapter. In addition, two fundamentally different approaches for integrating portfolio management into a firm's new product development process are provided. The "gates dominate" approach, where resource allocation decisions are integrated into NPD process gates, is best for larger firms in mature businesses where the portfolio of projects evolves only slowly over time. Fast-paced companies in fluid markets typically find the "portfolio reviews dominate" approach best, as it allows a constant shuffling of priorities across the project set.

In Chapter 14, "Assessing the Health of New Product Portfolio Management: A Metric for Assessment," two measures of portfolio health are presented and their use and utility demonstrated through examples. The depreciated product value metric calculates the sales value for each current product in the portfolio, depreciating each product's value year by year as time passes after introduction. This metric quantifies how little the firm's older products, with higher depreciation levels due to this year-by-year accrual, are actually contributing to the organization. A second metric, the danger index, is an indicator of how severe the threat to future sales is for each product in the portfolio. Analyzing the relative contribution of a firm's product lines using these metrics provides insight into which product lines and models the new product development team should be renewing or replacing in the portfolio.

While Chapter 8 presented a tool for managing risk at an individual project level, Chapter 15, "Risk Management: The Program Manager's Perspective," lays out a toolkit for managing risk from the program manager's perspective—managing risk across the full set of NPD projects in the firm or company. The three sections of the chapter describe the roles and responsibilities of program managers as gatekeepers and decision makers, defining risk within their context; how the focus of risk evolves according to the stage of development, explaining the six steps of effective risk management at gate meetings; and a set of templates to facilitate the dialogue around risk management.

The final chapter of Part 4, Chapter 16, "Process Modeling in New Product Development," describes a series of models of new product development that help NPD managers address several high-level issues, including strategic NPD planning, project selection, and project execution. Modeling the NPD process can help answer questions such as how many NPD projects are needed to meet the organization's growth targets; and are the NPD resources adequate for the desired number of projects. The chapter describes how to build a model of these processes for your firm, and presents a number of different templates as starting points. One overall purpose of modeling NPD is to create a steady commercial output of NPD projects. Another is to match the resources required to those actually available to increase the overall efficiency of the NPD process.

13 Portfolio Management: Fundamental to New Product Success

Robert G. Cooper, Scott J. Edgett,
and Elko J. Kleinschmidt

There are two ways for a business to succeed at new products: *doing projects right* and *doing the right projects*. Most new product prescriptions focus on the first route—for example, effective project management, cross-functional teams, and building in the voice of the customer. Portfolio management, the topic of this chapter, focuses on the second route, namely, doing the right projects.

A vital question in product innovation management is this: How should the corporation most effectively invest its R&D and new product resources? That is what portfolio management is all about: resource allocation to achieve corporate new product objectives. Much like a stock market portfolio manager, those senior executives who manage to optimize their R&D investments—to define the right new product strategy for the firm, select the winning new product projects, and achieve the ideal balance of projects—will win in the long run.

A ROADMAP FOR THE CHAPTER

This chapter first outlines the four goals in portfolio management together with the various tools and techniques for achieving each goal:

♦ The first goal is to *maximize the value* of the portfolio for a given resource expenditure. Various financial models, risk and probability models, and a scoring model approach are presented as ways to realize this goal.

♦ The next goal is *balance*—the right mix of projects. Here the emphasis is on visuals and graphics: bubble diagrams, including the popular risk-reward diagram and other variants, and more traditional charts, such as pie charts, which reveal the spending breakdowns in the portfolio.

♦ Achieving a *strategically aligned portfolio* is the third goal. Both bottom-up approaches (where careful selection of individual projects results in a strategic portfolio) and top-down methods, such as strategic buckets, where the business's strategy drives the portfolio, are described.

◆ The final goal is achieving the *right number of projects* for the limited resources available. While most techniques do deal with resource constraints, resource capacity analysis is presented as a possible solution here.

An assessment of popularity and results achieved reveals that the most popular portfolio methods aren't necessarily the best, and indeed, financial approaches yield the poorest portfolio.

Recommended approaches for portfolio management in your business are highlighted next. Two fundamentally different methods, the "gates dominate" approach and the "portfolio reviews dominate" approach, are described. Both use the same tools highlighted above, but the way the tools are applied is quite different. The pros and cons of the two approaches are outlined, along with some of the operational details of their use.

WHAT IS PORTFOLIO MANAGEMENT?

Doing the right projects is more than simply individual project selection; rather, it's about the entire mix of projects and new product or technology investments that your business makes. Portfolio management is formally defined as follows (Cooper, Edgett, and Kleinschmidt 1997a, 1999):

> Portfolio management is a dynamic decision process, whereby a business's list of active new product (and development) projects is constantly updated and revised. In this process, new projects are evaluated, selected, and prioritized; existing projects may be accelerated, killed, or deprioritized; and resources are allocated and reallocated to active projects. The portfolio decision process is characterized by uncertain and changing information, dynamic opportunities, multiple goals and strategic considerations, interdependence among projects, and multiple decision makers and locations. The portfolio decision process encompasses or overlaps a number of decision-making processes within the business, including periodic reviews of the total portfolio of all projects (looking at all projects holistically and against each other), making go/kill decisions on individual projects on an ongoing basis, and developing a new product strategy for the business, complete with strategic resource allocation decisions.

New product portfolio management sounds like a fairly mechanistic exercise of decision making and resource allocation. But there are many unique facets of the problem that make it perhaps the most challenging decision making faced by the modern business.

◆ First, new product portfolio management deals with *future events and opportunities*; thus much of the information required to make project selection decisions is at best uncertain and at worst very unreliable.

◆ Second, the decision environment is a very *dynamic* one: the status and prospects for projects in the portfolio are constantly changing as new information becomes available.

◆ Next, projects in the portfolio are at *different stages of completion,* yet all projects compete against each other for resources, so that comparisons must be made between projects with different amounts and "goodness" of information.

◆ Finally, *resources* to be allocated across projects are limited. A decision to fund one project may mean that resources must be taken away from another; and resource transfers between projects are not totally seamless.

Why So Important?

Portfolio management is a critical and vital senior management challenge, according to a study of best practices—see Figure 13-1 (Cooper, Edgett, and Kleinschmidt 1997b, 1998). Note how important the topic is rated by senior executives in the business as well as the senior technology people. Additionally, higher-performing businesses tend to rate portfolio management as much more important than poorer performers do.

Specific reasons for the importance of portfolio management, derived from the best-practices study, are:

1. Financial—to maximize return; to maximize R&D productivity; to achieve financial goals

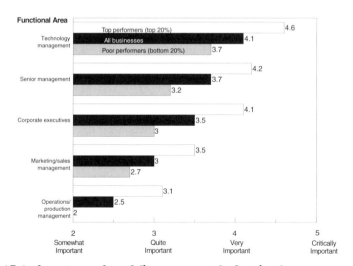

FIGURE 13-1. Importance of portfolio management by functional area.
This table shows how different functions rated portfolio management's importance. Note the significant differences between top and poor Performers—top-performing firms rate portfolio management to be much more important, as does the technology management function.
Source: Cooper, Edgett, and Kleinschmidt 1997b, 1998. Used with permission of Research-Technology Management, © 1997, 1998 Industrial Research Institute.

2. To maintain the competitive position of the business—to increase sales and market share

3. To properly and efficiently allocate scarce resources

4. To forge the link between project selection and business strategy—the portfolio is the expression of strategy and must support the strategy

5. To achieve focus—not doing too many projects for the limited resources available, and providing resources for the "great" projects

6. To achieve balance—the right balance between long- and short-term projects, and high- and low-risk ones, consistent with the business's goals

7. To better communicate priorities within the organization, both vertically and horizontally.

8. To provide better objectivity in project selection—to weed out bad projects

FOUR GOALS IN PORTFOLIO MANAGEMENT

There are four *high-level goals* common across businesses when it comes to portfolio management. The goal you wish to emphasize most will in turn influence your choice of portfolio methods. These four broad goals are:

Value maximization. Here the goal is to allocate resources so as to maximize the value of your portfolio. That is, you select projects so as to maximize the sum of the values or commercial worths of all active projects in your pipeline in terms of some business objective (such as long-term profitability, economic value added [EVA], return on investment [ROI], likelihood of success, or another strategic objective).

Balance. Here the principal concern is to develop a balanced portfolio—to achieve a desired balance of projects in terms of a number of parameters. For example, you might wish to seek the right balance of long-term projects vs. short-term ones; or high-risk vs. lower-risk projects; or across various markets, technologies, product categories, and project types (e.g., new products, improvements, cost reductions, maintenance and fixes, and fundamental research).[1]

Strategic direction. The main goal here is to ensure that, regardless of all other considerations, the final portfolio of projects truly reflects the business's strategy—that the breakdown of spending across projects, areas, markets, and so on is directly tied to the business strategy (e.g., to areas of strategic focus that management has previously delineated), and that all projects are "on strategy."

Right number of projects. Most companies have too many projects under way for the limited resources available (Cooper, Edgett, and Kleinschmidt 1997b, 1998, 1999, 2000). The result is pipeline gridlock, in which projects end up in a queue, they take longer and longer to get to market, and key activities within projects—for example, doing the up-front homework—are omitted because of a lack of people and time. Thus an overriding goal is to ensure a balance between resources required for the go projects and resources available.

What becomes clear is the potential for conflict between these four high-level goals. For example, the portfolio that yields the greatest net present value (NPV) or internal rate of return (IRR) may not be a very balanced one (it may contain a majority of short-term, low-risk projects, or is overly focused on one market); similarly, a portfolio that is primarily strategic in nature may sacrifice other goals (such as expected short-term profitability). Note that the nature of the portfolio management tool chosen indicates a hierarchy of goals. This is because certain of the portfolio approaches are much more applicable to some goals than others. For example, the visual models (such as portfolio bubble diagrams) are most suitable for achieving a balance of projects (visual charts being an excellent way of demonstrating balance), whereas scoring models may be poor for achieving or even showing balance but most effective if the goal is maximization against several objectives. Thus the choice of the "right" portfolio approach depends on which goal your leadership team has explicitly or implicitly highlighted.

What methods do companies find most effective to achieve the three portfolio goals? The next sections outline the portfolio management methods, complete with strengths and weaknesses.

Goal 1: Maximizing the Value of the Portfolio

A variety of methods can be used to achieve this goal, ranging from financial models through to scoring models. Each has its strengths and weaknesses. The end result of each method is a rank-ordered or prioritized list of go projects and hold projects, with the projects at the top of the list scoring highest in terms of achieving the desired objectives; the value in terms of that objective is thus maximized.

Net Present Value (NPV)

The simplest approach is merely to calculate the NPV of each project on a spreadsheet and then rank all projects according to their NPV. The go projects are at the top of the list, and projects continue to be added from further down the list until you run out of resources. Logically this method should maximize the NPV of your portfolio. Additionally, each project team usually determines the NPV for their project as part of their business case or capital appropriations request—so you're using a number that's already available.

Fine in theory . . . but: The NPV method ignores probabilities and risk; it assumes that financial projections are accurate (they usually are not); it assumes that only financial goals are important—for example, that strategic considerations are irrelevant; and it fails to deal with constrained resources—the desire to maximize the value for a limited resource commitment, that is, get the most bang for the buck. A final objection is more subtle: NPV assumes an all-or-nothing investment decision, whereas in new product projects, the decision process is an incremental one—more like buying a series of options on a project (Faulkner 1996).

Expected Commercial Value (ECV)

This method seeks to maximize the value or commercial worth of your portfolio, subject to certain budget constraints, and introduces the notion of risks and probabilities. The ECV method determines the value or commercial worth of each project to the corporation, namely, its *expected commercial value*. The calculation of the ECV is based on a decision tree analysis and considers the future stream of earnings from the project, the probabilities of both commercial success and technical success, and both commercialization costs and development costs (see Figure 13-2 for the calculation and definition of terms).

In order to arrive at a prioritized list of projects, the ECV of each project is determined. Next consider what resources are scarce or limiting. In the example in Table 13-1, R&D resources (people, but measured in terms of dollars) are thought to be the constraining or scarce resources—an R&D budget of $15 million. You may choose to use R&D people or work-months, or even capital funds, as the constraining resource.

Next, divide what you are trying to maximize, the ECV, by the constraining resource, which is the R&D costs per project (also in Table 13-1). Projects are rank-ordered according to this ECV/R&D cost ratio until the total R&D budget limit is reached. Those projects at the top of the list are go, while those at the bottom (beyond the total R&D budget limits) are placed on hold. The

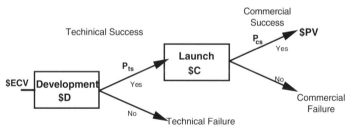

$$\$ECV = [(\$PV * P_{cs} - \$C) * P_{ts} - \$D]$$

$ECV	=	Expected commercial value of the project
P_{ts}	=	Probability of technical success
P_{cs}	=	Probability of commercial success (given technical success)
$D	=	Development costs remaining in the project
$C	=	Commercialization costs (capital equipment and market roll-out)
$PV	=	Present value of project's future earnings (discounted to today)

FIGURE 13-2. Determining the expected commercial value (ECV).
A model of a two-step investment decision process: First invest $D in development, which may yield a technical success with probability P_{ts}. Then invest $C in commercialization, which may result in a commercial success with probability P_{cs}. If successful, the project yields an income stream whose present value is $PV. More sophisticated versions would entail more than a simple 2-step model.

TABLE 13-1.
Expected Commercial Value Used to Prioritize Projects

1	2	3 Prob. Tech. Success	4 Prob. Comm. Success	5 Dev Cost	6 Comm. Cost	7 ECV	8 ECV/Dev. Cost	9 Rank and Status
Project Name	PV							
Alpha	30.00	80	.50	3	5	5.0	1.67	3 Go
Beta	63.75	.50	.80	5	2	19.5	3.90	1 Go
Gamma	8.62	.75	.75	2	1	2.1	1.05	6 On Hold
Delta	3.00	1.00	1.00	1	0.5	1.5	1.50	5 On Hold
Echo	50.00	.60	.75	5	3	15.7	3.14	2 Go
Foxtrot	66.25	.50	.80	10	2	15.5	1.55	4 On Hold

The ECV column shows the value of the ECV (column 7), determined from data in the table (columns 2–6), using the formula in Figure 13-2. This ECV is divided by the limiting resource, namely Development Costs (EVC/Development Cost in column 8). Projects are then ranked 1–6 according to this ECV/Dev cost metric, until one is out of resources (column 9). The development budget is $15 million. Projects Beta, Echo, and Alpha, totaling $13 million, are Go's. *Source:* Reprinted from Robert G. Cooper, Scott J. Edgett, and Elko J. Kleinschmidt, *Portfolio Management for New Products*, Second Edition (Cambridge: Perseus Publishing, 2001), with permission from the publisher.

method thus ensures the greatest bang for the buck; that is, the ECV is maximized for a given R&D budget.[2]

This ECV model has a number of attractive features. It recognizes that the go/kill decision process is an incremental one (the notion of purchasing options); all monetary amounts are discounted to today (not just to launch date), thereby appropriately penalizing projects that are years away from launch; and it deals with the issue of constrained resources, attempting to maximize the value of the portfolio in light of this constraint.

The major weakness of the method is the dependency on extensive financial and other quantitative data. Accurate estimates must be available for *all* projects' future stream of earnings, commercialization (and capital) expenditures, development costs, and probabilities of success—estimates that are often unreliable or, at best, simply not available early in the life of a project. A second weakness is that the method does not look at the balance of the portfolio—at whether the portfolio has the right balance between high- and low-risk projects, or across markets and technologies. A third weakness is that the method considers only a single financial criterion for maximization.

Productivity Index (PI)

The productivity index is similar to the ECV method, described above, and shares many of ECV's strengths and weaknesses. The PI tries to maximize the financial value of the portfolio for a given resource constraint (Evans 1996; Matheson, Matheson, and Menke 1994).

The productivity index is the following ratio:

$$PI = ECV \times P_{ts} / R\&D$$

Here, the definition of expected commercial value is different from that used above. In the productivity index, the ECV is a probability-adjusted NPV. More specifically, it is the probability-weighted stream of cash flows from the project, discounted to the present, and assuming technical success, less remaining R&D costs. There are various ways to adjust the NPV for risks or probabilities: employing a risk-adjusted discount rate, applying probabilities to uncertain estimates in calculating the NPV, or using Monte Carlo simulation to determine NPV. This risk-adjusted NPV is then multiplied by P_{ts}, the probability of technical success, and divided by R&D, the R&D expenditure remaining to be spent on the project (note that R&D funds already spent on the project are sunk costs and hence are not relevant to the prioritization decision). Projects are rank-ordered according to this productivity index in order to arrive at the preferred portfolio, with projects at the bottom of the list placed on hold.

Scoring Models as Portfolio Tools

Scoring models have long been used for making go/kill decisions at gates. But they also have applicability for project prioritization and portfolio management. Projects are scored on each of a number of criteria by management. Typical main criteria include:

◆ Strategic alignment
◆ Product advantage
◆ Market attractiveness
◆ Ability to leverage core competencies
◆ Technical feasibility
◆ Reward vs. risk

The project attractiveness score is the weighted addition of the item ratings, and becomes the basis for developing a rank ordered list of projects (Table 13-2 provides an illustration, using the six criteria listed above; projects are ranked until there are no more resources, in this case measured by FTE people). A sample scoring model is also shown in Table 13-3, with a more detailed list of criteria.

Scoring models generally are praised in spite of their limited popularity. Research into project selection methods reveals that scoring models produce a strategically aligned portfolio, one that reflects the business's spending priorities. They yield effective and efficient decisions and result in a portfolio of high-value projects (Cooper, Edgett, and Kleinschmidt 1997b, 1998).

TABLE 13-2
Prioritized Scored List of Projects: A Rank-Ordered List

Project	Leader	Strat.	Prod. Advtg.	Market Attract.	Core Comp.	Tech. Feasib.	Reward	Project Attract. Score	People FTE	Cum. FTE	Status
Epsilon	Peters	9	9	10	10	9	9	93.3	20	20	Active
Gamma	Cooper	10	10	7	7	7	7	80.0	20	40	Active
Alpha	Smith	8	7	7	8	8	9	75.0	15	55	Active
Delta	Scott	7	7	9	9	8	5	74.0	12	67	Active
Beta	Jones	7	7	6	6	8	6	66.7	20	87	HOLD
Omicron	Baily	8	6	6	8	7	5	66.7	20	107	HOLD

(resource limit: 70 FTEs)

1. Set up a spreadsheet—list your active, on-hold, and proposed projects.
2. Rank these projects according to some criterion (e.g., project attractiveness score or NPV).
   - In this example, six screening criteria are used (see text)—strategic fit, product advantage, etc.
   - The project attractiveness score—the average of these six criteria, but taken out of 100— is used as the ranking criterion.
   - All six projects are good ones, with scores over 65 points out of 100.
3. Include projects until you are out of resources (here measured by FTEs—full-time-equivalent people).
   - Here the first four projects are active (note the resource limit of 70 FTEs), and the last two are put on hold.

Source: Reprinted from Robert G. Cooper, Scott J. Edgett, and Elko J. Kleinschmidt, *Portfolio Management for New Products*, Second Edition (Cambridge: Perseus Publishing, 2001), with permission from the publisher.

Goal 2: A Balanced Portfolio

The second major goal is a balanced portfolio—a set of development projects balanced in terms of a number of key parameters. The analogy is that of an investment fund, where the fund manager seeks balance in terms of high-risk vs. blue-chip stocks, and balance across industries, in order to arrive at an optimal investment portfolio.

Visual charts are favored in order to display balance in new product project portfolios. These visual representations include portfolio maps or bubble diagrams (Figure 13-3)—an adaptation of the four-quadrant Boston Consulting Group (BCG) diagrams that have seen service since the 1970s as strategy models—as well as more traditional pie charts and histograms.

A casual review of portfolio bubble diagrams will lead some to complain that these new models are nothing more than the old strategy bubble diagrams of the seventies. Not so. Recall that the BCG strategy model and others like it (such as the McKinsey/GE model) plot business units on a market-attractiveness-vs.-business-position grid (Day 1986; Heldey 1977). Note that the unit of analysis is the strategic business unit (SBU)—an existing business whose performance, strengths, and weaknesses are all known. By contrast,

TABLE 13-3.
A Typical Scoring Model for Project Selection

Factor 1: Reward
- ◆ Absolute contribution to profitability (five-year cash flow: cumulative cash flows less all cash costs, before interest and taxes)
- ◆ Technological payback: the number of years for the cumulative cash flow to equal all cash costs expended prior to the start-up date
- ◆ Time to commercial start-up

Factor 2: Business Strategy Fit
- ◆ Congruence: how well the program fits with the strategy (stated or implied) for the product line, business, and/or company
- ◆ Impact: the financial and strategic impact of the program on the product line, business, and/or company (scored from minimal to critical).

Factor 3: Strategic Leverage
- ◆ Proprietary position
- ◆ Platform for growth (from "one of a kind" to "opens up new technical and commercial fields")
- ◆ Durability: the life of the product in the marketplace (years)
- ◆ Synergy with other operations/businesses within the corporation

Factor 4: Probability of Commercial Success
- ◆ Existence of a market need
- ◆ Market maturity (from "declining" to "rapid growth")
- ◆ Competitive intensity: how tough or intense the competition is
- ◆ Existence of commercial applications development skills (from "new" to "already in place")
- ◆ Commercial assumptions (from "low probability" to "highly predictable")
- ◆ Regulatory/social/political impact (from "negative" to "positive")

Factor 5: Probability of Technical Success
- ◆ Technical gap (from "large gap" to "incremental improvement")
- ◆ Program complexity (from "very high, many hurdles" to "straightforward")
- ◆ Existence of technological skill base (from "new to us" to "widely practiced in company")
- ◆ Availability of people and facilities (from "must hire/build" to "immediately available")

Sources: Cooper 1998; Cooper, Edgett, and Kleinschmidt 2001.

today's new product portfolio bubble diagrams, while they may appear similar, plot individual new product projects—future businesses, or what might be. As for the dimensions of the grid, here too the market-attractiveness-vs.-business-position dimensions used for existing SBUs may not be as appropriate for new product possibilities, so other dimensions or axes are extensively used.

Which Dimensions to Consider

What are some of the parameters that your business should plot on these bubble diagrams in order to seek balance? Different pundits recommend various parameters and lists, and even suggest the "best plots" to use. Table 13-4 provides

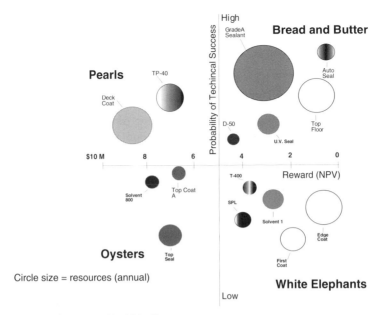

FIGURE 13-3. Risk-reward bubble diagram.
Source: Reprinted from Robert G. Cooper, Winning at New Products, *Third Edition (Cambridge: Perseus Publishing, 2001), with permission from the publisher.*

TABLE 13-4.
Popular Bubble Diagram Plots (Rank-Ordered by Popularity)

Rank	Type of Chart	First Dimension Plotted	Second Dimension Plotted	Percentage of Businesses Using Bubble Diagrams*
1	Risk vs. Reward	Reward: NPV, IRR, benefits after years of launch, market value	by Probability of success (technical, commercial, overall)	44.4%
2	Newness	Technical newness	by Market newness	11.1%
3	Ease vs. Attractiveness	Technical feasibility	by Market attractiveness (growth, potential, consumer appeal, life cycle)	11.1%
4	Strength vs. Attractiveness	Competitive position (strengths)	by Attractiveness (market growth, technical maturity, years to implementation)	11.1%
5	Cost vs. Timing	Cost to implement	by Time to implement	9.7%
6	Strategic vs Benefit	Strategic focus or fit	by Business intent, NPV, financial fit, attractiveness	8.9%
7	Cost vs. Benefit	Cumulative reward	by Cumulative development costs	5.5%

*Reads: Of all the businesses using bubble diagrams, 44.4% use the Risk vs. Reward version.
Source: Cooper, Edgett, and Kleinschmidt 1997b, 1998. Used with permission of *Research-Technology Management,* © 1997, 1998 Industrial Research Institute.

a list of the most popular bubble diagram plots (Cooper, Edgett, and Kleinschmidt 1997b, 1998).

Risk-Reward Bubble Diagrams

The most popular bubble diagram is a variant of the risk/return chart (see Figure 13-3 and Table 13-4). Here one axis is some measure of the reward to the company, while the other is a success probability.

- ◆ One approach is to use a qualitative estimate of reward, ranging from "modest" to "excellent" (Rousell, Saad, and Erickson 1991). The argument here is that too heavy an emphasis on financial analysis can do serious damage, notably in the early stages of a project. The other axis is the probability of overall success (probability of commercial success multiplied by probability of technical success).
- ◆ In contrast, other firms rely on very quantitative and financial gauges of reward, namely, the probability-adjusted NPV of the project (Evans 1996; Matheson, Matheson, and Menke 1994). Here the probability of technical success is the vertical axis, as the probability of commercial success has already been built into the NPV calculation.

A sample bubble diagram is shown in Figure 13-3 for an SBU of a major chemical company. Here the size of each bubble shows the annual resources spent on each project (dollars per year; it could also be people or work-months allocated to the project).

The four quadrants of the portfolio model in Figure 13-3 are:

- ◆ *Pearls* (upper left quadrant): These are the potential star products—projects that have a high likelihood of success and are also expected to yield a very high reward. Most businesses desire more of these. There are two such Pearl projects, and one of them has been allocated considerable resources (denoted by the sizes of the circles).
- ◆ *Oysters* (lower left): These are the long-shot projects—projects with a high expected payoff but a low likelihood of technical success. They are the projects where technical breakthroughs will pave the way for solid payoffs. There are three of these; none is receiving many resources.
- ◆ *Bread and Butter* (upper right): These are small, simple projects—high likelihood of success but low reward. They include the many fixes, extensions, modifications, and updating of projects, of which most companies have too many. More than 50 percent of spending is going to these Bread and Butter projects in Figure 13-3.
- ◆ *White Elephants* (lower right). These are the low-probability, low-reward projects. Every business has a few white elephants—they inevitably are difficult to kill—but this company has far too many. One-third of the projects and about 25 percent of spending falls in the lower right White Elephant quadrant.

Given that this chemical SBU is a star business seeking rapid growth, a quick review of the portfolio map in Figure 13-3 reveals many problems. There are too many White Elephant projects (it's time to do some serious pruning), too much money spent on Bread and Butter (low-value) projects, not enough Pearls, and heavily underresourced Oysters.

One feature of this bubble diagram model is that it forces senior management to deal with the resource issue. Given finite resources (e.g., a limited number of people or money), the sum of the areas of the circles must be a constant. That is, if you add one project to the diagram, you must subtract another; alternatively, you can shrink the size of several circles. The elegance here is that the model forces management to consider the resource implications of adding one more project to the list—that some other projects must pay the price.

Also shown in this bubble diagram is the product line that each project is associated with (via shading or crosshatching). A final breakdown is via color, to represent timing (not shown in our black-and-white map). Here hot red means "imminent launch," while blue is cold and means "early-stage project." Thus this apparently simple risk/reward diagram shows a lot more than simply risk and profitability data—it also conveys resource allocation, timing, and spending breakdowns across product lines.

Variants of Risk-Reward Bubble Diagrams: Dealing with Uncertainties

3M's ELLIPSES   One problem with the bubble diagram in Figure 13-3 is that it requires a point estimate of both the reward (the likely NPV) as well as the probability of success. Some businesses at 3M use a variant of the bubble diagram to effectively portray uncertain estimates (Tritle 1997). In calculating the NPV, optimistic and pessimistic estimates are made for uncertain variables, leading to a range of NPV values for each project. Similarly, low, high, and likely estimates are made for the probability of technical success. The result is Figure 13-4, where the sizes and shapes of the bubbles reveal the uncertainty of projects; here, very small bubbles mean highly certain estimates on each dimension, whereas large ellipses mean considerable uncertainty (a high spread between worst case and best case) for that project.

MONTE CARLO SIMULATION   Procter & Gamble uses Monte Carlo simulation to handle probabilities. P&G's portfolio model is a three-dimensional portfolio model, created by CAD software (Figure 13-5); the three axes are:

- ◆ NPV—a measure of the project's expected reward (probability-adjusted)
- ◆ Time to launch (the longer the time, the higher the risk and the more distant the reward)
- ◆ Probability of commercial success (as calculated from P&G's customized version of the NewProd model [Cooper 1987]).

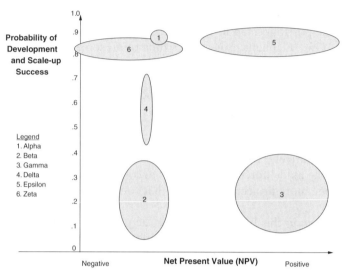

FIGURE 13-4. 3M's risk-reward bubble diagram showing uncertainties.
Source: Reprinted from Robert G. Cooper, Scott J. Edgett, and Elko J. Kleinschmidt, Portfolio Management for New Products, *Second Edition (Cambridge: Perseus Publishing, 2001), with permission from the publisher.*

In both firms, in order to account for commercial uncertainty, every variable—revenues, costs, launch timing, and so on—requires three estimates: high, low, and likely. From these three estimates, a probability distribution curve is calculated for each variable. Next, random scenarios are generated for the project,

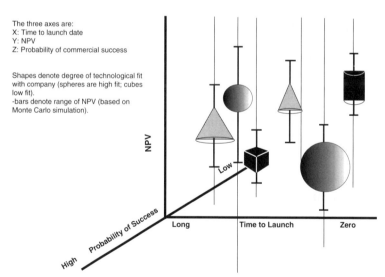

FIGURE 13-5. Procter & Gamble's 3-dimensional risk-reward bubble diagram.
Source: Reprinted from Robert G. Cooper, Scott J. Edgett, and Elko J. Kleinschmidt, Portfolio Management for New Products, *Second Edition (Cambridge: Perseus Publishing, 2001), with permission from the publisher.*

using these probability curves as variable inputs. Thousands of scenarios are computer-generated (hence the name Monte Carlo: thousands of spins of the wheel), and the result is a distribution of financial outcomes. From this, the expected NPV and its range are determined—an NPV figure that has had all commercial outcomes, and their probabilities, figured in. P&G shows this range of NPVs simply as an I-beam drawn vertically through the shapes.

Portfolio Maps with Axes Derived from Scoring Models

A combination scoring model and bubble diagram approach is employed by Speciality Minerals, a company spun off from Pfizer. A scoring model is used to make go/kill decisions on projects and also to rank order projects on a prioritization list. Here, seven factors are considered in the firm's scoring model (see Figure 13-6 for list). These same factors then provide the input data to construct the bubble diagram. For example:

♦ The horizontal axis, labeled "value to the company," consists of the weighted financial attractiveness and competitive advantage factors added together.

♦ The vertical axis is labeled "probability of success" and is made up of three weighted factors: customer interest, technical feasibility, and fit with technical/manufacturing capabilities.

The unique feature here is that this company's seven-factor scoring model does double duty: It is the basis for go/kill decisions at project or gate reviews. It also provides five of the factors (and data) to construct the two axes of the portfolio bubble diagram.

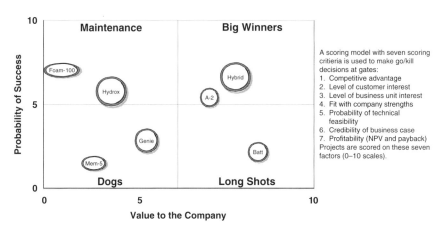

The two axes—Value and Probability—are computed from scoring model results (0–10 scores)
The computation is:
Value = .66 (Profitability) + .34 (Comp. Adv.)
Prob. of Success = (Cust. Int.) + .5 (Tech. Feasibilty) + .25 (Fit)

FIGURE 13-6. Risk-reward bubble diagram using scored axes.
Source: Reprinted from Robert G. Cooper, Scott J. Edgett, and Elko J. Kleinschmidt, Portfolio Management for New Products, *Second Edition (Cambridge: Perseus Publishing, 2001), with permission from the publisher.*

Traditional Charts for Portfolio Management

There are numerous parameters, dimensions, or variables across which one might wish to seek a balance of projects. As a result, there is an endless variety of histograms and pie charts that help to portray portfolio balance. Some examples:

Timing is a key issue in the quest for balance. One does not wish to invest strictly in short-term projects, nor totally in long-term ones. Another timing goal is for a steady stream of new product launches spread out over the years—constant "new news" and no sudden logjam of product launches all in one year. A histogram captures the issue of timing and portrays the distribution of resources to specific projects according to years of launch (not shown). Another timing issue is cash flow. Here the desire is to structure one's array of projects in such a way that cash inflows are reasonably balanced with cash outflows in the business. Some companies produce a timing histogram that portrays the total cash flow per year from all projects in the portfolio over the next few years (not shown).

Project types are yet another vital concern. What is your spending on genuine new products vs. product renewals (improvements and replacements), or product extensions, or product maintenance, or cost reductions and process improvements? And what should it be? Pie charts effectively capture the spending split across project types—actual vs. desired splits. Markets, products, and technologies provide another set of dimensions across which managers seek balance. The question faced is: Do you have the appropriate split in R&D spending across your various product lines? Or across the markets or market segments in which you operate? Or across the technologies you possess? Pie charts are again appropriate for capturing and displaying this type of data.

Goal 3: Building Strategy into the Portfolio

Strategy and new product resource allocation must be intimately connected. Strategy becomes real when you start spending money. Until you begin allocating resources to specific activities—for example, to specific development projects—strategy is just words in a strategy document.

The mission, vision, and strategy of the business are made operational through the decisions it makes on where to spend money. For example, if a business's strategic mission is to grow via leading-edge product development, then this must be reflected in the mix of new product projects under way—projects that will lead to growth (rather than simply to defense) and products that really are innovative. Similarly, if the strategy is to focus on certain markets, products, or technology types, then the majority of projects and spending should be focused on such markets, products, or technologies.

Linking Strategy to the Portfolio: Approaches

Two broad issues arise in the desire to achieve strategic alignment in the portfolio of projects.

♦ *Strategic fit*. Are all your projects consistent with your business's strategy? For example, if you have defined certain technologies or markets as key areas to focus on, do your projects fit into these areas? Are they in bounds or out of bounds?

♦ *Spending breakdown*. Does the breakdown of your spending reflect your strategic priorities? In short, when you add up the areas where you are spending money, are these totally consistent with your stated strategy?

There are two ways to incorporate the goal of strategic alignment:

1. *Bottom-up approach—building strategic criteria into project selection tools*. Here strategic fit is achieved simply by including numerous strategic criteria into the go/kill and prioritization tools.

2. *Top-down approach—strategic buckets method*. This begins with the business's strategy and then moves to setting aside funds—"envelopes" or "buckets" of money—destined for different types of projects.

Bottom-up Approach—Strategic Criteria Built into Project Selection Tools

Not only are scoring models effective ways to maximize the value of the portfolio, they can also be used to ensure strategic fit. One of the multiple objectives considered in a scoring model, along with profitability or likelihood of success, can be to maximize strategic fit, simply by building into the scoring model a number of strategic questions.

In the scoring model displayed earlier in this chapter (Table 13-3), two major factors out of five are strategic, and of the nineteen criteria used to prioritize projects, six, or almost one-third, deal with strategic issues. Thus projects that fit the business's strategy and boast strategic leverage are likely to rise to the top of the list. Indeed, it is inconceivable how any "off-strategy" projects could make the active project list at all, as this scoring model naturally weeds them out.

Top-Down Approach—Strategic Buckets Model

While strategic fit can be achieved via a scoring model, a top-down approach is the only method designed to ensure that the eventual portfolio of projects—that is, where the money is spent—truly reflects the stated strategy for the business.

The strategic buckets model operates from the simple principle that implementing strategy equates to spending money on specific projects. Thus setting portfolio requirements really means setting spending targets.

The method begins with the business's strategy and requires senior management to make forced choices along each of several dimensions—choices about how it wishes to allocate scarce money resources. This enables the creation of "envelopes" or "buckets" of money. Existing projects are categorized into buckets; then one determines whether actual spending is consistent with desired spending for each bucket. Finally projects are prioritized within buckets to arrive at the ultimate portfolio of projects—one that mirrors management's strategy for the business.

Sounds simple, but the details are a little more complex. Senior management first develops the vision and strategy for the business. This includes defining strategic goals and the general plan of attack to achieve these goals—a fairly standard business strategy exercise. Next, forced choices are made across key strategic dimensions. That is, based on this strategy, the management of the business allocates R&D and new product marketing resources across categories on each dimension. Some common dimensions are:

- ◆ *Strategic goals.* Management is required to split resources across the specified strategic goals. For example, what percentage should be spent on defending the base? On diversifying? On extending the base?

- ◆ *Product lines.* Resources are split across product lines. How much should be spent on product line A? On product line B? On C? A plot of product line locations on the product life cycle curve is used to help determine this split.

- ◆ *Project type.* What percent of resources should go to new product development? To maintenance-type projects? To process improvements? To fundamental research?

- ◆ *Familiarity matrix.* What should be the split of resources to different types of markets and to different technology types in terms of their familiarity to the business? You can use the "familiarity matrix" proposed by Roberts—technology newness vs. market newness—to help split resources (Roberts and Berry 1983).

- ◆ *Geography.* What proportion of resources should be spent on projects aimed largely at North America? At Latin America? Europe? The Pacific? Globally?

Now management develops strategic buckets. Here the various strategic dimensions listed above are collapsed into a convenient handful of buckets. For example, buckets might be product development projects for product lines A and B, cost reduction projects for all product lines, product renewal projects for product lines C and D, and so on (see Table 13-5).

Next, the desired spending by bucket is determined: the "what should be." This involves a consolidation of desired spending splits from the strategic allocation exercise above.

After that comes a gap analysis. Existing projects are categorized by bucket, and the total current spending by bucket is added up (the "what is"). Spending gaps are then identified between the "what should be" and "what is" for each bucket.

Finally, projects within each bucket are rank-ordered. You can use either a scoring model or financial criteria to do this ranking within buckets (Table 13-5). Portfolio adjustments are then made, either by immediate pruning of projects or by adjusting the approval process for future projects.

TABLE 13-5.
Projects Prioritized within Buckets

Four of Twelve Buckets with Spending Targets in Each Buckets			
Bucket #1: New Products for Product Line A Target Spending: $8.7M	Bucket #2: New Products for Product Line B Target Spending: $18.5M	Bucket #3: Maintenance of Business for Product Lines A and B Target Spending: $10.8M	Bucket #4: Cost Reductions for All Product Lines Target Spending: $7.8M
Project A 4.1	Project B 2.2	Project E 1.2	Project I 1.9
Project C 2.1	Project D 4.5	Project G 0.8	Project M 2.4
Project F 1.7	Project K 2.3	Project H 0.7	Project N 0.7
Project L 0.5	Project T 3.7	Project J 1.5	Project P 1.4
Project X 1.7	Gap = 5.8	Project Q 4.8	Project S 1.6
Project Y 2.9		Project R 1.5	Project U 1.0
Project Z 4.5		Project V 2.5	Project AA 1.2
Project BB 2.6		Project W 2.1	

Twelve "buckets" or categories of project types are defined in this example; only four buckets are shown here. Projects are sorted according to bucket, and then are rank-ordered within columns according to a maximization method (ECV), a financial criterion such as NPV × probability of success, or better yet, a scoring model. Numbers within columns show resources required to do each project ($Mil.). Note that Bucket 1 runs out of resources after Project L; whereas in Bucket 2, there is a shortage of good projects. *Source:* Cooper, Edgett, and Kleinschmidt 1998. Used with permission of *Research-Technology Management,* © 1998 Industrial Research Institute.

You may find that the number of dimensions and splits outlined above becomes too complex and onerous. A somewhat simpler breakdown can be found in Honeywell-AlliedSignal's "Mercedes-Benz star" method of allocating resources (Figure 13-7). The leadership team of the business begins with the business's strategy, and uses the Mercedes-Benz emblem (a three-pointed star) to help divide up the resources. There are three buckets: fundamental research and platform development projects, which promise to yield major breakthroughs and new technology platforms; new product developments; and maintenance—technical support, product improvements and enhancements, and so on. Management divides the R&D funds into these three buckets and then rates and ranks projects against each other within each bucket. In effect, three separate portfolios of projects are created and managed. The spending breakdown across projects mirrors strategic priorities.

The major strength of the strategic buckets model is that it firmly links spending to the business's strategy. Over time, the portfolio of projects, and the spending across strategic buckets, will equal management's desired spending targets across buckets. Another positive facet of the strategic buckets model is the recognition that all development projects that compete for the same resources should be considered in the portfolio approach. Finally, different criteria can be used for different types of projects. That is, you are not faced with comparing and ranking very different types of projects against each other—for

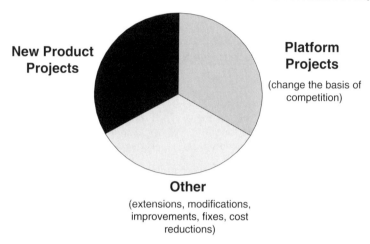

New Product Projects

Platform Projects

(change the basis of competition)

Other

(extensions, modifications, improvements, fixes, cost reductions)

FIGURE 13-7. Strategic buckets method of portfolio management.
The business's strategy dictates the split of resources into buckets; projects are rank-ordered within buckets, but using different criteria in each bucket (method used by Honeywell-AlliedSignal). Source: Reprinted from Robert G. Cooper, Scott J. Edgett, and Elko J. Kleinschmidt, Portfolio Management for New Products, *Second Edition (Cambridge: Perseus Publishing, 2001), with permission from the publisher.*

example, comparing major new product projects to minor modifications. Because this is a two-step approach—first allocate money to buckets, then prioritize like projects within a bucket—it is not necessary to arrive at a universal list of scoring or ranking criteria that fits all projects.

Goal 4: The Right Number of Projects

Superimposed across all three goals above, of course, is resource constraints. That is, management must try to achieve these three goals but always remain wary of the fact that if too many projects are approved for the limited resources, pipeline gridlock is the result.

The problem of too many projects and too few resources can be partly resolved by undertaking a resource capacity analysis. This analysis attempts to quantify your projects' demand for resources (usually people, expressed as person-days of work) vs. the availability of these resources—see Figure 13-8 (Cooper 1999; Cooper, Edgett, and Kleinschmidt 2000).

1. *Do you have enough of the right resources to handle projects currently in your pipeline?* Begin with your current list of active projects. Determine the person-days each month required to complete them according to their timelines. Then look at the availability of resources. You usually find major gaps and hence potential bottlenecks.

2. *Do you have enough resources to achieve your new product goals?* Begin with your new product goals. What percent of your business's sales will come from new products?

3. *Now, determine the person-days required to achieve this goal.* Again, you will likely find a major gap between demand based on your goals and capacity available. It's time to make some tough choices about the realism of your goals or whether more resources are required.

This capacity analysis is a beginning and usually highlights key problems.

- ♦ It detects far too many projects in the pipeline, resulting in an immediate prioritization and pruning effort.
- ♦ It causes senior management to rethink its fairly arbitrary new product revenue and profit goals for the business.
- ♦ It identifies the functional areas that are major bottlenecks in the innovation process, leading to decisions to increase or shift personnel.

The capacity analysis chart shown in Figure 13-8 can also be used at portfolio and gate review meetings to show the impact of adding a new product project to the active list—what the addition means to resource commitments and constraints.

POPULARITY AND EFFECTIVENESS OF PORTFOLIO METHODS

In practice, financial methods dominate portfolio management, according to the best-practices study cited above (Cooper, Edgett, and Kleinschmidt 1998). Financial methods include various profitability and return metrics, such as NPV, ECV, ROI, EVA, or payback period—metrics that are used to rate, rank-order, and ultimately select projects. A total of 77.3 percent of businesses use such an approach in portfolio management—see Figure 13-3. For 40.4 percent of businesses, this is the dominant method.

Other methods are also quite popular:

- ♦ *Strategic approaches.* Having decided the business's strategy, money is allocated across different types of projects and into different envelopes or buckets. Projects are then ranked or rated within buckets. A total of 64.8 percent of businesses use this approach; for 26.6 percent of businesses, this is the dominant method.
- ♦ *Bubble diagrams or portfolio maps.* A total of 40.6 percent of businesses use portfolio maps; only 8.3 percent use this as their dominant method. The most popular map is the risk-vs.-reward map in Figure 13-3, but many variants of bubble diagrams are used.
- ♦ *Scoring models.* Scaled ratings are added to yield a project attractiveness score, which becomes the criterion used to make project selection and/ or ranking decisions. These models are used by 37.9 percent of businesses; in 18.3 percent, this is the dominant decision method.
- ♦ *Checklists.* Projects are evaluated on a set of yes/no questions. Each project must achieve a certain number of yes answers to proceed. The number

Method 1. Resource Demand Created by Your Active Projects:

Determine resource demand:

- Begin with your current list of active development projects, prioritized from best to worst (use a scoring model to prioritize projects, or a financial approaches, such as NPV or ECV). Develop a prioritized project list table, as below (here, Alpha is the best project; Foxtrot is the least attractive).
- Then consider the detailed plan of action for each project (use a timeline software package, such as Microsoft Project).

- For each activity on the timeline, note the number of person-days of work (or work-months), and what group (or what department) will do the work. These are shown in the "Person-days" column.
- Record these person-day requirements in the prioritized project list table—one column per department. In other columns, note the cumulative person-days by department.
- Develop such a table for each month.

Resource Demand Vs. Capacity Chart—Example

Project	Product Management		Marketing		Research Group A		Research Group B	
	Persondays	Cumulative	Persondays	Cumulative	Persondays	Cumulative	Persondays	Cumulative
Alpha	3	3	2	2	10	10	5	5
Beta	4	7	2	4	10	20	5	10
Gamma	3	10	2	6	15	35	5	15
Delta	5	15	3	9	15	50	8	23
Epsilon	6	21	3	*12*	5	*55*	8	31
Foxtrot	6	27	2	*14*	5	60	5	36
Demand		27		14		60		36
Available Persondays		20		10		60		40
% Utilization		135.00%		140.00%		100.00%		90.00%

What is your resource capacity?

- Next, look at the capacity available—how many person-days each department (or group) has available in total. (These person-days look at all people in that group or department, and what proportion of their time they have available for new products. Be sure to consider their "other jobs" in this determination—for example, the fact that a Marketing group likely has 90% of their time consumed by day-to-day assignments).

- Then mark the point in your prioritized-list-of-projects table where you run out of resources—where demand exceeds capacity. (Numbers in bold and italics show where we run out of resources in the sample table above).
- Determine the Percent Utilization—cumulative resources demanded divided by resources available—last row.

In the sample table above, note that two departments have over-committed resources, and two projects—Epsilon and Foxtrot—are the reasons why.

Results:

You will likely learn three things from this exercise:

♦ You really do have too many projects, often by a factor of two or three

♦ You can see which department or group is the constraining one, and

♦ You also begin to question where some departments spend their time (and why such a small proportion is available to work on new products!).

Method 2. Resource Demand Generated by Your Business's New Product Goals:

Determine resource demand:

♦ Begin with your new product goals—what sales or percentage of sales you desire from new products.

♦ Translate these goals into numbers of major and minor new product launches annually.

♦ Then, using your attrition curve—how many Stage 1, Stage 2, Stage 3, etc. projects does it take to yield one successful launch?—determine the number of projects per year you need moving through each stage.

♦ Next, consider the person-days requirements in each stage, broken down by function or department. The numbers of projects per stage combined with the person-days requirements yield the resource demand—namely, the person-days and personnel requirements to achieve your business's new product goals, again by department.

What is your resource capacity?

♦ Now turn to availability—how many person-days are available per department (same as the second part of Method 1 above).

Results:

♦ Again you'll likely find a major gap between demand versus capacity.

♦ At this point, you either modify your goals, making them a little more realistic; or make tough choices about adding resources or reassigning people in order to achieve your goals.

These two exercise can be done either with person-days (people × days) or dollars as the measure of resources.

FIGURE 13-8. Two ways to undertake resource capacity-versus-demand analysis.
Source: Adapted from Cooper 1999.

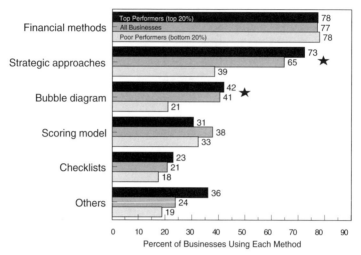

FIGURE 13-9. Popularity of portfolio methods employed.
Chart shows percentage of businesses using each portfolio method. Numbers are percentages, and add to more than 100 due to firms using multiple methods. Note that top performers tend to use strategic approaches, while poor performers do not use bubble diagrams (denoted by stars).
Source: Cooper, Edgett, and Kleinschmidt 1997a, 1998. Used with permission of Research-Technology Management, © 1997, 1998 Industrial Research Institute.

of yeses is used to make go/kill and/or prioritization (ranking) decisions. Only 21 percent of businesses use checklists; in only 2.7 percent is this the dominant method.

Popularity does not necessarily equate to effectiveness, however. When the performance of firms' portfolios were rated on six metrics in this study, companies that relied heavily on financial tools as the dominant portfolio selection model fared the worst (Table 13-6). Financial tools yield an unbalanced portfolio of lower-value projects and projects that lack strategic alignment. By contrast, strategic methods produce a strategically aligned and balanced portfolio. And scoring models appear best for selecting high-value projects, and also yield a balanced portfolio. Finally, firms using bubble diagrams obtain a balanced and strategically aligned portfolio.

It is ironic that the most rigorous techniques—the various financial tools—yield the worst results, not so much because the methods are flawed, but simply because reliable financial data are often missing at the very point in a project where the key project selection decisions are made.

PUTTING THE PORTFOLIO TOOLS TO WORK*

How should you use these various portfolio tools? Here we make the assumption that you already have a new product process in place—a gating or Stage-

*Parts of this section are taken from Cooper, Edgett, and Kleinschmidt, 2000.

TABLE 13-6.
Strengths/Weaknesses for Each Portfolio Method

Performance Metric	Performance Ratings for Each Portfolio Method (1–5 Scores)			
	Financial Methods	Strategic Methods	Scoring Model	Bubble Diagrams
Projects are aligned with business's objectives	3.76✗	4.08	3.95	4.11★
Portfolio contains very high-value projects	3.37✗	3.77	3.82★	3.70
Spending reflects the business's strategy	3.50	3.72★	3.59	3.00✗
Projects are done on time—no gridlock	2.79✗	3.22★	3.13	2.90
Portfolio has good balance of projects	2.80✗	3.08	3.20★	3.20★
Portfolio has right number of projects	2.50	2.93★	2.70	2.25

★ = Best method on each performance criterion
✗ = Worst method on each criterion
Ratings are 1–5 mean scores for each method, when used as dominant portfolio method.
Here 1 = poor and 5 = excellent
Source: Cooper, Edgett, and Kleinschmidt, 1998. Used with permission of *Research-Technology Management,* © 1998 Industrial Research Institute.

Gate™* process, as in Figure 13-10. (Note: A PDMA best-practices study reveals that the great majority of PDMA members have such processes in place [Griffin 1997]; if not, implementation of such a process is your first step.)

There are two fundamentally different approaches to integrating portfolio management tools into your new product process:

1. The "gates dominate" approach is best for larger firms in mature businesses where the portfolio of projects is fairly static. A solid gating process, where resource allocation methods are integrated into the gates, is likely best here. There is simply no great need to reprioritize the entire set of projects every few months; rather, the focus is more on in-depth reviews on individual projects and making sound go/kill decisions on each. Portfolio management is simply added to the process by modifying the gates somewhat (e.g., displaying portfolio lists and charts at gates) and holding several portfolio reviews annually to make any necessary course corrections.

2. The "portfolio reviews dominate" approach is best suited to fast-paced companies in fluid markets, whose portfolios are likely to be more dynamic. Here a constant reprioritization of the portfolio of projects is essential, simply because things change so fast in the marketplace. What

*Stage-Gate is a trademark of R. G. Cooper & Associates Consultants Inc., a member company of the Product Development Institute Inc.

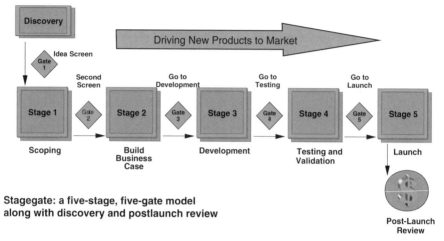

Stagegate: a five-stage, five-gate model
along with discovery and postlaunch review

FIGURE 13-10. The typical Stage-Gate™ model—from discovery to launch.
Source: Reprinted from Robert G. Cooper, Winning at New Products, Third Edition (Cambridge: Perseus Publishing, 2001), with permission from the publisher.

was a great project several months ago suddenly is not so good any-more—the whole market has changed. In this method, all projects are up for auction about four times per year. Portfolio reviews are the key decision meetings and amount to an all-project, mass gate meeting, where all projects and all resources are on the table.

Approach 1: The Gates Dominate

Here, the philosophy is that if your gating or new product process is working well, the portfolio will take care of itself. Therefore, make good decisions at the gates! The emphasis of this approach is on *sharpening gate decision making* on individual projects. (See Chapter 7, "Decision Making.")

In this approach, senior management or gatekeepers make go/kill decisions at gates on individual projects. Also at gates, the project is prioritized and resources are allocated. Gates thus provide an in-depth review of projects, one project at a time, and project teams leave the gate meeting with committed resources—with a check in hand. This is a real-time decision process, with gates activated many times throughout the year. By contrast, the periodic portfolio review, held perhaps once or twice a year, serves largely as a check to ensure that real-time gate decisions are good ones.

This "gates dominate" approach is often used by companies that already have a well-functioning Stage-Gate™ process in place. They then add portfolio management to their gating process, almost as a complementary decision process. This approach is used most often in larger companies, in science-based industries, and where projects are lengthy.

Here's how it works: Projects proceed through a gating process, as portrayed in Figure 13-10. Projects are rated and scored at gates, usually by senior management, especially at more critical gates (gate 3 and beyond).

To introduce portfolio management, gates become two-part decisions (Figure 13-11). The first part or half of the gate is a pass/kill decision, where individual projects are evaluated using the financial, checklist, and scoring model valuation tools described above.

The second half of the gate meeting involves prioritization of the project under discussion relative to all other projects (Figure 13-11). In practice, this means making a go/hold decision and, if the decision is to go, allocating resources to the project. A rank-ordered list of projects is displayed to compare the relative attractiveness of the project under discussion to the other active and on-hold projects (Table 13-2). Here, projects can be ranked on a financial criterion (for example, NPV or, better yet, ECV) or on the project attractiveness score derived from the scoring model.

Additionally, the impact of the proposed project on the total portfolio of projects is assessed. The question is: Does the new project under discussion improve the balance of projects (or detract from balance), and does the project improve the portfolio's strategic alignment? Bubble diagrams and pie charts are the tools used for visualizing balance and alignment, as outlined above.

Note how the gates dominate the decision process in this approach: go/kill, prioritization decisions, and resource allocation decisions are made in real time, right at the gate meeting. But other projects are *not* discussed and reprioritized at the gate; only the project in question is given a relative priority level compared to the rest.

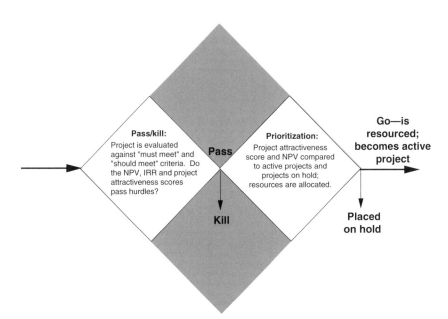

FIGURE 13-11. The two-part decision process at gates.
Source: Cooper, Edgett, and Kleinschmidt 1998. Used with permission of Research-Technology Management, © *1998 Industrial Research Institute.*

Portfolio Reviews in Approach 1

What about looking at all projects together? That's the role of portfolio reviews. In this approach, the portfolio reviews serve largely as a check that the gates are working well. Senior management meets perhaps twice a year to review the portfolio of all projects using the various pie charts, bubble diagrams, and lists described (Figure 13-7 and Table 13-2):

♦ Is there the right balance of projects?

♦ Is the mix right?

♦ Are all projects strategically aligned (fit the business's strategy)?

♦ Are there the right priorities among projects?

If the gates are working, not too many decisions or major corrective actions should be required at the portfolio review. Some companies indicate that they don't even look at individual projects at the portfolio review, but only consider projects in aggregate.

Recap: Approach 1

To recap, the gates are where the day-to-day go/kill decisions are made on projects in this approach. Gates focus on individual projects—one at a time—and are in-depth reviews. At gates, each project is evaluated and scored before moving on to the next stage—a real-time decision process. At gates, poor projects are spotted and weeded out, and good ones are identified and prioritized accordingly. Note that resource decisions—committing people and

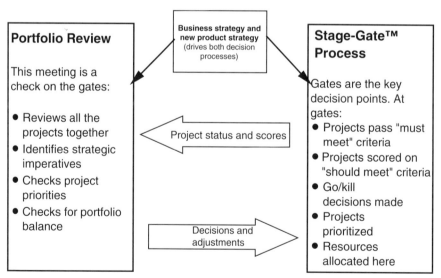

FIGURE 13-12. Method 1—An integrated portfolio management process (Stage-Gate™ with portfolio reviews).
Source: Cooper, Edgett, and Kleinschmidt 2000. Used with permission of Research-Technology Management, © 2000 Industrial Research Institute.

money to specific projects—are made right at these gate meetings. Thus the gates become a two-part decision process, with projects evaluated on absolute criteria in the first part (pass/kill decisions in Figure 13-11), followed by a comparison with other active and on-hold projects in the second part (go/hold decisions).

Portfolio reviews, by contrast, are periodic meetings, held perhaps twice per year. They serve as a check on the portfolio and oversee the gate decisions being made. If the gates are working well, the portfolio reviews are largely a rubber stamp.

Note that the portfolio reviewers and the senior gatekeepers are most often the same people within the business. The result of the gating process working in tandem with the portfolio reviews is an effective, harmonized portfolio management process (Figure 13-12).

Approach 2: Portfolio Review Dominates

The philosophy of the second approach is that every project must compete against all the others. A single decision on all projects replaces one of the gates in the gating process.

Here, senior management makes go/kill and prioritization decisions at the portfolio reviews, where all projects are up for auction and are considered on the table together. This portfolio review typically occurs four times a year. The gates in the Stage-Gate™ process serve merely as checks on projects—ensuring that projects remain financially sound and are proceeding on schedule.

The result of this "portfolio review dominates" approach is a more dynamic, constantly changing portfolio of projects. The method may suit faster-paced companies, such as software, information technology, and electronics firms, but it requires a much stronger commitment by senior management to the decision process, spending the time to look at all projects together and in depth several times a year.

Approach 2 uses many of the same portfolio tools and models described above, but in a different way. The result is a more dynamic portfolio of projects. In this approach, the project enters the portfolio process typically after the first stage (at gate 2 in Figure 13-10) when data are available.

The main difference from the first approach is that early in the life of projects, a combined gate 2 and portfolio decision meeting takes place. All new gate 2 projects, together with all projects past gate 2, are reviewed and prioritized against one another. Every project at gate 2 and beyond is thus in the auction, and all these projects are ranked against each other. Active projects, well along in their development, can be killed or reprioritized here, and resources are allocated here rather than at gates.

The role of gates in this approach is very different from their role in the first approach. Successive gates (after gate 2) are merely checkpoints or review points.

- They check that the project is on time, on course, and on budget.
- They check quality of work done—the quality of deliverables.
- They check that the business case and project are still in good shape.

If the answers to these are no, the project could be killed at the gate, recycled to the previous stage, or flagged for the next portfolio review/gate 2 meeting.

An "All Projects" Gate 2 Meeting

The major decisions, however, occur at the combined gate 2/portfolio decision point, which is a more extended, proactive meeting than the portfolio reviews in the first approach. And although this is a periodic process, it is almost real-time because this portfolio/gate 2 meeting is usually held every three months.

The format of this vital, quarterly portfolio/gate 2 decision point is typically this: All gate 2 and beyond projects are on the table. The portfolio managers (senior management) first identify the "must do" projects—the untouchables. These are projects that are either well along and still good projects or strategic imperatives. Then management votes on and identifies "won't do" projects, which are killed outright.

Next, the projects in the middle are evaluated. There are different methods here:

- Some firms use the same criteria they use at gate meetings, and in some cases, the most recent gate 0–10 scores; that is, the project attractiveness score from each project's most recent gate meeting is used to rank-order the projects.
- Other managements rescore the projects right at the portfolio/gate 2 meeting (using a shorter list of criteria than the list found in the typical scoring model).
- Forced ranking on criteria is also used. Here management ranks the projects against one another on each criterion. Again, a handful of major criteria are used, such as those used by Kodak at its portfolio review (Patton 1999):
  - Strategic fit
  - Product leadership (product advantage)
  - Probability of technical success
  - Market attractiveness (growth, margins)
  - Value to the company (profitability based on NPV)

We recommend the forced ranking method because it yields better discrimination than a traditional scoring model, forcing some projects to the top of the list and others to the bottom. One of the weaknesses of a scoring model is that projects tend to score middle-of-the-road—every project scores 60 out of 100. But any of these three methods yields a list of projects, rank-ordered according to objective scores. Projects are ranked until one runs out of resources. This ranked list is the first cut or tentative portfolio.

Following this, it is necessary to check for portfolio balance and strategic alignment: The proposed portfolio is displayed using some of the bubble dia-

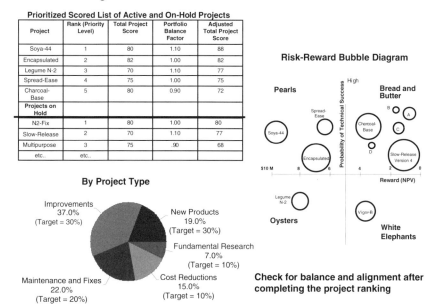

FIGURE 13-13. The portfolio display after completing the ranking.
Source: Cooper, Edgett, and Kleinschmidt 2000. Used with permission of Research-Technology Management, © *2000 Industrial Research Institute.*

grams, prioritized lists, and pie charts described above (summarized in Figure 13-13). The purpose here is to visualize the balance of the proposed portfolio and also to check for strategic alignment. If the tentative portfolio is poorly balanced or not strategically aligned, projects are removed from the list and other projects are bumped up. The process is repeated until balance and alignment are achieved.

A Recap: Approach 2

To recap, the portfolio/gate 2 decision meeting is where the key decisions are made in approach 2. The portfolio review is really a gate 2 and portfolio review all in one; these are held about four times a year. It is here that the key go/kill decisions are made, and consequently this is a senior management meeting. With all projects at or beyond gate 2 on the table, the meeting does the following:

◆ Spots "must do" and "won't do" projects
◆ Scores (via a forced ranking) the ones in the middle
◆ Checks for balance and strategic alignment (using various portfolio charts and bubble diagrams)
◆ Decides the portfolio: which projects, what priorities, how much resources

The gates serve mainly as a check. Projects are checked as they progress from stage to stage to ensure that they are on time and on budget and that they

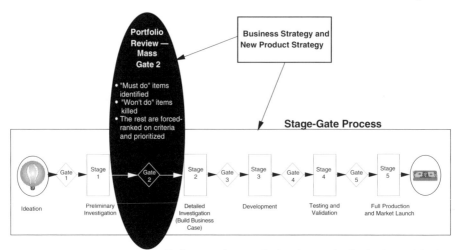

Projects are force-ranked against each other in the combined Gate 2/Portfolio Review. Prioritization is established, and resource are allocated here. Subsequent gates serve as checks.

FIGURE 13-14. Method 2—Portfolio management intersecting with the new product process.
Source: Cooper, Edgett, and Kleinschmidt 2000. Used with permission of Research-Technology Management, © *2000 Industrial Research Institute.*

remain good projects. Go/kill decisions are still made at gates to weed out poor projects. Gates rely on criteria, and the scores at these gates are often used as inputs to the portfolio review meeting.

Approach 2 thus lashes together the two decision processes: the gating process and the portfolio review. Gate 2 is really the integrative decision point in the scheme, and the point where the two decision processes intersect (Figure 13-14).

Pros and Cons: Approach 2 vs. Approach 1

The second approach has some advantages (and disadvantages) compared to the first. Management indicates that it is easier to prioritize projects when looking at all projects on the table together (rather than one at a time at real-time gates). Additionally, some people have difficulty with the two-part gate approach in the first approach and Figure 13-11; for example, how does one find resources for a good project when that is the only project being considered at the meeting? Finally, some managers like the notion that prioritization of all projects is redone regularly—no project is sacred.

There are also disadvantages to the second approach and areas in which the first approach is superior. Many managements believe that if projects are to be killed, then the project team should be there to defend the project (or at least to provide updated information), such as happens at an in-depth gate meeting. Another criticism is that the second approach requires a major time commitment from senior management; often taking several days every quarter to conduct this portfolio/gate 2 decision meeting.

A final advantage of the first approach is that gate reviews provide a much more in-depth assessment than is possible when all the projects are considered at a single meeting.

JUST DO IT!

New product portfolio management has become a vital concern, particularly among leading firms. Although a number of tools have been described that help to select projects and visualize a portfolio, the choice of tool may not be that critical; indeed, the best performers use an average of 2.4 tools each—no one tool can do it all.

Two different approaches to portfolio management—where the gates dominate, and where the portfolio review dominates—have also been outlined. Both have their merits, and both are recommended. Regardless of which portfolio method or which specific tools you favor, do move ahead: Choose a method and implement it. Our research shows clearly that those businesses that feature a systematic portfolio management process—regardless of the specific approach—outperform the rest.

NOTES

This chapter is based on a number of books and articles by the authors, listed in the bibliography. See also www.prod-dev.com.

1. Although the focus here is on portfolio management for new products, to the extent that technology resources used in new products are also required for other types of projects, portfolio management must consider the fact that new product projects compete against process developments, product maintenance projects and even fundamental research projects.
2. This decision rule of rank order according to the ratio of what one is trying to maximize divided by the constraining resource seems to be an effective one. Simulations with a number of random sets of projects show that this decision rule works very well, truly giving maximum bang for the buck.

BIBLIOGRAPHY

Cooper, R. G. 1987. "Selecting Winning New Products: Using the NewProd System." *Journal of Product Innovation Management* 2: 34–44.

Cooper, R. G. 1998. *Product Leadership: Creating and Launching Superior New Products*. Reading, MA: Perseus.

Cooper, R. G. 1999. "The Invisible Success Factors in Product Innovation." *Journal of Product Innovation Management* 16, 2: 115–33.

Cooper, R. G. 2001. *Winning at New Products: Accelerating the Process from Idea to Launch,* 3rd ed. Reading, MA: Perseus.

Cooper, R. G., S. J. Edgett, and E. J. Kleinschmidt. 1997a. "Portfolio Management in New Product Development: Lessons from the Leaders—Part I." *Research Technology Management,* September-October, 16–28; Part II, November-December, 43–57.

Cooper, R. G., S. J. Edgett, and E. J. Kleinschmidt. 1997b. *R&D Portfolio Management Best Practices Study.* Washington, DC: Industrial Research Institute.

Cooper, R. G., S. J. Edgett, and E. J. Kleinschmidt. 1998. "Best Practices for Managing R&D Portfolios." *Research Technology Management* 41, 4: 20–33.

Cooper, R. G., S. J. Edgett, and E. J. Kleinschmidt. 1999. "New Product Portfolio Management: Practices and Performance." *Journal of Product Innovation Management* 16, 4: 333–51.

Cooper, R. G., S. J. Edgett, and E. J. Kleinschmidt. 2000. "New Problems, New Solutions: Making Portfolio Management More Effective." *Research Technology Management* 43, 2: 18–33.

Cooper, R. G., S. J. Edgett, and E. J. Kleinschmidt. 2001. *Portfolio Management for New Products.* 2d ed. Reading, MA: Perseus.

Day, G. 1986. *Analysis for Strategic Marketing Decisions.* St. Paul, MN: West Publishing.

Evans, P. 1996. "Streamlining Formal Portfolio Management." *Scrip,* February.

Faulkner, T. 1996. "Applying 'Options Thinking' to R&D Valuation." *Research Technology Management,* May-June, 50–57.

Griffin, A. 1997. *Drivers of NPD Success: The 1997 PDMA Report.* Chicago: Product Development and Management Association.

Heldey, B. 1977. "Strategy and the Business Portfolio." *Long Range Planning.*

Matheson, D., J. E. Matheson, and M. M. Menke. 1994. "Making Excellent R&D Decisions." *Research Technology Management,* November-December, 21–24.

Patton, E. 1999. "The Strategic Investment Process: Driving Corporate Vision Through Portfolio Creation." *Proceedings: Product Portfolio Management: Balancing Resources with Opportunity.* Boston: The Management Roundtable.

Roberts, E., and C. Berry. 1983. "Entering New Businesses: Selecting Strategies for Success." *Sloan Management Review,* spring, 3–17.

Roussel, P., K. Saad, and T. Erickson. 1991. *Third Generation R&D: Managing the Link to Corporate Strategy.* Boston: Harvard Business School Press and Arthur D. Little.

Tritle, G. L. 1997. "New Product Investment Portfolio." Internal 3M document.

14 Assessing the Health of New Product Portfolio Management: A Metric for Assessment

Robert J. Meltzer

It was Oliver Cromwell who, in the seventeenth century, said, "He who stops being better stops being good." Cromwell's words apply as well today as they did then. And his words certainly apply when it comes to the new product development (NPD) process. Unless a company continually strives to improve its NPD process, its ability to survive in a competitive environment will "stop being good" enough to survive.

But how is a company to know if its NPD process is getting better? One way to look at the health of a company's NPD process is to look at an analogous process, the process by which the financial health of a company is diagnosed. To assess the financial health of a company an accountant will audit the books. The audit is a process of discovery by which the accountant attempts to discover the reality of the company's financial health. To this end accounting auditors apply tests to management-supplied data. An audit of the new product development (NPD) process is also a means of discovery, but the goal is different. Its goal is the continuous improvement of the process, to make sure the company is not "he who stops getting better." If there is to be continuous improvement, there must first be a metric that allows comparison of the current performance of the NPD process with performance in the past.

Anyone responsible for the NPD portfolio, someone such as the NPD manager, will find it valuable to distinguish between a metric for the success of individual product developments and a metric for the success of the entire NPD process. The distinction is valuable because some measures may show an individual product to be a smashing success. At the same time the entire process by which a portfolio is established and ultimately brought to market may be failing. This chapter will explore methods by which the success of single product developments can be assessed and show a metric by which the entire NPD process can be measured.

METRICS FOR SINGLE PRODUCT DEVELOPMENT

Many different metrics have historically been applied to assess the success of a single product development. One of those metrics may signal success and may indeed be a fair measure, but it is essential to be sure the metric has validity, that is, it is applied uniformly so as to allow valid comparisons to be made among developments.

No single metric has been shown to accurately compare the performance of developing today's new product with the development of some similar product in the past. There is the question of similarity—is this new product sufficiently similar to a new product developed in the past to warrant comparison? But that question aside, we should be aware that many suggested metrics for single product development can suffer from built-in ambiguities.

Time to market is one suggested metric, but measuring time to market, unless defined in the same way for all product developments, can be ambiguous. The start date for any particular project may not be at all clear unless the start date has a uniformly applied definition. Does the project start when senior management gives the go-ahead and allocates funds? Or does it start when R&D and marketing get together and spend time to produce a product concept, prepare a preliminary evaluation of market potential, and assemble a proposal to be presented to management? And does the end of the time-to-market period come when the development is complete, or when there is inventory of the product on the shelf, or when the first revenue unit is sold to a real customer? If time to market is to be a metric, then a good definition, applied uniformly, of start and end is essential. The exact definition of start and end is of less importance than that it be applied uniformly to all new product developments.

New product revenue contribution has also been suggested as a metric. But this too may be ambiguous without detailed definition. If the new product is an accessory to an existing product and increases sales of the existing product, then how much new product revenue is to be ascribed to the new accessory? If expansion of the product line increases sales of an existing product, how much profit expansion is to be credited to the new product? Or if a new product cannibalizes an existing product, is the loss of existing product sales to be deducted from the new product revenue? These accounting questions can be difficult to answer quantitatively. But this metric too can be useful if there a consistent rule by which new product revenue contribution is to be measured. Consistency is more important than correctness. For the measure of success to be a useful tool for improvement of the NPD process, a comparative measure is needed, not an absolute one. Comparison is the basis for deciding if the company is getting better at new product development.

The ratio of new product profit to R&D expenditure is another suggested metric for the efficiency by which a single product is developed. But that metric supposes that the R&D organization alone is responsible for the profitability of a new product in the market. One might, with equal validity, construct a ratio of new product profit to marketing expenditure or manufacturing cost.

Anyone in R&D knows examples of products developed, with marketing approval, through pilot production, only to have the product canceled by marketing. The reasons for cancellation may be excellent, but cancellation will certainly have a negative effect on this metric if applied to overall R&D expenditure, inappropriately faulting R&D. But this metric can have value if the R&D expenditure is considered only for new products that reach the market. Applied in this way, the metric measures the validity of the estimates made by R&D of the cost of its efforts, of estimates made by marketing for product profit, and of estimates by manufacturing for manufacturing cost.

Product goal attainment as a metric may also be ambiguous. A comparison with other product developments is possible only if the goals were the same and had the same priority. A measure for product goal attainment can depend on who is measuring and which of many possible goals has been attained and which has not. Product performance goals may have been achieved, but at the expense of losing schedule integrity. Projected time to market may have been achieved, causing momentary celebration. But if sales of the newly developed product are less than projected, there will be much gloom among the marketing, sales, and finance people. If this metric is to be valid, then prioritizing the goals must be part of the measure. Prioritizing the goals for a development is the responsibility of the product portfolio manager and is most appropriately part of a product requirements document.

CHARACTERISTICS OF A USEFUL NPD PROCESS METRIC

Any metric that might be applied to the development of a single product will often focus on one function or another of the NPD process. But no one function, not R&D or marketing or manufacturing, is the sole contributor to the process that produces new products. Some metric for the productivity of, for example, the R&D organization, may show constant improvement. In spite of the improvement in the R&D organization, there may be no improvement in the rate at which new products reach the market. If the total process is to improve, some method for measuring overall effectiveness is needed.

To be a useful metric for the effectiveness of a company's total NPD process, the metric should possess several characteristics:

1. *It uses historical data.* The metric should be able to compare the effectiveness of the practice of the present NPD process with the process as practiced in years past using existing historical data. What we want to measure is the effectiveness of the process as it is being practiced, not as described in some written company development process manual.

2. *It does not place blame.* The goal of the metric is to improve the process, not to heap blame or praise on a particular individual or organizational function. How the process is to be improved is a matter that must be

separated from the need to improve it. It will be up to the individual contributors or senior management to find the means for improving the process.

3. *It is widely applicable.* The metric should be equally applicable to commoditylike products and to products that embody rapidly moving technology. For some product lines success is measured by market share, while for others it is measured by the rate at which new products are introduced to the market. Measuring improvement in the rate at which new products are introduced to the market is of importance for a company strategy that emphasizes new product introduction. But even a company with a strategy directed primarily toward new product introduction may have commoditylike products that are cash cows. For these cash cows little conventional development effort is appropriate. A good metric for a development process should be equally adaptable to both situations.

4. *It need have no absolute meaning.* The metric need only be a means for year-to-year comparison for internal use.

5. *It is easy to apply and to understand.* All functions in the company should be able to employ the metric.

THE DEPRECIATED PRODUCT VALUE METRIC (DPVM)

The metric for NPD process described here is called the depreciated product value metric. The metric operates by depreciating the value of each sales dollar of old product as time passes. The value of each dollar of sales depreciates even though the total dollar sales of the product increases, and even though profit from the old product remains high.

The value of each dollar of sales depreciates because the older the product, the more it is in danger of losing sales volume and becoming less profitable. The danger may come from competitive product introduction or technology change, from saturation of the market, from a customer perception that it is "old stuff," or from changes in any of the economic, social, or political drivers of customer preference.

All products will not depreciate in value at the same rate. The value of products in an area of rapidly changing technology will depreciate rapidly. Computer technology is an obvious example. A computer product can easily be moribund in eight months and dead in twelve. But cornflakes are a technically stable product, the value of which can be expected to depreciate slowly.

The DPVM calculates the value of a product by applying Equation 14-1:

Depreciated sales = (This year's sales)

$$\times \text{ (Depreciation rate)}^{\text{(Years since introduction)}} \quad (14\text{-}1)$$

An example of the calculation is given in Table 14-1. The example applies the metric to a product that had first-year sales of $120 million. The depreciation

TABLE 14-1.
Example: Calculating a Depreciated Value

	Year 0	Year 1	Year 2	Year 3
Annual sales (millions)	120	160	170	130
Sales depreciated @ 0.6/yr.	120	96	61	28

rate applied to this product is 0.6. In year 2 the sales for year 2 will be multiplied by $0.6^2 = 0.36$. The depreciated value of the sales in year 2 is then $170 million multiplied by 0.36, or $61 million. Although the table shows the age of the product in whole years, there is no reason there cannot be a snapshot at any point. For example, a product introduced two years and three months ago would have an age of 2.25 years.

Determining the Depreciation Rate

There will always be some question about how to set the depreciation rate for any given product or product line. One approach would be to just pick a plausible number and see how it works out. But that's not necessary.

The advantage of the DPVM method is that it is possible to use historical data to see what the depreciation rate should have been to prevent declining sales in the past for products taken to be similar. A plausible depreciation rate might be arrived at by asking how many years will pass before the value of each sales dollar is reduced to, say, one-tenth of its estimated value at product introduction. A product whose value will be reduced to 10 percent in five years will carry a depreciation rate of 0.65. Some other product, whose value will be reduced to 10 percent in ten years, will have a depreciation rate of 0.8.

If there are no internal historical data, as for a product new to the company, then a look at the product life cycle of potential competitors, or a look at companies in analogous businesses, will give a clue. How often have those companies with analogous products changed the technology of their products or changed the marketing/advertising strategy?

Some have suggested the possibility of taking a more quantitative approach. One suggestion for a more quantitative approach is to pick some value of the negative slope on a declining sales curve as a marker to define a suitable depreciation rate. That much arithmetical effort hardly seems worthwhile. The depreciation rate is, after all, an inexact number. It is useful as a tool, but it is only a tool.

The Danger Index: Another Application of the Depreciation Rate

Even for those products that seem unlikely to be challenged in the near term there are competitive risks. The older the product is, the more likely it is that the competition will have had an opportunity to discover the product's weaknesses and to respond with a superior product. Or a competitor may have found a way to reduce manufacturing cost and so lower the price. And the older the product, the more likely it is that there will be new technology or that customer desires will have changed. It is also true that the higher the sales volume of a product, the more likely it is to attract competition, just as more rabbits will attract more foxes. The depreciation rate can be combined with the current sales figures and product age, using Equation 14-2, to produce a danger index, an indicator of how severe the threat is to future sales volume of the product. In Table 14-2 is an example of how the danger index can be calculated for four of a company's products. Product C, which enjoys the highest sales, is also in the most volatile market. It is already two years old and, unsurprisingly, is in the greatest danger.

$$\text{Danger Index} = \text{(This year's sales)} \div \text{(Depreciation rate)}^{\text{(Years since introduction)}} \quad (14\text{-}2)$$

Applying the DPVM: Assessing the Health of New Product Portfolio Management

The DPVM is determined by applying Equation 14-3. The inputs for Equation 14-3 are readily available.

Effectiveness of portfolio management year to year = (Depreciated sales for this year) ÷(Depreciated sales of the previous year)   (14-3)

1. The dollar sales of each product of the company for the current year and the year before are available to marketing, sales, and finance.
2. The depreciation rate for every new product is most appropriately set at the beginning, as part of the product requirements document negotiated by marketing and R&D. If there is no existing depreciation rate, historical sales data can be analyzed to see what the depreciation rate should have been to ensure the timely introduction of new products and prevent declining sales.

TABLE 14-2.
Example: Calculating a Danger Index

Product	Depreciation Rate	Current Sales	Years Since Introduction	Danger Index
A	0.6	160	2	444
B	0.8	220	3	429
C	0.5	350	2	1,400
D	0.9	180	1	200

3. The sum of this year's depreciated sales of company products is to be divided by the sum of last year's depreciated sales, as in Equation 14-3. The higher this ratio, the more effective is the management of the new product portfolio. Equally important is whether this ratio is moving up from one year to the next. Certainly, a declining ratio indicates a need for action, but the ratio, of itself, does not specify the action to be taken.

Still another way to apply the DPVM is to construct a ratio of the total depreciated sales to the actual sales, as in Equation 14-4, and compare this year's ratio to the ratio of previous years. The higher the ratio determined by Equation 14-4, the better the management of the new product portfolio. By comparing the ratio of Equation 14-4 this year with the ratio for previous years the health of the product portfolio management can be assessed.

Effectiveness of portfolio management in a given year =
$$(\text{Depreciated sales}) \div (\text{Total sales}) \quad (14\text{-}4)$$

Access to the metric, once generated, is a matter of company policy. Some companies are more open than others. There is no immediately evident reason why the total of any year's discounted sales should not be available to all, nor is there an evident reason why the ratio of this year's depreciated sales to last year's should not be made known; sensitive sales data of any one product are not discoverable from the total or from the ratio. But the total and the ratio do convey to all hands that the company is doing a better job of introducing new product or that changes must be made. On the other hand, the depreciation ratio applied to a given product may well be considered to be proprietary company information. That ratio would indicate to competition the rate at which the company plans to replace existing product.

ADVANTAGES OF THE DPVM—AND A DISADVANTAGE

◆ The DPVM avoids the possible misuse of metrics associated with measures more applicable to individual projects than to the overall development process.
◆ The start date for the DPVM method is unambiguous. It is the date of the first revenue sale of the new product.
◆ If the product is not profitable, it will be withdrawn. Absence of revenue from the product reduces the effectiveness measure. Some deficiency of portfolio management resulted in a product undesired by customers.
◆ No finger of responsibility is pointed for failure of portfolio management. The measure is of the process as a whole, not of the shortcomings of any particular contributor. The measure indicates that it is the process that needs improvement, but it takes no position on how improvement is to be achieved.

CASE STUDY: APPLYING THE DPVM AT GENERIC PRODUCTS INTERNATIONAL

Generic Products International (GPI) is a company with sales in the $500 million range. In spite of increasing sales, the CEO became uneasy about the management of the company's new product portfolio. He had no direct evidence that portfolio management was in trouble. He said, "It's a gut feeling, and I trust my gut."

The CEO conveyed his unease at a staff meeting. Sales and marketing tried to reassure him. After all, sales were up. The R&D people were less sanguine. Technology was moving rapidly in the area of some company products. The COO had heard about the DPVM from a colleague and suggested that implementing the DPVM might be a way to show the CEO that matters were better than his gut was telling him. The CEO, willing to try anything, appointed a committee to provide an assessment of how well the new product portfolio was being managed using the DPVM. The committee comprised a member from marketing, a member from R&D, and a member from finance. The CEO scheduled a report in two weeks

About 90 percent of GPI's sales came from five product lines. The committee easily agreed that they would apply the metric to only those five.

Sales data for each of the lines were immediately available. The only contentious issue was choosing the depreciation rate for each of the five. Total sales were increasing, and so the marketing member argued for a rate that would depreciate sales value slowly. Rapidly changing technology was the argument made by the R&D member for more rapid depreciation. There seemed to be no way to reconcile the two viewpoints. The scheduled report date was in danger.

In an effort to resolve the issue, the committee sought the advice of the COO who had originally presented the idea. The COO, in turn, went to the colleague from whom he had first heard of the DPVM. From the colleague he learned that the issue was a common one and that a facilitator was often added to the committee. The COO's colleague recommended a facilitator, one who had worked with other companies to resolve this issue. With the CEO's permission the facilitator was brought on board.

The facilitator held separate discussions with marketing and R&D and found no real surprises. R&D's view of marketing was that they had no vision of the future, had no understanding of where technology was at the present and where it was likely to go, and could never decide on product requirements before a new product development was begun. They were all standard complaints, no different from the complaints the facilitator had heard in other organizations. Marketing, for its part, had a litany of complaints about R&D. Chiefly, these had to do with failing to meet project schedule and budget goals. R&D was also accused of being more

TABLE 14-3.
Sales and Age of GPI Products

Product	Sales Previous Year	Sales Current Year	Current Age
A	117	124	4
B	79	81	3
C	131	141	1
D	96	101	2
E	99	109	2
Total	522	556	

interested in pursuing new technology than in the sales that would result from satisfying customer desires. It was all familiar stuff to the facilitator. He also recognized that forcing either organization to adopt the other's choice of depreciation rate would only exacerbate the existing adversarial relationship—which was a separate problem that needed to be addressed.

The first suggestion made by the facilitator was to have marketing choose one product line for which they most wanted to set the depreciation rate. Then R&D would get a choice, then marketing, and so on. R&D did not like having second choice, and they claimed the suggestion made no sense at all. Marketing, for its part, didn't want R&D choosing the depreciation rate for any product. Since this first idea had never flown anywhere else, the facilitator was not surprised. Moreover, the facilitator didn't think much of the idea either; he'd have had a problem if R&D and marketing had agreed. On the bright side, at last the facilitator had found a place where R&D and marketing were on common ground—neither thought much of this way to resolve the issue. The gambit had worked.

"Well, since no one wants to be second, why don't both marketing and R&D be first, and second, and third, and so on? Let's have both set the depreciation rate they think most appropriate for each product line. Do it independently. Then we'll do the math and see what happens."

And in this way the issue was resolved. Both marketing and R&D would use the same sales data supplied by finance, but the committee member from marketing and the one from R&D, after consulting with others in their respective organizations, would separately choose a depreciation rate for each product line. The results of the metric would be compared. If the metric, as calculated by both, showed the same trend in portfolio management effectiveness, irrespective of the difference in the chosen depreciation rates, there would be clear indication if there were need for action. But if the metric calculated by one showed improvement in effectiveness and the measure of the other showed effectiveness was declining, and if the difference was significant, then the reasons for one's complacency and the other's anxiety would need close examination.

Table 14-3 gives the sales data, rounded to the nearest million, for the current year and the year just past for the five most important product lines of GPI. Also shown is the number of years, with respect to the current year, since each product was introduced (age).

The depreciation rates chosen by marketing and R&D for each product line are shown in Table 14-4. Note that while marketing generally depreciated the value of sales more slowly than did R&D, product B is a small exception. Also R&D and marketing agreed that product D is the one to be depreciated most rapidly, albeit at different rates.

Marketing calculated the effectiveness of new product portfolio management, as shown in Table 14-5. R&D's calculation is shown in Table 14-6. Despite the differences in depreciation rates, the conclusion was the same: New product portfolio management has been less effective in the current year than it was in the previous year.

TABLE 14-4.

Marketing and R&D Depreciation Rates

Product	R&D Rate	Marketing Rate
A	0.6	0.8
B	0.7	0.65
C	0.55	0.75
D	0.35	0.5
E	0.4	0.7

CASE STUDY (continued)

TABLE 14-5.
Marketing Calculates Effectiveness

Product	Rate	Previous Sales	Age	Dep. Sales	Current Sales	Age	Dep. Sales
A	0.8	117	3	60	124	4	51
B	0.65	79	2	33	81	3	22
C	0.75	131	0	131	141	1	106
D	0.5	96	1	48	101	2	25
E	0.7	99	1	69	109	2	76
Total		522		341	556		280

Effectiveness (year to year) = 280 ÷ 341 = 0.82
Effectiveness (previous year) = 341 ÷ 522 = 0.65
Effectiveness (current year) = 280 ÷ 556 = 0.5

The CEO's gut was right. The DPVM enabled senior management to recognize that product portfolio management needed an overhaul to ensure a steady stream of new products. And, in the present instance, differences in the chosen depreciation rates had no effect on the conclusion.

Just where the new product development process at GPI was inadequate was information not supplied by the DPVM. It might have been that project schedules were not met because of constant changes in the product requirements. Or it might have been that scheduling tools were inadequate. Or the efforts of the R&D organization might have been starved for resources, or its efforts diluted by nondevelopment tasks. Possibly the root cause was the present adversarial relationship between R&D and marketing. Or there might have been any of a number of other managerial, procedural, or cultural inhibitors standing in the way of new product development. In all cases it is the task of senior management to discover the cause(s) and implement the remedies.

TABLE 14-6.
R&D Calculates Effectiveness

Product	Rate	Previous Sales	Age	Dep. Sales	Current Sales	Age	Dep. Sales
A	0.6	117	3	25	124	4	16
B	0.7	79	2	39	81	3	28
C	0.55	131	0	131	141	1	78
D	0.35	96	1	34	101	2	12
E	0.4	99	1	40	109	2	17
Total		522		269	556		151

Effectiveness (year to year) = 151 ÷ 269 = 0.56
Effectiveness (previous year) = 269 ÷ 522 = 0.51
Effectiveness (current year) = 151 ÷ 556 = 0.27

♦ The higher this year's revenue contribution from a new product is, compared to the revenue from an old product, the higher will be the measure of process effectiveness. An accessory that increases the sales of an old product will go toward an increase in effectiveness. If the accessory has not sufficiently increased the sale of the old product so as to overcome the loss associated with the depreciation rate, then development of the accessory may not have been a worthwhile project.

If there is a disadvantage to the DPVM, it is that the depreciation rate is subject to manipulation at the time the product is defined in the product requirements document. At that time in project life the DPVM can be a whip applied by marketing to R&D by setting the depreciation rate unrealistically. As with other product attributes, there is no escaping negotiation between R&D and marketing. At the same time, as shown by the case study, the metric is not extremely sensitive to the rate chosen.

15 Risk Management: The Program Manager's Perspective

David J. Dunham

This chapter lays out a toolkit for managing risk in new product development (NPD) from the program manager's perspective. It is divided into three sections that:

1. Describe the roles and responsibilities of program managers as gatekeepers and decision makers, and defines risk in their context
2. Explore how the focus of risk evolves according to the stage of development, and explain the six steps of effective risk management at gate meetings
3. Provide a set of templates, based on those used by the program manager of one corporate development team, to facilitate the dialogue around risk management

It should be stated at the outset that the toolkit outlined in this chapter is just that, a set of aids to facilitate discussion of risk between the program manager and development teams. Like any toolkit, its value is dependent on the behavior of the participants. Poor preparation by the development manager can confuse rather than focus the discussion. Disregard of the development team's risk analysis by the program manager, coupled with arbitrary decision making, can be equally counterproductive. In respect to the behavior of program managers, there appears to be a governance trend among many senior management teams toward taking the risk management process more seriously than might have been the case in the past. While preparing this chapter, the author conducted a highly qualitative survey among program managers (also known as senior management) in fifteen companies ranging from Fortune 100 firms to start-ups. From these interviews, it can be hypothesized that there is a growing involvement by senior management in the NPD process in general (perhaps characterized as more hands-on behavior), and in the management of risk in particular. Reasons for this closer attention are predictable, and include bigger investment bets, fuzzier market spaces, accelerated development cycles, multiple alliances and partnerships, and young development teams.

In short, development is getting riskier. While many executives would

FIGURE 15-1.

argue that this involvement is not new behavior for them, anecdotal evidence from the interviews suggests a heightened awareness and a need to proactively manage the tough realities of product development. Bottom line: The "graybeards" are paying increasing attention to risk management in NPD.

By getting more involved, they are changing the dynamic of risk management within NPD. Historically, program managers and project managers could have been described as being on opposite sides of the risk management table—even, at times, on opposing sides. Now, as in the model defined by the venture capital community, the dialogue is increasingly collaborative (see Figure 15-1).

THE PROGRAM MANAGER'S ROLE

You are a gatekeeper and you are sitting in a review meeting receiving a presentation from the project leader on a new product opportunity. You received the briefing package the day before the meeting, and there is a short section in the plan about "risks and contingencies." The presentation will run an hour, and you are expected to decide on funding the next phase of development at this meeting, or soon after. How do you go about thinking about risk management in this context? What questions will you ask of the development team to get past the numbers and into the real risk issues? What do you need to see in the presentation to get a real understanding of the project risk? How do you contribute to the team's insight on the risk management program? How do you accelerate the development process, rather than create a bottleneck? (For a more systematic description of the program manager's information needs at decision meetings, see Figure 15-2.) To answer these questions one must first define the various roles of the program manager as they apply to NPD risk management.

You're in the gate review meeting. Think about what you need to hear in order to participate productively in a risk management discussion. Some suggestions:

♦ You want a clear understanding up front about the new idea. It has to make sense to you as a business proposition. What is it, why will the customer pay for it, how do we make money from it?

♦ You want the facts separated from the team's opinions. Objectively, what do we know about customers' needs, our ability to make it and sell it, etc. versus what are we assuming?

♦ You want to avoid team bias and conflict. Am I hearing the views of an engineer that believes in the product but doesn't understand the market? Am I hearing the views of a marketer that understands the customer's needs but doesn't see the technical difficulty of delivering the solution?

♦ You want to hear the financial story certainly, but also the thinking behind the numbers. Have the numbers been engineered to meet our new product hurdle rates or is this truly such a great opportunity?

♦ You want to track the risk management from one meeting to another. Have we addressed the risk that was identified in the last meeting and what has changed in our learning since the last team presentation? As one executive said: "Is this what we agreed to in our last meeting?"

FIGURE 15-2. Program manager's needs.

Three Potential Roles

A program manager may have the title of marketing director, business unit general manager, VP or head of R&D, or VP of business development. Regardless of the title, these executives, as program managers, share a common responsibility. They are accountable for developing and bringing to market new products and services that will meet certain targets for revenue and profit growth. Further, they have to deliver this top- and bottom-line growth in the context of strategic business priorities. These priorities are likely to include such objectives as filling gaps in the existing product line, entering new market space for the business unit, and developing a less expensive solution to a customer's needs. With this accountability in mind, program managers can be said to have three roles that are relevant to NPD risk management.

Role 1: Investor

As stated, program managers are investment gatekeepers and portfolio managers. They seek to achieve financial and strategic growth goals from a group of investments. They are responsible for achieving a certain return on the investment in NPD. They have to balance, or rebalance, the portfolio according to new information and new priorities that emerge during the year—for example, urgent requests from senior management, or termination of a project already in the development pipeline.

The primary tool to ensure the productivity of the development portfolio is the allocation of resources. Program managers decide what projects to fund, when to fund them, and how much to allocate to them. They also use the allocation of resources to balance the direction and speed of the portfolio and specific development projects. Program managers do not have the same role as project managers. Their roles are interdependent and complementary. The program manager is the investment decision maker, while the project manager is the leader of the entrepreneurial development team. The project manager is, one hopes, the advocate for the new opportunity, the standard-bearer rallying the troops and displaying the required conviction in the opportunity to drive the project forward. Project managers are the proponents of the project by definition. They have to maintain an energy level that stems from a real belief in the importance of the opportunity. They have to prove the case to the investor and request the resources for successive stages of the project.

This advocacy role makes it particularly tough for project managers to quickly come to terms with risks that can threaten the continuation of the project. Indeed, it is because many project managers are such ardent advocates that program managers frequently have trouble killing no-go projects. Program managers can be loath to play the role of bad cop, and rather than make a decision to terminate a project cleanly, they allow it to limp along without formally committed resources. Here's how one executive at Eastman Kodak expressed her attitude toward effective risk management. A training package on risk management was being developed for incorporation into the overall new product development process. Part of the training was an actual case study of a project that had been terminated relatively early in development. The senior executive, acting as program manager, was asked whether the case should be adapted to show it going all the way to launch, thus illustrating risk management across the entire development cycle. "Absolutely not," she replied. "Show the case being killed after the risk analysis at the second gate. That's the message I want to get across to the teams—it's okay to recommend a no-go on a project!"

Instinctively, it is the program manager who will focus first on the risk in a new product project. He or she must know when to fish and when to cut bait. Program managers need a heads-up if projects are going wrong. They need to see yellow flags before they turn red. Their concern is that they get a clear, honest perspective on risk as it evolves, and on the risk mitigation program as it is executed. They need that information from project managers, who are in the trenches.

Role 2: Coach

In addition to the portfolio manager and investment manager, program managers are coaches to development teams. Their role as coach is based on the broader perspective they have on the business opportunity and their broader business experience. They bring this experience to bear on the identification of risk. For example, they will have insight on the likelihood of customers paying 30 percent more for a proposed new product. Equally important, they apply their experience on how to best manage the risk in the project—for example, by resolving an internal, corporate risk caused by a lack of resources. Program managers, as senior managers, also have greater insight into the strategic, political, and cultural issues surrounding the development process.

It is this insight that must be shared with development teams in order for project mangers to have the relevant context as they push their project forward. They can identify possible risks that might have eluded the team. For example, many development teams do not have strong sales representation, especially in the early stages of development. Consequently, it may not occur to a team that is focused on customer and technical risks to examine the fact that the sales force has never sold this kind of product before. Yet the absence of a qualified sales force may turn out to be a project showstopper for reasons of strategy, expense, or missed market windows. So the program manager should be looking down the road on behalf of the development team and coaching on the dangers to be anticipated.

However, the required flow of information is not a one-way street. Project managers have their own inside track on information. They are more intimate with the new product concept as it is developed. They have more immediate access to fresh data—for example, the results of customer research. They are also more likely to be the first to experience an emerging risk to the project.

The mutual exchange of information between those around the table at a gate meeting is central to successful risk management in NPD.[1] Ultimately, risk management is a collaborative effort between the project manager and program manager. It is a learning process to be undertaken together, each side contributing their particular perspective. Too often risk management is made more difficult because it is not managed as a team effort. Both sides need to be able to put the discussion of risk on the table and agree on how best to reduce it. Another senior manager described the issue as one of trust: "If I have been working with a development manager for years, I know that he knows where the risk is likely to lie in a new product project. I don't have to spend a lot of time probing potential risks because I trust the manager to have taken an unbiased view of the risk. The reason you need a risk management process," he added, "is that typically bias will creep into the development team's view of risk." A risk management process can alleviate bias by helping to check the validity of the risk analysis.

Role 3: Facilitator to the Gatekeeping Team

The third role that the program manager plays is that of facilitator to the gatekeeping team. The program manager is at the center of a multifunctional decision-making group of executives, frequently consisting of functional departmental heads. The program manager's job is to help that team make quality investment decisions.

These functional team members are needed to help resolve inevitable trade-offs in the project plan during decision meetings. If the key people are not present and their participation is not sought by the program manager, disconnects occur, contingency plans break down, and the project is at risk. As an example, a story reported in the *Wall Street Journal* described the challenge faced by a developer before Hewlett-Packard overcame the silo culture of its business units. "H-P lost a Chinese banking-system deal worth roughly $50 million to rival IBM. The main stumbling block: a midlevel H-P manager in the U.S. declined to spare $50,000 to build a prototype of the system for the would-be customer."

Therefore, at a gate meeting the program manager should have around him or her those senior functional managers who can address and resolve these internal risks and, additionally, contribute to the discussion of external, market-based risks. For suggestions on how to make the dialogue on risk more productive at gate meetings, see Figure 15-3.

Risk Discussion Challenges

Discussing risk in new product development certainly seems to be a difficult thing to do. Despite the fact that the high-risk nature of new product development is built into the corporate psyche, many corporations still take an almost fatalistic approach toward managing the risk. Reasons for not being anxious to dwell on risk differ depending on the chair in which you are sitting.

Program Manager

- ◆ Spending time on risk assessment and management is counter to the action culture of many corporations. "Risk management does not create an asset," to quote one executive.
- ◆ Management feels that the learning can/should be done in the market.

Project Manager

- ◆ There is a natural aversion among developers to focus on the downside.
- ◆ Highlighting risk is counterintuitive for development teams who want to promote the opportunity when competing for NPD funding.

Risk management is also time-consuming to execute, and it requires skills that many developers don't have (financial modeling skills, for example). However, despite resource, time, and skill constraints, the benefits to the program manager of a process that systematically identifies, sizes, and mitigates NPD risk are apparent:

- ◆ Lower chance of redos and crisis management in later development stages
- ◆ Higher probability that what gets to market will be a success
- ◆ Quicker kill decisions, faster resource reallocation
- ◆ Better focus of development resources on the few critical things that matter and can accelerate development

In a review session, the discussion of risk is made the more difficult by factors that frequently have little to do with the opportunity being considered:

- *Personalities*. The gatekeeper team may include champions of the concept who are as intent on selling the concept as the development team presenting. Alternatively, the gatekeeper team may include the "devil's advocate" who makes a point of doubting any idea presented, and sees risk everywhere.

- *Hidden Agendas*. Heads of different functional groups or different business units may have a bias towards the new product concept that is based on political, organizational or strategic concerns. These concerns are usually not apparent to the development team presenting and can cloud an objective exploration of risk.

- *Lack of expertise*. A senior manager responsible for reviewing new services for a joint venture confessed that since he didn't have a technical background, sometimes "he lacked the conviction to press a development team about a risk even though (his) gut told him something was wrong".

- *Lack of dialogue time*. Presentations consume the available meeting schedule leaving time only for a rushed or postponed gatekeeper decision.

To facilitate a quality discussion about risk and to arrive at a more productive outcome, one telecom Program Manager has used the following techniques.

- *Limit the presentation time*. The development team has to make their presentation in no more than 50% of the scheduled time. The remainder is left for discussion.

- *Build risk into the presentation*. Each team must include a slide addressing a) what are the major perceived risks, b) what is the mitigation plan, and c) what resources (people and dollars) are needed to execute the plan (e.g., customer research, prototype testing, etc.)

- *Separate management questions*. Encourage clarification questions during the team's presentation, even if they sound naïve. In the subsequent dialogue, ask gatekeepers to separately discuss concerns of, and their suggested improvements for, the risk management program presented by the development teams.

- *Get everyone's opinion*. Regardless of whether they are a champion of the project or devil's advocate, ask each gatekeeper his or her view on the risks and the best way to address them.

- *Use a facilitator*. As Cooper notes in *Winning at New Products*, best-practice companies have someone act as a director of proceedings during gate meetings to ensure due process and, in this case, an open discussion on risk. The Program Manager can act as facilitator.

- *Record the decision*. Get the project manager to repeat the decision and action items that she or he is taking away from the meeting. Specifically, have them confirm the risks that need managing next, the agreed mitigation plan, and the authority to spend the necessary resources to execute the plan. That recorded decision becomes the starting point for the risk management discussion at the next meeting.

FIGURE 15-3. Discussing risk at gate meetings.

Risk: The Program Manager's Perspective

NPD risk can be defined as an event that can negatively impact the ability of a new product or service to meet its stated business objectives in the market.

The key concept underlying NPD risk is that it is defined as an occurrence that poses a threat to the viability of the business proposition as it was conceived. In other words, how will the proposed business opportunity be threatened by specific and collective risks? It follows that all NPD risk can ultimately be defined as a threat to either revenue (e.g., revenue is not enough or doesn't come soon enough) and/or expense (e.g., development, marketing, sales expense, or capital investment are too high).

The program manager would do well to build this business orientation into every dialogue with the development team. Ask a development team, "How exactly will this concern/risk affect the business opportunity? Point to the line item in the P&L" (see Figure 15-4). For example, a development team may express a concern such as "We hear rumors that the competition will cut prices on their product when we introduce our new offer." This might translate into "We won't get the market share we forecast originally." Now the risk discussion can move to a more specific and productive point. "How much market share would we lose?" asks the program manager. "It depends on how much they cut prices," responds the project manager. "Okay," says the program manager, "how much can they afford to cut prices and still achieve a reasonable margin? Let's do an analysis of that." The conversation has moved from the vague to the specific and the quantitative. The risk has been defined, and the next step can be a particular action item for the team. So program managers should ask the development team to convert generally stated concerns into business risk

Risk Event	Business Impact
Not enough customers want our solution at the price	Reduced revenue
The vendor developing the product is overrunning on cost development expense	Development expense
It can't be developed in time to hit the planned launch date	Delayed revenue
The quality specs aren't being met at volume runs	Delayed revenue
We have underestimated the cost of building awareness	Marketing expense
We need more feet on the street	Sales expense
We are going to be late to market	Delayed revenue (and perhaps market share)
We don't have the marketing resources we need	Delayed revenue or incremental marketing expense

FIGURE 15-4. How risks affect the P&L.

Drill-down on Assumptions

FIGURE 15-5.

statements (P&L line items) and then to drill down using the technique of repeating, "Why is that?"

If risk should be expressed in terms of revenue or expense impact, it also follows that projects labeled "strategic" should not be immune to risk scrutiny. The temptation to classify a new product project as "strategic" and therefore to be pursued regardless of apparent risk is common in corporations. Senior management and program managers have been known to override concerns that were surfaced by the development team. A possible response by management to such concerns may be expressed as "We need this product to remain competitive" and "We must have it in the market in nine months."

It is the job of the program manager to identify strategic imperatives that are truly outside the parameter of the development team's responsibility and to reduce them to assumptions that can be managed. For example, the claim that the product "must be in the market in nine months" may be made as a factual statement by management, but it rests on a set of assumptions that might or might not prove to be the case (Figure 15-5).

The program manager should provide "air cover" for the development team in these situations by first untangling any strategic imperatives that come from above, then sorting them into their underlying assumptions and, if necessary, confirming with top management that they are comfortable with the underlying assumptions.

RISK MANAGEMENT STAGES AND GATES

The Stage-Gate™ approach to managing development programs divides the NPD process into development activities that are interspersed with decision points, typically held as gate meetings. At these gate meetings, the project's progress is reviewed and is redirected or stopped if it is going off track. For simplicity, four decision gates are shown, though some companies employ six or more gates within the development cycle (Figure 15-6).

Gate Meetings in Development

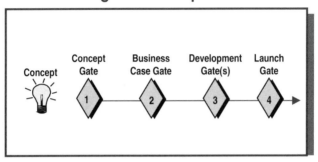

FIGURE 15-6.

The gates are more than arbitrary checkpoints to tell project managers about the completed activities and to get an okay to proceed to the next stage. They are forward-looking and represent investment milestones for program managers and the gatekeeping team.

As explained in *Corporate Venturing*, there are a number of milestones along the path from raw idea to commercial success: "Each of these milestones gives the company the chance to decide whether to continue the business as planned, redirect the effort, or bow out" (Block and McMillan 1993: 176). Indeed, in the world of venture capital, the discount rate applied to the investment in a new venture drops proportionately at each milestone to reflect the reduction in risk. "At each stage [milestone]," the authors continue, "venture managers [program managers] should ask the following questions about the business:

◆ What assumptions did we make?

◆ How have they changed?

◆ In light of these changed assumptions, what actions do we need to take (including abandoning the idea)?" (Block and McMillan 1993: 176)

This idea, that milestones are points at which previous assumptions are tested, the risk is thus reduced, and direction and investment are provided to reach the next milestone, is also reflected in the concept of option pricing theory (OPT) in product development. As Cooper and colleagues write in *Portfolio Management for New Products*, "OPT recognizes that management can kill a project after each incremental investment is made" (Cooper, Edgett, and Kleinschmidt 1998, p. 29).

Because OPT allows management to commit resources on a milestone-by-milestone basis, the absolute investment on the line at any given time is limited, and thus risk is managed by limiting financial exposure. Whether or not program managers choose to apply OPT in their investment strategy, the principle of investing one milestone at a time is a valuable one. For example, an executive was bemoaning that the R&D budget seemed to be an entitlement program. In his company, budgets were allocated annually for R&D projects, and it was tough to redirect those resources during the year. There was little flexibility in the R&D portfolio. To avoid the rigidity of annual development budgets, program managers should adopt milestone resource allocation.

Risk Management by Stage

As the development process progresses, the nature of risk management also progresses (see Figure 15-7). Program managers will want to change their focus and toolkit accordingly.

	Concept	Business Case	Development	Launch
Business Risk	◆ Not in strategic ballpark (e.g., too small to make a difference; & will consume resources and clutter up portfolio)	◆ Concept is flawed, unacceptable level of risk ◆ Opportunity-cost (ie., stronger options)	◆ Business case weaker than forecast ◆ Late to market	◆ Launch doesn't live up to projections
Risk Metric	◆ Strategic fit	◆ Discounted cash flow (NPV, IRR)	◆ Discounted cash flow ◆ Schedule, performance, cost	◆ Time to break-even
Next Investment	◆ Resources to conduct due diligence and prepare business case	◆ Development expense and capital	◆ Production ramp-up ◆ Test market	◆ Sales and marketing expense
Risk Management	◆ Clarify the concept ◆ Identify high level risks ◆ Fill information gaps	◆ Reduce uncertainty around "show-stopper" risks	◆ Continuous test programs ◆ Contingency plans (trade-offs)	◆ Launch readiness program (e.g., product, channel)
Program Manager Risk Management Role	◆ Provide strategic criteria ◆ Assess strategic fit ◆ Facilitate company buy-in ◆ Allocate resources	◆ Coach team on risk management ◆ Allocate resources	◆ Coach team (MBWA) on risk ◆ Help create contingencies ◆ Allocate resources ◆ X-Functional facilitation	◆ Coach team on risk ◆ X-functional facilitation

FIGURE 15-7. Risk management by stage gates.

Gate 1: Concept

A new product concept is considered for inclusion in the development portfolio at this gate. The concept is considered on its own merits, judged against strategic and high-level financial criteria, and against other projects in the portfolio. Questions asked are "Should we even get started on this idea given what else we are, or could be, working on?" and "Does this concept help improve our overall development portfolio and get us to our growth goals?"

Risk here is typically viewed in broad terms. For example, a financial risk may be whether the concept is going to satisfy the company's goal of only investing in business opportunities of >$100 million. A strategic risk may relate to whether the company feels it has the capability to succeed in the new market. Decisions at the concept gate meeting are likely to be based on limited facts, interwoven with management experience.

At this stage there will be a tendency for decision making to be based as much on heat as light. This magnifies the inherent risk in the concept. A sense of strategic necessity may drive management's desire to do the project, for example, while initial, overly optimistic revenue projections may excite the development team. (To the latter point, one program manager who has overseen a portfolio of projects over several years observed that from the first rough-cut revenue forecasts at the concept gate to actual entry into the marketplace, new product revenue projections had a habit of shrinking by 50 percent at every gate.)

For the program manager there may be a temptation to skip a systematic risk analysis at this early stage. One reason is to think that the relatively low level of investment required for the next stage (i.e., business analysis, financial modeling, and customer research) implies that there is little cost in letting the idea move forward. However, there is a real danger that this decision may create a bottleneck in the development program. For example, the resources to do the next stage of business case development and due diligence are scarce in most corporations. Marketing and finance people are usually in great demand. Overloading these functions can result in less careful risk analysis. Another reason for postponing a risk discussion at the concept gate is that program managers tend to take a wait-and-see approach to ideas at this early stage. "It could be too risky, but let's look at the business case and see some numbers and then we can decide whether to proceed." This leads to a postponement of risk management, but there are real issues that need to be addressed at the front end before committing to the next stage. (A classic example of an internal company risk would be establishing the buy-in of the relevant business units. Indeed, at some companies, the explicit commitment of all business units involved in a cross-business unit initiative is required before proceeding to the business case stage.)

A program manager should take a coaching role at the concept gate and help the development team focus on the critical issues for the next stage. As a program manager, ask yourself what concerns and risk you would focus on if you were the project leader who had little time and fewer resources to develop a business case.

Gate 2: Business Case

Gatekeepers now consider the opportunity with a more comprehensive description in front of them. The discussion of risk at the business case gate can take place with an understanding of its financial impact on the business (P&L). A more quantitative assessment of the opportunity and its attendant risks can be reviewed against financial criteria, for example, "Does it give us a sufficient net present value (NPV) in five years?" Risk management includes specific plans to reduce uncertainties and prepare contingencies. Of course, the degree of due diligence will be in proportion to the investment at stake. At a leading semiconductor company, a full risk management program with a quantitative risk assessment is reserved for about 50 percent of its business initiatives. Clearly, the more innovative the concept is to the business unit or corporation and the higher the investment, the more rigor should be applied.

A comprehensive list of uncertainties are documented and sized. The experience of the program manager is important in helping bring reality to these uncertainties. At a Fortune 100 company an executive responsible for the electronic components business described his technique for drilling down on the forecasts prepared by development teams: "Tell me exactly what the customer said [about this new product offer]." By asking this question, he explained, "I'm looking for tangible evidence that the forecasts are based on actual customer contact."

Gate 3: Development

Development is a complex stage made up of activities that range from prototype development to volume ramp-up and test marketing. From the program manager's perspective the focus on risk shifts from "Does this opportunity warrant investing the necessary development resources?" to "We have committed the development resources; now let's make sure we execute according to plan." The interaction between program and project manager is no longer one of selling or buying the concept, but rather one of bringing the product to market on time, within budget, and to the required specifications. Risk management in this environment is one of regular updates and reviews, perhaps done as a series of gate meetings or as more informal exchanges. The risk management process centers on the constant trade-offs that need to be made as reality challenges the opportunity as it was originally conceived. Customer needs become clearer and specifications therefore change. Suppliers can't deliver as expected. Regulatory approval doesn't happen when expected. The existing channel isn't as comfortable selling the new product as was first thought.

For the program manager, risk management at development gate/review meetings is conducted at two levels. First, as an investor, the program manager needs to ensure that the business proposition continues to be viable. This requires that as new learning is generated about previously identified risks and about emerging risks, it is fed into the business equation. For the Health Imaging new business development group at Kodak, this meant updating the P&L

and a sensitivity analysis and reviewing it at monthly board meetings (assuming there is sufficient new learning to warrant it).

Second, as coach, the program manager becomes actively involved in applying past experience to anticipate potential risks and thus head them off. As mentioned at the start of this chapter, this anticipatory approach to risk management appears to be a change in management attitudes that started to take place in the late 1990s as Wall Street raised the value put on high-growth companies. Seasoned program managers working in industries such as software, electronics, chemicals, telecommunications, and medical devices describe the shift as one from crisis management to preemptive risk management. No longer, it appears, is senior management averse to discussing and addressing risk; rather, many are taking a proactive approach to managing it. This active involvement can take the form of strict participation in scheduled review meetings. For example, at a leading wireless service provider, key functional heads always attend review meetings, and sending delegates is strongly discouraged. Or it can follow the less formal process of managing by walking around.

Gate 4: Launch

At the conclusion of the development stage a new kind of risk will likely emerge as the development team prepares to hand off to product management and to the channel. These risks reflect the potential misalignment that occurs when one group transfers a huge amount of data to another group. Assumptions are perhaps not successfully shared, and the risk of a disappointing execution in the market is real. The program manager again has to perform the role of facilitator around these issues and facilitate the dialogue between marketing and the sales force, for example. As a representative of the investor group that has sponsored the new product opportunity from its inception, the program manager now worries about actual adoption rates by the customer, actual supply problems, and actual channel performance. The risk can be summarized as "Will we meet revenue projections at launch and make break-even on schedule?"

A vice president at a European communications equipment manufacturer illustrated her predicament with managing risk at launch as follows:

◆ Customers are nervous. They are facing increased risk because of the complexity of their communications purchases. They need a lot of hand-holding by the sales force, which in turn needs to understand the principles of risk management.

◆ But we in turn face the risk of long-term customer service contracts and a 30 percent turnover in the direct sales force.

◆ So we have to institutionalize risk management in a sales channel that is young and somewhat inexperienced.

◆ Coming up with a predefined set of criteria—a checklist of risks—to educate the sales force doesn't work because no list is a substitute for live situations.

♦ We've found the most effective risk management approach is to select a "team of people with scars" (battle-seasoned managers) and transfer their knowledge to the field through constant, direct interaction.

The Six Steps of Risk Management

For effective risk management the program manager will want to ensure that the following six steps have been taken by development teams as they prepare to discuss risk at gate meetings (step 1 is applicable to the concept gate; all six steps apply to the remaining gates).

1. Be Clear on the Concept

Strangely, this first step in risk management is one that is often poorly executed. New product projects get off the ground based on a description of a technology capability or a set of features. Yet without a clear statement of the business concept, the greatest risk in a project is likely to be about miscommunications and misunderstandings between development team and management. The program manager should demand a concise description of the concept as a business opportunity. Specific questions to be asked include: "What is the value proposition for the customer?" "How do we get paid?" "Why should we be doing this now?"

2. Identify a Comprehensive List of Uncertainties

Has the development team identified a reasonably complete set of uncertainties? For the program manager, the concern is that the project team will overlook key risks. Tools can help this process (see the toolkit described later in this chapter), but many program managers have also developed their own set of questions to probe the completeness of the analysis by the development team. For example, management at a small medical device company uses four basic questions when exploring the risk in a new opportunity:

♦ Market feasibility (will they buy it?)
♦ Technical feasibility (is it doable?)
♦ Regulatory approval (what is the path to FDA approval?)
♦ Reimbursement (how much of the price will be reimbursed by public and private insurers?)

Experience has taught this management team that more than likely there will be major risks associated with one or more of these areas, and they know to address them systematically when reviewing new opportunities with project managers. The general partners of a major U.S. telecommunications equipment manufacturer's new-ventures group use a checklist of twenty to thirty criteria when conducting an assessment of a new venture. The list includes such categories as technical and market risk as well as intellectual property (IP) risk. They also consider the quality of the new venture team as a category of potential risk (as do venture capitalists) when considering whether to invest in a start-up.

These program managers are using customized checklists to systematically explore where the risk lies in the new opportunity. These questions are applied at the front end to get the development team thinking comprehensively about the possible risk. If suggested areas of risk are put on the table at the outset of the project, project teams are more likely to address them during the development process.

3. Identify "Root Cause" Uncertainties

Program managers need to know that project teams are getting beyond the superficial uncertainties and drilling down to the core issues that need managing. So they might ask, "Do we have the risk correctly stated?" For example, development teams may state a concern as, "We are not comfortable that we will reach $30 million in revenue in year 3." The program manager recognizes that this concern is expressed at too high and vague a level to be effectively managed. It needs breaking down. The program manager can coach the team by asking it to "peel the onion." For example, the risk of missing the revenue forecast in year 3 can be broken down into two discrete assumptions. The first assumption relates to whether sales will build to $30 million fast enough; that is, will customer adoption occur as planned? The second concern relates to the absolute size of the opportunity; is there $30 million of demand out there? The program manager's experience can help this drill-down process.

4. Assess the Impact of Those Uncertainties

The next question is "Which risks are the ones we must address most urgently?" To know how dangerous a risk might be, it is necessary to measure its impact on the business. The first step is to set the high and low values for the uncertainty. For example, if the assumption is that the price of the new product will be $1,200, a low value might be $600 and a high value $1,400. The second step is to assess the probability of each value occurring, and the third step is to measure the financial impact of the uncertainty surrounding the assumption. The program manager's role is to inject a dose of reality and experience into the discussion. As many program managers using these tools have observed, watch out for assumptions where the base case value ($1,200) and high case value ($1,400) are close, while the gap between the base case value ($1,200) and low case value ($600) is much wider. These ratios usually indicate an overly optimistic base case assumption.

5. Plan to Reduce the Uncertainties

To manage risk we must first discover what we can do to reduce the uncertainty component. The question posed by the program manager to the development team is framed as "What could we find out quickly to reduce the uncertainty range surrounding this assumption and thereby raise our comfort level with the base case number?" Program managers develop their own approaches to testing the assumptions and uncertainties that they find are repeated in one project after another. For the Fortune 100 executive mentioned earlier, reducing risks associated with customers' acceptance of the new product means having a customer partner on the development team. In effect, this manager has standardized the test program for customer-based assumptions. The only proof of market demand that he accepts is a paying customer expressing intent to purchase the product as soon as it is developed. Thus, working with a customer partner throughout development becomes a proxy for broader market acceptance. Then there's the more entrepreneurial approach. It was reported that a regional competitor to UPS wanted to prove its assumption that it was worth opening a new "spoke" location out of its region. Rather than commit to an expensive research study among corporate clients, they opted for less expensive and more direct competitive intelligence. They parked an observer at the edge of the regional airport and for a week counted packages as they were transferred from UPS vans to the plane on the runway.

Program managers would do well to invoke this same entrepreneurial spirit in their development teams. Ask project managers to create quick, inexpensive tests to encourage multiple learning cycles and fast turnaround of programs to test assumptions.

6. Plan Contingencies

Once a plan is in place to reduce the uncertainty, the second action in managing risk is to have the mitigation plan in your back pocket. For the program manager this means coaching the team with questions such as "If we are wrong about the assumption, where do we go next with the development program?" This is a tough exercise for the average project manager, who is not focused on what can go wrong. As one telecommunications executive points out, the gatekeepers have the experience and are in a position to know what a viable business trade-off might be—for example, whether to go outside and acquire a technology or continue to go it alone with internal development.

The stage has now been set. The roles and responsibilities of program managers in the risk management process have been defined, their evolving perspective on risk has been outlined, and the principles of a productive risk discussion have been described. Let us now review a toolkit that can help facilitate the dialogue on risk between program manager and development teams at gate meetings.

THE TOOLKIT

The toolkit is presented as a set of templates that can be prepared by project managers for presentation to program managers. The intent of every template is to facilitate the risk dialogue between all stakeholders. The toolkit is dynamic. Individual templates are used iteratively throughout the development process, being updated to capture the new learning for each gate meeting. In this way the templates capture the process of continuous learning that takes a concept from maximum uncertainty at the outset to reasonable confidence in the business proposition by the time of launch.

The toolkit is broadly based on an approach to risk management that was introduced by Eastman Kodak's Health Imaging[2] new business development group. It does not contain any components that would be considered proprietary, and many companies use similar tools. We have selected only those templates that would be used by development teams at gate meetings. That is, each template is designed to stimulate a discussion about the risk in a new product development project.

Any tools like the templates on the following pages should carry a warning. In this case it is "Caveat program manager!" The caution is that once development teams master the art of filling in such templates, they can become an exercise in presentation preparation and the thinking gets short-circuited. Tools can't do the thinking for people; as the vice president of a major health care product manufacturer noted, "If I had to choose between good people and good process, I'd choose people." It is important to remember that the purpose of the toolkit is to capture the thinking of the development team about the risks and to allow program managers and their gatekeeper colleagues to see that thinking, and thus be able to join the risk dialogue quickly and effectively.

Concept Gate

Template 1: The Concept Statement

The concept statement is the first of several templates designed to capture a comprehensive picture of the new opportunity. It is created on day one during the concept stage. But, like several other templates, it gets updated throughout the development process and is presented at every gate meeting. The thrust is to get an overview of the idea as quickly as possible so that whoever is developing the concept has a 360-degree look at the risk and can then focus on the risks that need the most urgent attention.

The template (Figure 15-8) requires a description of the idea covering the key aspects of the concept. At this early stage, only major headers are filled in (e.g., target market identification versus the three subquestions that are filled in as more is learned). Developers (i.e., whoever is responsible for this initial concept statement) may respond that they don't know enough to fill in portions of the template at the outset. It is true they do not *know* the answers to all the line items, but they can take an informed guess. By doing so they are really capturing their assumptions about why the company is pursuing the opportunity (strategy), who the target is (customer), why the product answers the cus-

Business Objective

How does the project fit with business strategic priorities?

How does the company make money from this new product? Why should this project be pursued?

What impact will the project have on other business opportunities in the NPD portfolio

Target Market

Which customer segments are we targeting?

What are the targeted geographic regions/countries?

Who will use and who will buy this solution?

Customer Needs and Values

What are the customer needs being addressed?

What is the current solution to the customer's need

Market Potential

What is the approximate market size, definition and growth rate?

Competition

Who is the competition and what threat do they pose?

What advantages does our product have over those of our competitors?

Key Product Features and Benefits

Give a short description of what the product is (hardware, software etc.) and what it does

What are the key features and benefits of the product?

What will the product offer as a solution to customer needs?

What are our development costs?

Delivery and Support

How will we produce our product/service offer?

Marketing strategy

Value Proposition

Price

Sales channel

Awareness

Alliance Candidate (if relevant)

Who are they?

What are they selling?

What do they want?

FIGURE 15-8. Concept statement.

tomer's needs (product), how it is superior to alternatives (competition), and so on. Just the process of putting down this summary of facts and assumptions permits a dialogue to take place between program managers and developers.

RISK DISCUSSION   At the concept gate meeting the risk is managed at two levels. The first risk is that program manager and project manager may not be aligned on the strategic purpose that the new product serves. For example, does the developer understand how the product fits into the broader product portfolio—for instance, to build market share without cannibalizing the core business? Also, does the project manager have a clear sense of how the corporation is going to make money with the new product? For example, for a software product, is the company to be paid per user, per transaction, or a flat fee? The program manager has the portfolio perspective and the responsibility to clarify any strategic assumptions up front for the development team.

Second, the concept statement provides a tool by which future risk can start to be identified. For example, a program manager can ask the developer which line items in the description intuitively raise the greatest concern. This kind of question begins the discussion of risk, though the formal, quantitative risk analysis has not yet started.

Template 2: The Screening Matrix

The screening matrix (Figure 15-9 on pages 398–399) provides a structured method for developers to identify areas of risk early in the development process. It also is prepared for the concept gate. It shows which areas of the concept are solid and which are risky or unknown to the team. Because there is little data to back up the screening matrix scores at the concept stage, categories such as technical feasibility or customer need are the focus of discussion, rather than the particular scores for individual line items. Program managers should expect the scores in the matrix to change as the team learns more about the product and the market. In this sample template each line on the matrix is scored a 1, 4, 7, or 10. Descriptions are given for each cell, and these are customized to match business unit priorities.

RISK DISCUSSION   Don't put too much emphasis on the numbers at this point because they are probably based on a developer's best guess. However, the line items and scores are a jumping-off point for a discussion about the significant areas of risk, and allow the program manager to focus the development team on the particular areas that will need the most urgent analysis. For example, the program manager will want to probe line items with scores of 1 or 10 to confirm the team's belief that the concept is very weak or strong in those areas.

Business Case Gate

For illustrative purposes we have used a case study to describe the content of several of the following templates used at this gate meeting. The case involves an actual development project terminated before commercialization.[3] Figure 15-10 shows the detailed concept statement as it might have been presented at the business case gate meeting.

Template 3: Assumption Generation

Using the tools already described (concept statement and screening matrix), a development team is ready to separate the known facts from assumptions. One question they must ask themselves is "How do we know if the most critical risks have been identified?" A development team will list assumptions under each of the template headings as the first approach to risk management in the business case stage (Figure 15-11). They may generate a set of twenty to thirty individual assumptions. Obviously, they are not all equally risky, but at least the team has thought through risk from every angle of the proposition.

RISK DISCUSSION   This list of assumptions can be presented at a gate meeting to illustrate the breadth of the team's risk analysis, or it becomes a backup worksheet. An assumption should be presented as a statement. For example, under "Target Market," the assumption would read "The two thousand companies that service the automotive aftermarket." An assumption stated this specifically avoids confusion and helps the development of focused test programs later.

(For another productive approach to identifying risky development issues, we recommend the concerns-mapping tool described in Chapter 8 of this volume and also in Smith and Reinertsen 1995, page 230. This tool allows a development team to assess the importance of different concerns by mapping the concerns against probability and impact axes. Concerns mapping is an effective tool for development teams to narrow down the number of key concerns and risks before measuring their potential financial impact, as indicated in the next template.)

Template 4: Sensitivity Analysis (Tornado Chart)

The tornado chart (Figure 15-12) is the graphical representation of the uncertainty ranges for the most important assumptions. It measures the NPV impact of the low and high ends of the uncertainty range for each assumption. The impact on NPV is measured in terms of individual assumptions and deliberately avoids measuring multivariable scenarios (e.g., "What if the low case for price, the low number of customers, and the highest cost of COGS all turn out to be the actual case?"). The tornado chart shows which assumptions are most dangerous according to two indicators:

◆ The degree to which the low value of an assumption will result in a negative NPV.

◆ The width of the NPV range resulting from the spread between the low and high range of uncertainty. This indicates a priority need for further research to narrow the uncertainty.

Key Factors	1	4	7	10	Row Score
Strategic Fit					
Product demonstrates company leadership	No product leadership demonstrated	Minor product leadership attributes are apparent	Major product leadership attributes are demonstrated	Demonstrates sustainable leadership in the market segment	4
Proprietary Position	Easily copied	Protected, but not a deterrent	Solidly protected with trade secrets, patents	Position protected with trade secrets, raw material access	4
Platform for growth	Dead end/one of a kind	Other opportunities for business extension	Potential for diversification	Opens up new technical and commercial fields	1
Synergistic with company	No direct synergy with current company offerings	Limited synergy with company	Adds value to one or more offerings in company	Synergy with existing and future offerings	7
Technical Feasibility					
Technology skill base	Technology new to the company: (almost) no skills	Some R&D experience, but probably insufficient	Selectively practiced within the company	Widely practiced within the company	7
Technology "GAP"	Multiple inventions required	Single invention required	Technically challenging	No technical hurdles or engineering problems	7
Probability of manufacturing success	Low; new to us: need to acquire or build	Modest; doable, but means major modification	Good; doable, with minor modifications only	High, can be done with existing people and facilities	10

	Existing technology	Novel	Novel—difficult to imitate	Novel—breakthrough	
Novel technology					4
Customer Need					
Clear target market	Not clear who will buy	Some sense of target segment	Strong reason to believe the target market can be defined	Specific customers have already identified themselves	4
Clear market need	No evidence yet of need	Believe there is some need	Believe there is broad and urgent need	Customers are asking for the concept/product/service	4
Price/value proposition	Low	Moderate	High	Very high	4
Financial					
Estimated revenue to company	<$50M	$50–100M	$100–200M	>$200M	4
Time to commercial start-up	>5 years	3–5 years	1–2 years	<1 year	10
Attractive business model	Contains no recurring revenue stream or service components	Product is disposable	Product is disposable with possible add-ons	High recurring revenue from source components	4

FIGURE 15-9. Concept screen.

Background

It's the early 1990's and the telephone is still the dominant communications device on the desktop. PCs are not yet standard for office workers. There is some data communications via PC/modems and the fax is ubiquitous in the office as a shared device "down the hall". The former ATT Global Communications Systems business unit is committed to maintaining control of telecommunications on the desktop and is exploring a number of different telephony-based approaches to voice/data transmission.

Docfon (document sharing by phone) allows simultaneous, interactive communication to discuss the content of documents. Coming out of the Concept stage, initial, tentative customer research indicates that engineering firms would be the first market interested in the Docfon concept.

Concept Statement for Business Case Gate

Business Objective

Alignment with business participation and investment strategies
- The Docfon concept is in accordance with the company's strategy to "own telephony on the desktop"
- Docfon is one of the several complementary projects in early stage development that explores desktop voice/data telephone integration
- There is no significant impact on the company's existing businesses from this project
- We make money selling the hardware device. Secondarily, we expect to stimulate incremental long-distance telephony usage

Market Analysis

Target Market Identification
- The target customers are engineers that work for firms with 10 or more professionals and their customers
- Docfon will be released in the US in year 1 and in Europe in year 2

Customer Needs and Values
- The need is for a simultaneous (time-saving), interactive way to see, discuss, and change documents when working at two different locations
- Currently, the customer must first use fax or express mail service to send a document, call recipient, discuss changes, and then send changes

Market Potential and Growth Projections
- There are approximately 400,000 engineers in target firms with 10 or more professionals
- Other professional groups (e.g., advertising agencies) will be addressed in year 2 and beyond

FIGURE 15-10. Docfon Case Study

RISK DISCUSSION   The Health Imaging team at Kodak, as well as gatekeeper teams at other corporations, has found this to be a valuable tool in facilitating the program/project manager discussion. Management asks two questions on seeing this template.

- "Do I see any obvious missing assumptions? For example, the team has not mentioned COGS as a major concern, but last time we ordered these VLSI chips, they were 20 percent more expensive than expected."

Product Features and Performance

Product Conceptualization
- Docfon is a stand-alone, desktop device that allows simultaneous speech and data over a standard analog phone line. Specifically, it includes a document scanner, 14" LCD display, electronic pen, speaker, and modem

Fit with customer need
- Docfon allows users to simultaneously discuss and share the same document while making changes with the pen

Product Features and Performance

Key Features and Benefits
- Simultaneous interactive voice and document communication between two or more users at different locations regarding a design, drawing, layout, or text document.

Technical Feasibility
- Docfon can be developed in less than one year and uses current DSP technology
- The key technical challenge is confirming data transmission speed using the proposed protocol
- The product leverages ATT (Bell Labs) strengths in analog telephony
- ATT's I.P. at launch is the proprietary software protocol, but this is only expected to give the company nine to twelve months lead time over competition
- All development work will be done internally but the LCD screen will be purchased externally
- Development costs are estimated to be modest (largely software) as a percent of sales
- Repair and maintenance will be managed by AT&T

Competition

- Competitive products include document collaboration and conferencing systems. These units use two telephone lines or an ISDN line and range in price from $5,000–$10,000
- Docfon's competitive advantage is that no current, alternative, or proposed product adequately meets the customer's need and price range
- Our competitive advantage will come from Bell Labs' expertise in telecom technologies and from AT&T's market leverage

Commercialization Plan

- To generate sufficient profit margin the product will be priced at $3,000+
- Our current direct sales force will sell the product to customer firms
- Advertising expenses are estimated at 25% of sales for year one, declining to <15% over three years

- "Am I comfortable with the uncertainty range the team has put on the assumption?" For example, the team estimates FDA clearance could be delayed six months. Experience says if the FDA requires additional clinical trials, it could be delayed twelve months or more.

Program managers reviewing such tornado charts might want to keep a general observation in mind. In contrast to the Docfon tornado chart (Figure

	Assumptions

1) Business Objective
    ◆ Fit with strategic priorities
    ◆ Fit with other businesses
    ◆ Fit with portfolio priorities
	1 ATT will accept a gross margin business of less than 60%
	2 We must be in the market in 12 months to preempt similar products
	3 We will not compete for desktop space with other ATT products
	4 ATT is not interested in pursuing PC solutions to this customer need

2) Market Analysis
    ◆ Select addressable market
    ◆ # of accepting engineering firms
    ◆ Customer need
    ◆ Market potential
    ◆ Competition
    ◆ Education process
    ◆ Standards
	1 20% of prospect engineering firms will buy Docfon
	2 The customer will accept another device on the desktop
	3 Customers will be available for real-time interactivity
	4 20% of prospects that sales force calls on will buy Docfon
	5 Engineering professionals can persuade their customers to adopt Docfon
	6 On average, each customer will buy 5 units

3) Product Features & Performance
    ◆ Technical/IP feasibility
    ◆ Customer Role
    ◆ Development Costs
    ◆ Manufacturing
	1 The technology is available to meet the technical performance specs.
	2 Product will meet customer requirements at COGs of $1,200 in year 1
	3 Our technical performance specs will match/exceed user requirements
	4 COGs will decrease by 20% per year

4) Product Positioning ◆ Differentiation	1 New technology will not make our product obsolete within 5 years 2 Customers prefer a pen-based versus mouse-based (PC) solution 3 The AT&T brand name gives us an advantage over competitors 4 Customers perceive our solution to be better value than competitors'

5) Commercialization
    ◆ Pricing
    ◆ Channel
    ◆ Delivery & Support
    ◆ Advertising & Promotion
	1 Customers are willing to pay $3,000 per unit in year 1
	2 Each sales person will call on 500 new prospects in year 1
	3 Our existing channels are the most efficient way to launch the product
	4 The existing channel will be willing to divert time to the new product

6) Alliance (if applicable)
    ◆ Partner's Expectation
    ◆ Partner's Reliability
    ◆ Schedules
	Not applicable

FIGURE 15-11. Docfon assumption generation.

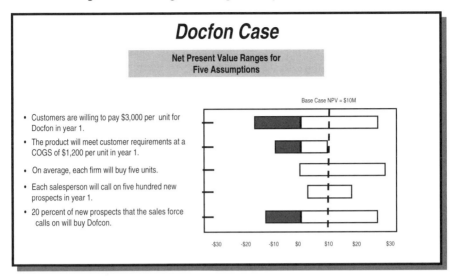

FIGURE 15-12.

15-12), which shows the negative NPV impact of assumptions, the two tornado charts shown in Figure 15-13 should be considered suspect.

On the left, the all-white tornado chart implies that the development team considers none of the risks (assumptions) identified to be dangerous enough (i.e., could have a sufficiently low value or worst case) to cause the project to generate a negative NPV. This is possible in theory, for example, if a new prod-

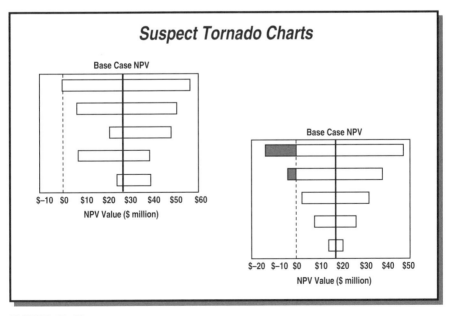

FIGURE 15-13.

uct's costs have been heavily absorbed in a prior development project so that the investment in the new product is minimal. However, more typically tornado charts like this indicate that the development team thinks the new product is bound to be a slam-dunk, no-risk, guaranteed commercial success!

The chart on the right shows a set of assumptions with equally distributed high and low values. This typically occurs when a standard ± 10 percent distribution has been applied to the uncertainty range surrounding each assumption, and it communicates that little real thinking has gone into considering the actual uncertainty ranges and risks. Again, this is worth challenging.

These tornado charts are built on a very simple sensitivity analysis using a P&L spreadsheet. Each assumption is considered as an independent variable, and the NPV impact of its uncertainty range is calculated on a stand-alone basis, without worrying about the accumulated risk of all assumptions. Obviously, there are times when a more sophisticated sensitivity analysis is warranted. For a major development program or acquisition, teams will want to consider different what-if scenarios that deal with possible outcomes from combinations of the assumptions. The advantage of this simplified sensitivity analysis is that it is a user-friendly device that does the job of engaging both sides in a discussion about the risks that matter, without overwhelming the team with numbers and number crunching.

Template 5: Assumption Uncertainty Ranges

This template (Figure 15-14) shows the uncertainty ranges for the assumptions that are thought to be dangerous and whose impact on the business NPV is worth measuring. The uncertainty ranges are fed into the P&L to measure their impact, and then into the tornado chart. This template is created as a worksheet and is used as an input to generate the tornado chart. However, like the assump-

Assumptions	Link to P&L Model	Base Case Value	Uncertainty Range	
			Low Value	High Value
Customers' willingness to pay for Docfon	Price	$3,000	$1,200	$4,000
Unit Materials Cost (1st year)	COGS	$1,200	$2,000	$1,000
# of sales calls per rep	# customers visited in year 1	500 calls	200	500
5 Docfon units per customer	Docfon units	5	2	10
20% of sales calls successful	# customers buying in year 1	20%	5%	25%

FIGURE 15-14. Docfon assumption uncertainty ranges.

Assumption	On average, each location will buy 5 units.		
NPV Range	$46.3 MM		
Test Program	**Alternative #1**	**Alternative #2**	**Alternative #3**
Description	Project manager conducts phone survey with 5 customers	30 one-on-one customer interviews with brochure	Loan 100 prototype units to customers for 3 months
Required Resources	$2,100	$65,000	$150,000
Time	2 weeks	4 weeks	4 months

FIGURE 15-15. Docfon test program alternatives.

tion list (Figure 15-11), it may also be used to show the team's thinking during a gate meeting because it describes the uncertainty ranges for each assumption, and gatekeepers may want to discuss these.

RISK DISCUSSION   Should the gatekeepers choose to explore the uncertainty ranges for the assumptions with the development team, the focus of the dialogue will tend to be about their downside potential. For example, they may ask, "Has the Docfon team really considered how low the price of the product may have to go in order to attract a broad customer market?" On occasion, the upside potential of an assumption can also constitute a risk, such as when a development team has badly underestimated demand for its new product, resulting in back orders, frustrated customers, and a disastrous product introduction.

When a multifunctional team is presenting, management should take the opportunity to check whether there are any dissenters on the uncertainty ranges by asking the respective functional representative for a particular assumption whether he or she is comfortable with the range.

Template 6: Test Programs

In order to reduce the level of uncertainty surrounding the critical assumptions, teams should design test programs (Figure 15-15). The objective of a test program is to minimize uncertainty as economically (in terms of both money and time) as possible. When designing test programs, it is important to consider alternative approaches (e.g., speed, cost, level of uncertainty reduction) to find the most effective way to proceed. In the Docfon example, the team recommended alternative 2. The team thought it was important to show the customer the product, in the form of a brochure, in order to give the product a sense of reality in the customer's eyes.

RISK DISCUSSION   Test programs such as those illustrated for Docfon will often not be mutually exclusive. That is, all three tests could be run sequentially over time since each test researches the uncertainty in the assumption to a differing degree of rigor (and cost and time). For the program manager, who is going to sign off on the funds to conduct the test program, the question is which test is most relevant given the product's stage of development. For example, a program manager may have said to the Docfon team, "I agree that you will want to do a decent-size study with customers down the line, and that developing a brochure may make sense in that context. But we are early in development and I am uncomfortable with the assumption that the average customer is going to buy as many as five units each. So, as an immediate next step, I want you to get on the phone with a handful of customers and do a quick reality check. Then I will consider spending $65,000 for your recommended alternative."

In general, program managers should encourage teams to incorporate more than one assumption within the same test program—in effect, being as economical and entrepreneurial as possible in their approach. Also, challenge them to consider less predictable approaches with questions such as "If we had to test this in a week, how would we do it?" Finally, establish the decision rules before conducting the test—for example, "Unless 20 percent of physicians indicate a clear interest in our product at $200 we will not recommend continuing with the business opportunity as it is currently defined." Or, as in the Docfon case, "Unless at least 20 percent of customers are willing to pay at least $3,000 per unit, we will not recommend continuing with the business opportunity as it is currently defined." Stating the required test results in quantitative terms ahead of time will decrease the danger that the results of the test program will be subject to interpretation when they come in.

Template 7: Contingency Plans

For each assumption, the program manager needs to ensure that development teams know where to go and what to do next. This template (Figure 15-16) is helpful because it lays out systematically the cost (trade-off) of any contingency plan.

RISK DISCUSSION   Contingency planning is about anticipating new learning and creating alternative solutions. Program managers are essential to this pro-

Contingency Analysis			
Original Risk	Contingency Plan	Trade-off	Acceptable Business Result
Price of $3,000 too high	Revisit cost and eliminate features	Loss of customer appeal	10% reduction in customer interest
Cannibalize existing business	Raise price to deter automatic switching	Drop in market share for new product	5% market share reduction by year 2
Internal development program moving too slowly	Contract out development	Higher development costs	10% increase in R&D expense

FIGURE 15-16. Contingency analysis.

cess because the trade-off required will frequently involve reallocating resources. For example, deciding to double the number of engineers on a project for the next two months, or to restructure sales force compensation in order to support a struggling new product launch, are contingencies that likely require senior management involvement. In reality, more discussion occurs around contingency planning than any other topic at a typical gate meeting review.

SUMMARY TIPS

At the start of this chapter we asked you, the program manager, to imagine sitting in a gate meeting and being presented with a new product opportunity. We posed a set of questions such as: What questions will you ask of the development team to get past the numbers and into the real risk issues? What do you need to see in the presentation to get a real understanding of the project risk? How do you contribute to the team's insight on the risk management program?

Our hope is that the toolkit described above will be helpful in addressing these questions. Along with a well-engineered toolkit, however, having a clear sense of your role as the decision maker and gatekeeper in the development process is essential. In sum:

1. The key is to get the "graybeards" and "people with scars" contributing to the risk dialogue. Whether in meetings or less formal forums, pull the management team together to talk about and resolve risk issues in each other's presence.

2. Remember your role as investment manager. Don't be afraid to say no to a project, and don't be bullied into hasty decisions. When you do say yes, allocate the resources to do risk management properly. The development team of a telephone service provider was well into the development phase of a project without any financials or quantitative risk assessment completed. The reason? Finance was consumed with doing an acquisition valuation analysis for the corporation, and finance was the only group able to run NPV calculations. (Companies who have had a formal development process in place for some time have learned that it is essential to allocate a full-time finance person once the project has been acknowledged by the company as a high-risk/high-reward opportunity.)

3. Remember your second role as coach. Help teams to identify risk, figure out how to manage it, and think through contingencies. The program manager's experience is invaluable to the risk management process. Take a leaf from the venture capitalist's manual and roll up your sleeves to help identify and resolve critical issues.

4. Remember your third role as facilitator to the gatekeeping team. Escalation of issues is inevitable in risk management. You are the bridge to your functional peers, who are needed for their experience and resources. If an opportunity requires the cooperation of different business units within the company, the chances are it won't fly without you providing "air cover."

5. Use the risk management process wisely. Not all risk in NPD is equally

risky. Any serious risk analysis is demanding on a development team. As a manager at the semiconductor company mentioned, "When they hear we are going to do a full risk management analysis on a project, development teams groan." Don't demand equal risk analysis on product improvements and line extensions, for example. The rigor required for effective due diligence does consume time and talent. There is always a need to balance speed and due process. On the other hand, don't sidestep the rigor of proper risk management just because of time constraints. To quote one executive interviewed: "Formal risk management may be a decelerator in terms of speed to initial product launch, but it is not to final success."

NOTES

1. At many companies gate meetings convert to monthly board meetings for a high-profile project, once it passes through the business case gate into development. The are called board meetings because senior management is involved and takes an active role in the discussion and mitigation of risk.
2. The author wishes to thank Kodak Health Imaging Business for permission to use templates from their risk management process.
3. The Docfon case study was first described by Hollister (Ben) Sykes and David Dunham in *Visions,* April 1998. The author wishes to thank AT&T and Lucent Technologies for permission to reproduce the Docfon new product development case.

BIBLIOGRAPHY

Block, Zenas, and Ian McMillan. 1993. *Corporate Venturing.* Boston: Harvard Business School Press.

Cooper, R. G., S. J. Edgett, and E. J. Kleinschmidt. 1998. *Portfolio Management for New Products.* Reading, MA: Addison-Wesley.

Smith, Preston, and Donald Reinertsen. 1995. *Developing Products in Half the Time.* New York: John Wiley and Sons.

16 Process Modeling in New Product Development

Paul Bunch and Gary Blau

Bringing new products to market is a challenging process involving many complex decisions. Issues that make the process challenging include long timelines from concept to product launch, high attrition of projects throughout the development process, and high development costs. For example, the time required to develop a new pharmaceutical compound typically exceeds twelve to fifteen years, while the associated costs can exceed $500 million per compound (Pisano 1996; Drews 1998). Fewer than 25 percent of the molecules that begin clinical testing actually make it to the marketplace. In addition to discovering and developing new products internally, many companies engaged in new product development (NPD) attempt to diversify their development portfolios by relying on external partners to co-develop and market their products. Furthermore, many companies are attempting to market products for multiple uses in targeted or niche markets. While these diversification strategies create a richer set of opportunities to pursue, they also create demand for scarce NPD resources required to bring products to market. Clearly, firms that are able to use their NPD resources most productively will be rewarded in the marketplace.

In this chapter, we will be primarily concerned with the use of models to help NPD managers address several high-level issues, including strategic NPD planning, project selection, and project execution. A steady-state strategic planning model will be presented that can be used to estimate how projects and resources should be distributed along the NPD process to achieve a steady output to the marketplace. The steady-state model assumes that the projects are completely balanced along the R&D chain. While this is clearly an idealization, it is useful to help determine what fraction (or multiple) of resources currently available is necessary to achieve a given output from R&D.

With regard to project selection and execution, steady-state models do not go far enough. In particular, such models fail to consider resource constraints and the impact of time-variant factors that influence NPD performance over time. For example, one clearly must account for the real (unbalanced) composition of any given set of projects. In general, this distribution will not be balanced so that a steady output can be reached from NPD. Furthermore, one must account for the addition and deletion of projects over time. Projects can be eliminated due to undesirable product qualities such as unsatisfactory safety

409

or quality results. Projects can be added from internal development efforts, in-licensing efforts, or initiation of a line-extension project. We will present models that account for these issues and can be used to guide project selection, predict how a given portfolio will perform over time, and help improve the results.

BACKGROUND

Process Description

In this section, we provide concepts that will be useful throughout the rest of the chapter.

We consider the R&D process to consist of a set of tasks, each of which consumes resources and whose function is either to filter diverse opportunities or eliminate uncertainty in some feature of the product, such as safety or efficacy in the case of pharmaceuticals. This view of the R&D process is a stage gate process as described by Rosenau (1996) and is depicted in Figure 16-1. Note that each stage in this abstraction of the R&D process could consist of any number of more detailed stages. In simple terms, as projects proceed from concept to commercialization, either they are terminated due to some undesirable feature or they progress to the next stage of development. We assume, then, that all projects are either launched as commercial products or are terminated. Determining the resource requirements for projects in this process at any instant in time simply consists of summing the requirements across all projects in the process. This approach requires that one know the stage of development for each project and the resources required to execute a single project in each stage of the development process. When possible, one should use project plan data to estimate the resources required in each stage of development for each project in the process. When this is not possible, it may be appropriate to use average historical resource requirements in each stage of development for each project in the process.

The abstraction of the R&D process shown in Figure 16-1 serves as a good starting point, but it does not go far enough. Specifically, in addition to working on projects that will produce new products generated from internal develop-

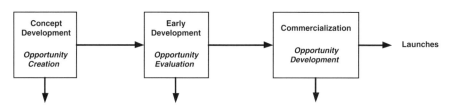

Attrition (Terminated Projects)

FIGURE 16-1. NPD projects flow through a series of process steps from concept development to commercialization.

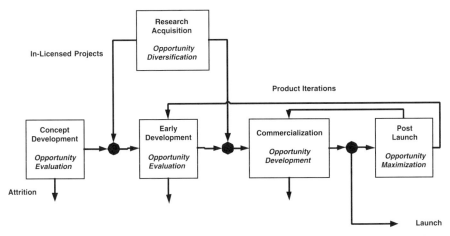

FIGURE 16-2. The NPD portfolio consists of projects from a range of sources, including new internal efforts, research partnerships, and product iterations. Each type of project may require different levels of resources.

ment efforts, many R&D programs undertake efforts that extend the use of existing products. In the case of pharmaceuticals, extensions could include efforts to pursue using a single product (molecule) to treat multiple disease states or to administer the product in multiple dosages or delivery forms (capsules and intravenous), for example. Furthermore, R&D programs may add projects to their set by licensing from partner companies projects or technologies that then must be further developed. Thus an R&D program could have in its set of projects various types of projects originating from several sources. This is depicted in Figure 16-2. As before, each stage in this diagram could consist of any number of more detailed stages. Note that the characteristics of projects will most likely depend upon their type. That is, the demand for resources by product iterations may differ from (be less than) the demand for resources from new products developed internally.

Critical functional resources are required to complete work at each stage of the development cycle. In the pharmaceutical R&D process, these functional resources include clinical pharmacologists, toxicologists, chemists, statisticians, and clinical research physicians, among others. In most instances, these resources are drawn from functional areas in which areas of technical expertise are concentrated. Completion of work in a given stage requires a pool of resources, which will generally comprise a variety of functional areas. Any given function then can be called upon to support multiple projects over a range of development stages in the R&D process. This situation is depicted in Figure 16-3, in which solid lines represent the flow of projects and dotted lines indicate the input of resources to complete a project at a given step. In the remainder of this section, we describe models that can be used to estimate resources required to execute a set of NPD projects.

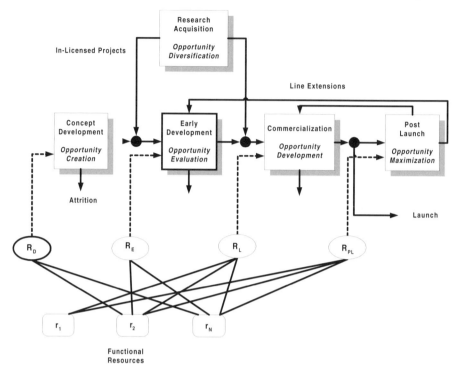

FIGURE 16-3. Sets of resources are required to support work in each of the stages of NPD. *Resource sets often consist of teams of experts from functional areas.*

Algebraic Models

The power of algebraic models is their simplicity and ease of implementation. Additionally, if the parameters in the models are uncertain, the algebraic models can easily be subjected to Monte Carlo simulation.[1] If one is willing to assume the existence of an average project, then the resource requirements for executing a set of projects at any instant in time can be estimated. Specifically, by knowing the number of projects by type in each stage of the NPD process and the resources required per project by type and stage, total resources requirements can be determined as

$$r_{i,Tot}(t) = \sum_{S}\sum_{T} WIP_{S,T}(t) \times r_{i,S,T} \qquad (16\text{-}1)$$

where

$r_{i,Tot}(t)$ = total number of resources of type i required at time t
S, T = indices referring to development stages (e.g., concept development, early development, and commercialization) and project types (e.g., new products, product iterations), respectively

$WIP_{S,T}$ (t) = work in process (number of projects) of type T in development stage S at time t

$r_{i,S,T}$ = number of resources of type i required to complete a single project of type T in development stage S

Assuming resource requirements can be determined, the essence of estimating the resources required at time t is the determination of $WIP_{S,T}$ (t). Note that Equation 16-1 can be used to determine the resource requirements at any instant in time t or over some time period t over which the quantity $WIP_{S,T}$ (t) remains constant.

If one is not willing to live with the assumption of average projects, then resource requirements must be estimated based on individual projects. Wheelwright and Clark (1992) present the aggregate project plan as a way of estimating and allocating resources to individual projects over time. Their approach is to present a list of projects under consideration and their resource requirements over some future time horizon. Since projects may be terminated throughout the NPD process, one should consider a scheme for modifying the expected resource requirements. We suggest that such an approach can be expressed compactly as

$$r_{i,Tot} (t) = \sum_{np\varepsilon NP} \sum_{u\varepsilon U} \bar{r}_{i,np,u} \times d_{np,u} (t) \times p_{np,u} \qquad (16\text{-}2)$$

where

$r_{i,Tot}(t)$ = total number of resources of type i required during time period t

NP, U = indices referring to the set of new projects and tasks performed by the new projects, respectively

$\bar{r}_{i,np,u}$ = number of resources of type i required per unit time to work on task u of project np

$d_{np,u} (t)$ = duration of project np on task u during time period t

$P_{np,u}$ = probability that project np will progress from its task at time t = 0 to task u

This formula simply states that one should scale the resources to perform a task by the probability that the task will actually occur and that resources should be estimated only for the duration during time period t which they are in use. To demonstrate the use of Equation 16-2, consider Figure 16-4, showing three tasks: concept ($u1$), prototype ($u2$), and scale-up ($u3$). We consider estimation of resources during the time period t = 12 months. The starting task for this time period is $u1$. During this time, the time spent on $u1$ is two months, the time spent on $u2$ is six months, and the time spent on $u3$ is four months. Furthermore, there is a 75 percent chance that task $u2$ will actually be performed and only a 25 percent chance that task $u3$ will be performed ($\frac{3}{4} \times \frac{1}{3}$). During the twelve months that are of interest, if we neglect the probability of project failure, we would estimate resource requirements (in person-months) as $(2 \times 3) + (6 \times 4) + (4 \times 10) = 70$. If we acknowledge that the project could fail between tasks, however, we estimate that the resources requirements

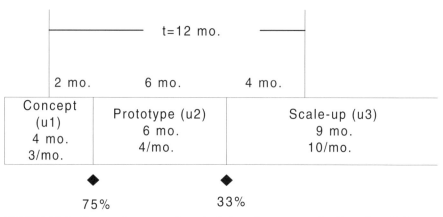

FIGURE 16-4. Process demonstrating calculation of resources using Equation 16-2.

would be $(2 \times 3) + (0.75 \times 6 \times 4) + (0.25 \times 4 \times 10) = 34$ person-months. To make full use of Equation 16-2, one simply needs to apply this approach for each critical resource to each project in the portfolio. That is, to determine the resource requirements of a particular resource for a portfolio of projects, apply this approach to each project in the portfolio and sum across all projects. This approach can be implemented in a straightforward manner using a spreadsheet and as demonstrated with the following example. Consider an NPD portfolio consisting of ten projects, as shown in Table 16-1. This is the state of the portfolio at the current time. In addition to the projects shown in Figure 16-1, we assume that three projects are added into the concept stage every quarter. The purpose of the model is to predict the state of the portfolio over some future time horizon. In this case, we predict the state of the portfolio over four future time periods. Since each time period represents a quarter (three months), we make resource predictions for our portfolio over the next year. During each time period, we predict the time spent in each stage by each project, the time remaining in each stage for each project at the end of the time period, whether a transition occurred from one stage to the next, and the resources consumed. Time profiles of projects by stage of development and FTEs predicted by the model are shown in Figure 16-5. A spreadsheet implementation of this model is included with this volume.

Simulation Models

While the power of algebraic models is their elegance and simplicity, their shortcomings lie in their inability to deal with events that change over time. To address these shortcomings, one can use more sophisticated modeling methods. These methods, which will be broadly referred to as *process simulation models,* can include discrete event models as well as system dynamics models (Sterman 2000; Pritsker 1986). The mathematical details of these models are beyond the

TABLE 16-1.
Project Portfolio with Ten Projects in Various Stages of the NPD Process

Project	Parameter	Stage		
		concept	prototype	scale-up
1	CT	4	9	12
	P(TS)	70%	50%	90%
	FTEs	1	2	4
2	CT	4	6	8
	P(TS)	60%	70%	90%
	FTEs	1	3	6
3	CT	2	9	10
	P(TS)	50%	60%	80%
	FTEs	1	2	4
4	CT	0	12	18
	P(TS)	100%	60%	90%
	FTEs	1	4	6
5	CT	0	4	10
	P(TS)	100%	70%	80%
	FTEs	3	3	7
6	CT	0	2	8
	P(TS)	100%	80%	90%
	FTEs	1	2	5
7	CT	0	1	10
	P(TS)	100%	80%	90%
	FTEs	2	2	6
8	CT	0	0	10
	P(TS)	100%	100%	80%
	FTEs	2	3	8
9	CT	0	0	5
	P(TS)	100%	100%	90%
	FTEs	2	4	6
10	CT	0	0	8
	P(TS)	100%	100%	90%
	FTEs	1	3	5

scope of this chapter, but we believe that it will be quite useful to make the reader aware of the types of models that can be constructed and how they can be used.

Process simulation models are particularly good at tracking and controlling the movement of individual projects through the steps of a process over time. These models also allow one to control or limit the movement of projects through the process based on the availability of resources. In addition to providing a more realistic picture of NPD capacity, this analysis can help identify bottlenecks in the NPD process. Furthermore, if one is able to estimate the extent to which resource shortages affect NPD performance, one can investigate the impact of various resource allocation strategies on overall NPD perfor-

Projects By Stage Profiles

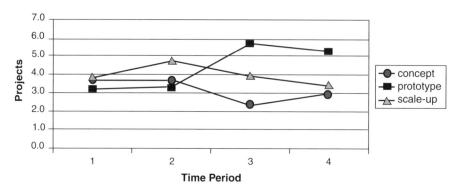

FTE Profiles

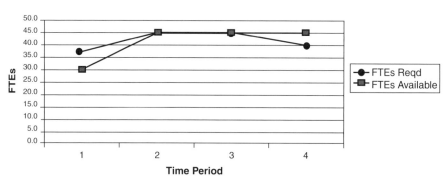

FIGURE 16-5. Projects by stage and FTE profiles over four time periods predicted using Equation 16-2.
The portfolio initially contained ten projects, shown in Table 16-2. Three projects per quarter enter the NPD process.

mance. Although process simulation models are more difficult to build than simple spreadsheet models, they can provide significantly more insight.

In later sections, we discuss several specific uses of these models. In addition to tracking the movement of projects, these models can track the implied flow of resources. That is, they can track resource consumption of projects during development and while on the market as well as resource generation (revenue) of projects that achieve launch.

The simulation framework we consider consists of two high-level models. The first model tracks the movement of individual projects through the NPD process and onto the market. The second model tracks the movement of resources. Conceptual process diagrams representing the basic structures of the project flow model and resources model are shown in Figure 16-6 and Figure 16-7, respectively. The flow diagrams use the convention that shaded variables originate in other models and thus capture interactions between models. Variables that are underlined and italicized are model input parameters. As an

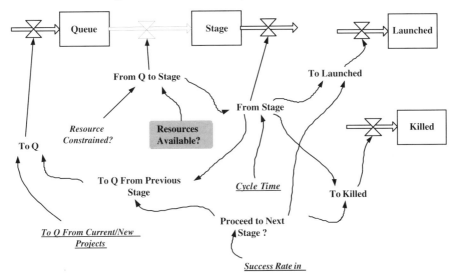

FIGURE 16-6. There are a number of variables that affect the flow of projects through a staged NPD process.

These factors are shown here. In a resource-constrained model, projects progress from the queue to the stage only when there are sufficient resources available.

example, consider the resource model that is shown in Figure 16-7. The box named "Stage" is shaded and represents the aggregate of the stages from the project flow model that track the number of projects in each stage of NPD. Furthermore, the variable "Resources Reqd. per Project/Stage" is input data read from an exogenous data file.

Figure 16-6 is a conceptual representation of the process flow model. The variables labeled "Stage," "Queue," "Launched," and "Killed" are used to track the location of projects in NPD. Specifically, projects enter the flow scheme at the queue to the first stage. For any given stage, a project will move from the queue into the stage when resources are available. Once a project has been completed in a given stage, it may be terminated, in which case it moves to the "Killed" variable. If a project is not terminated, then it either moves to the queue of the succeeding stage or it is launched if the current stage is the last stage. A project moves from a queue to a stage only when there are sufficient resources to support it. To determine if there are sufficient resources, the value of the variable "Resources Available?" is monitored. This variable is distinct for each resource type and project. Since this variable is shaded, it originates in another model. Since the variables "Cycle Time" and "Success Rate in Stage" are bold and italicized, they represent model input parameters. These parameters are uniquely specified for each project and each stage of NPD. They are used to specify the amount of time each project will spend in each stage and the probability that each project will be successful in each stage of NPD, respectively. At any point during the simulation, then, one knows the location of each project and thus its resource requirements.

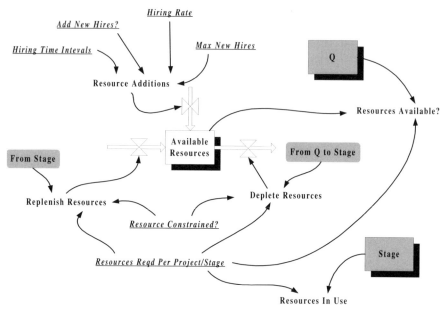

FIGURE 16-7. If the movement of projects is controlled by available resources, one must track the flow of resources.
Here we consider depletion and replenishment from project work as well as resources from new hiring.

The purpose of the process flow model is to simulate the movement of individual projects through the NPD pipeline to their ultimate destination. That is, each project will be tracked until it exits as a launched product or is terminated due to some unfavorable finding. The projects tracked could consist of those currently in the pipeline and those yet to be added. All projects will either exit as a launched product or be terminated due to some unfavorable finding. At each simulation time step, the location of each project is tracked. This information is needed to determine the resources required by projects or revenue generated by each project.

The resource model accounts for resource requirements and availability at all times during the simulation. Resources can be added to the available pool when work in a given stage has been completed and when new hires are brought on board. Note that we include only net hires—that is, we do not explicitly model resources that leave the pool due to non-project-related reasons. We can account for hiring freezes and maximum allowable hiring levels, and include explicit recruiting goals. Resources are depleted from the available pool when projects are allowed to pass from a queue to a development stage. They are returned to the available pool when projects are completed.

Note that one can consider instances of the model in which resources either limit or do not limit the flow of projects through the development process. Although resources always limit the flow of projects, one may want to determine the resource requirements in both the limiting and nonlimiting cases. When run in the resource-limiting mode, the resource model serves as a feedback mechanism, since common resources are required at various stages along the R&D process flow. It is this feedback that allows us to get a more realistic estimate of our throughput capabilities.

Regardless of whether one uses average project data or project-specific data, estimated resource requirements can be compared to the available resources and planned additions to quantify utilization levels over time. Significant differences in resource utilization from target levels indicate that interventions might be necessary. These could include increasing or reducing head count, reengineering key business processes, modifying the project portfolio by the level of in-licensing, terminating projects, or outsourcing selected activities of key functional resources.

In the following sections, we consider applications of both steady-state and simulation models.

NPD PLANNING MODELS

Steady-state Planning Model

The aim of the steady-state planning model is to determine the number of milestones required as well as the level of work activity and resources required to support a constant desired output from R&D. That is, the steady-state model could help one determine the R&D budget as well as the number of people needed to support two product launches per year, for example. The data required to drive the steady-state model consist of task-specific data (cycle time, success rate, and resource requirements for each resource) and global process data (number of new discovery projects required over a time horizon, and project mix(new products, co-development products, and product iterations).).

Calculation of the quantity $WIP_{S,T}(t)$ (defined previously) for the steady-state model is straightforward. Hopp and Spearman (1996) present Little's Law, which relates the throughput (TH) of a process to its work in process (WIP) and cycle time (CT). We use a generalization of this relationship to determine WIP at each stage of the R&D process. Specifically, the relationships used to calculate WIP at stage j are:

$$WIP_j = TH_j \times CT_j \tag{16-3}$$

$$TH_j = TH_{j-1} \times SR_{j-1} + \sum_{k=External\ Inputs} TH_{k,j} \tag{16-4}$$

In this context, we define the throughput of stage j, TH_j, to be the number of projects entering stage j for further development. Equation 16-4 shows that projects entering stage j may have originated either in an upstream stage of development or in another external input source such as product iterations or an in-licensing opportunity. Thus for the N-staged process shown in Figure 16-8, one can calculate WIP for all stages by setting the desired input rate from discovery, TH_1, and recursively applying Equations 16-3 and 16-4.

To demonstrate the use of the model just presented, consider a simple example based on the process shown in Figure 16-2. Suppose that we wish to have two product launches per year and that we want 50 percent of these launches to be new product introductions and the remaining 50 percent to be from product iterations. Note that this example will not include in-licensed projects. In this example, product iterations (line extensions) only participate in the early development and commercialization steps. Furthermore, the data for this process are shown in Table 16-2. Based on the total number of product launches and the fraction of launches from new products and product iteration, we calculate the number of launches from new products and product iterations. These values correspond to the number of projects exiting the commercialization stage. By applying Equation 16-4 (with no external inputs) to the commercialization stage, we can calculate the number of projects exiting the early development stage. Note that this value is identical to the number of projects entering the commercialization stage. By applying Equation 16-3, we can calculate the WIP (work in process) in the commercialization stage. Finally, we apply Equation 16-1 to calculate the required resources (cost) of work in this stage. Now we can apply the same approach to the early development and concept development stage. Using this model, we are able to determine the total number of projects required in the NPD pipeline and the total annual R&D budget. Furthermore, we can see how the number of projects and budget should be distributed by stage in the NPD process or how they should be distributed to new products vs. product iterations.

The steady-state planning model is a useful aggregate tool to assess the resource requirements under a range of conditions to achieve a given output from R&D. Furthermore, it can be easily implemented in a spreadsheet and its concepts are easily communicated. Also, one can easily assess the impact of uncertain parameters using standard Monte Carlo simulation add-on tools. The shortcomings of this model lie in its simplifying assumptions. In particular, the steady-state model assumes at all times that the R&D value chain is balanced with respect to work in progress and as a result does not predict fluctuations in requirements over time. In practice, however, R&D project portfolios are rarely balanced, since there is uncertainty whether or when a given project will move to the next stage or be terminated. At times more projects than expected will progress; at other times fewer than expected will progress. Furthermore, the steady-state model treats the flow of projects through the process as continuous when the reality is that partial projects do not move from one stage to another—a project either moves to the next stage or it is terminated. Additionally, many factors affecting resource requirements change over time and are difficult to effectively track. These factors include the inclusion of specific projects in the project set, head count on board, and the number and mix of projects entering the R&D process.

TABLE 16-2.
Example Algebraic Planning Model Using Equations 1, 3, and 4

Launch Distributions				
New Products	50%			
Line Extensions	50%			

Launches				
Total Launches	2			
New Product Launches	1			
Line Extension Launches	1			

	Concept Development	Early Development	Commercialization	Total
New Products				
Cycle Time	2	3	3	
Probability	50%	67%	75%	
Cost Per Project*	15	25	150	
Projects Exiting	6	4	1	
Projects Entering	12	6	1.3	
WIP	24	18	4	46
Cost**	360	450	600	1410
Line Extension				
Cycle Time	—	3	3	
Probability	—	80%	90%	
Cost Per Project*	—	10	50	
Projects Exiting	—	1.1	1	
Projects Entering	—	1.4	1.1	
WIP	—	4.2	3.3	7.5
Cost**	—	42	167	208
Totals				
Projects Exiting	6	5	2	
Projects Entering	12	7	2	
WIP	24	22	7	
Cost*	360	492	767	
Total WIP	54			
Total Cost*	1618			

*$million/project/yr
**$million/yr

KEYS FOR SUCCESS

The following ideas should be helpful as you begin using models to better manage your NPD process.

◆ Before building your models, get agreement from decision makers regarding how the model should be used. That is, will the model be used to set future performance targets, predict the resource requirements of projects, or determine which process improvements to pursue?

◆ While the model is being constructed, get input regarding model assumptions and data from those who will be impacted by the model's output. After the model has been constructed, make the details of the model available to anyone who is interested. This gives the model more credibility and minimizes second-guessing. The details should not be kept a secret.

◆ Make the model as accessible to others as possible. To help with this, use standard, easy-to-use tools if possible. Aside from helping with the credibility of the models, this minimizes the need to rely on an expert to evaluate model scenarios.

◆ Build models of your process at the highest level of abstraction possible. Add detail to your models only when necessary. Models with less detail often represent your process adequately, require less data, and are easier to build and communicate to others.

◆ Validate your model with historical data to give yourself and others confidence that the model adequately represents the key features of your process.

◆ Reconsider the assumptions and data in your model on a routine basis to account for changes in your business. The frequency with which you update your model will depend on how fast your process changes.

◆ Include variability in your models. Everyone knows that parameters such as cycle times and resource requirements vary. You can account for variability by using Monte Carlo analysis or by using a mix of project types.

◆ Explore the sensitivity of your model to the input parameters. Do your conclusions about your process change if key input parameters are changed only slightly?

Dynamic Simulation Planning Models

One of the most straightforward uses of process simulation models is to make the resource estimation methods of the steady-state model more realistic. The project flow model is initialized with information that includes a list of the current projects in the portfolio, the current location (stage of development) of each project, the estimated time each project will spend in each subsequent stage, and the likelihood that each project will progress through each subsequent stage of development. If we are not aware of the identity of projects in the future projects set, we make assumptions about the number of these projects that might enter the pipeline from various sources. Furthermore, we assume a range of values for the cycle times, success rates, and resource requirements for these projects. These estimated or assumed values can be based on performance objectives, historical data, industry benchmark data, or reasonable assessments of what is achievable.

Note that we represent the flow of projects through our process recursively. Consider the N-staged process shown in Figure 16-8, for example. When work is completed on a project at stage j, the project is either terminated (killed) or proceeds to stage $j+1$. If a project is successful after completing work in stage

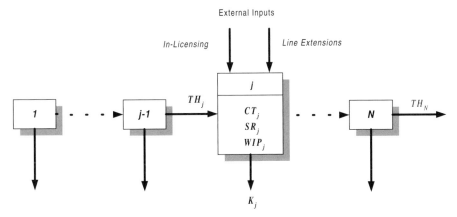

FIGURE 16-8. A general staged process model.

N, then it exits the process as a successfully launched product. Note that projects may enter the NPD process at any stage of development. In particular, product iterations of existing products and in-licensed projects only participate in a subset of the tasks in the entire NPD process flow. Furthermore, projects currently in the pipeline complete only the subset of steps remaining in the NPD process.

Bunch and Schacht (2001) provide a detailed explanation of how simulation models can be used to predict resource requirements associated with different initial pipeline distributions. They considered time-based resource requirements of various project pipelines. They showed how simulation models could be used to estimate resource requirements for various NPD pipelines. In their simulations, each pipeline was initialized with the same total number of projects. The distribution of projects within the stages of development varied, however. The time profiles of required resources under the various initial conditions were then tracked and compared.

To demonstrate these results, consider a staged process as shown in Figure 16-8 where N = 5. Consider pipelines with the same number of initial projects and initial distributions as shown in Figure 16-9. Case 1 represents a portfolio that is initially perfectly balanced. That is, the projects are distributed so that a steady output will be achieved. Cases 2 and 3 represent pipelines in which all projects begin in the first and last stages, respectively. Case 4 represents a pipeline in which projects are distributed throughout but are not properly balanced. The cycle times, success rates, and resource requirements vary by stage of development but are assumed to be the same for each project. Note the assumption that later stages of development consume more resources than earlier stages. The resource requirement profiles corresponding to these initial project distributions are shown in Figure 16-10. Note that since the current projects list for case 1 is perfectly balanced, it follows the steady-state line. After a transition period, the resource profiles of the other cases follow the steady-state requirements. This is a result of adding new projects in at a constant rate. These results

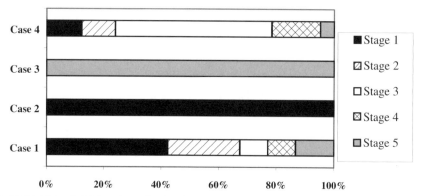

FIGURE 16-9. Initial project distributions for the same number of total projects.
*Case 1, steady state; case 2, all projects in stage 1; case 3, all projects in stage 5; case 4,
disproportionate number of projects in middle (stage 3).*

are intuitive. Since projects in earlier stages require less resources, we expect
the resource requirements for case 2 to start low and gradually increase as
projects progress to the later stages, where resource requirements per project
are much higher. The resource requirements for case 3 start high since all projects start at the last stage of development. As projects initially in the portfolio
exit the process, the pipeline is depleted of high-resource-consuming projects.
Therefore resource requirements drop off dramatically and gradually reach
steady-state as new projects make their way through the process. Case 4 is

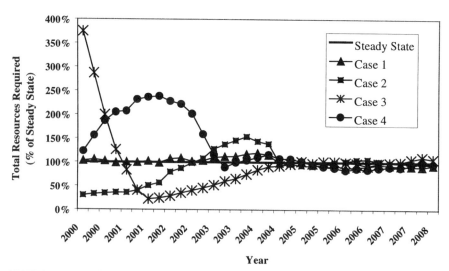

**FIGURE 16-10. Time-variant resource requirements to execute the same number of
total projects with different initial distributions of projects.**
*The amount of resources required to execute the same number of projects can vary dramatically and is
profoundly affected by the distribution of projects within the NPD process.*

essentially a hybrid of the previous cases. Note that resource requirements increase dramatically as the large number of projects initially in stage 3 progress to the later stages of the pipeline and decrease as these projects exit the system.

Simulation models can be used to estimate the resource requirements over time for a set of projects planned for execution. Using these models, one can compare available resources with required resources to predict time periods over which resource shortfalls might be expected. This might suggest management interventions to minimize the shortfalls. The interventions might include hiring additional resources, changing the set of projects to work on, or changing the timing of project execution. If management interventions are difficult to implement or take time to put in place, simulation models can be used to limit the flow of projects based on resource constraints to provide a more realistic estimate of pipeline performance.

PROJECT PRIORITIZATION AND PLANNING USING SIMULATION

Blau and colleagues (2000) have suggested a framework in which project selection and planning with simulation are combined. Using their approach, one develops a rank-ordered list of projects that are being considered for addition into the NPD pipeline. Next, one specifies scheduling rules so that projects in the priority list can be moved through the NPD process (in rank order) as fast as possible. Once the priority list and scheduling rules are developed, the process simulation model is run many times to predict how the portfolio will perform over time. The performance is measured by net present value (NPV) and required resources distributions. The objective is to maximize NPV while utilizing resources in a reasonable way. One can easily change the priority list and scheduling rules to find project prioritization and scheduling rules that improve pipeline performance. This method will be reviewed in the following paragraphs and an example will be provided.

Consider a set of projects under consideration in an NPD portfolio. We assume that reasonable point estimates and possibly ranges of important properties are available for each project. These properties include both development and market information. Development information includes required resources (capital and human), probability of technical success, and completion date for each stage of development. Market information includes mature sales as well as selling and marketing costs.

Using this information, one can use a process simulation model to track the movement of each candidate project individually over time. The predicted estimates of required resources and revenue generated will vary from run to run for two reasons. First, the input parameters can be specified by distributions rather than by point estimates. Second, the final destination of the project will change from run to run since there is a nonzero chance that each project will be terminated at each step of the NPD process. By running these models many times, distributions of important financial parameters—including total resources consumed, total revenue generated, and NPV—can be determined for each project. These parameters are then used along with other criteria to develop the rank-ordered list of projects.

As a result of using simulation to estimate financial parameters, one finds that the distribution of NPV for individual projects is bimodal. The first distribution is associated with simulation runs in which the project is discontinued prior to launch. We refer to the mean value of this distribution as the *risk* (Blau and Sinclair 2000) of the project. The risk will always be negative. The second distribution is associated with simulation runs in which the project is launched on the market. We refer to the mean value of this distribution as the *reward* of the project. The reward will be positive in most cases. An example of the NPV distributions generated using simulations for a specific project is shown in Figure 16-11. It is obvious from this figure that it is futile to use a single measure for the NPV of this development project.

Project prioritization has received a significant amount of attention in the literature. Cooper, Edgett, and Kleinschmidt (1997a, 1997b) suggest many of the common methods in use. We simply note here that most methods consider multiple project attributes. The major attributes include financial returns, resource requirements, time to market, probability of success, strategic fit, and portfolio balance, among other things. Cooper, Edgett, and Kleinschmidt (1997a) suggest several methods used in portfolio management and project selection, including financial methods, scoring models, and bubble diagrams or portfolio maps.

Financial methods often make use of a range of financial metrics, including NPV, return on investment (ROI), or payback period. Using scoring models, each project is given a rating on each of several dimensions. Next, the scores are multiplied by weightings and then summed to yield an overall project score. This score is then used to rank the projects. Using bubble diagrams, one plots projects on an XY scatter plot in which one axis represents the financial reward of the projects and the other axis represents the success probability of the projects. The size of the symbol representing each project depends on some other

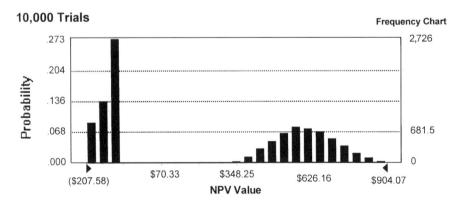

FIGURE 16-11. Bimodal NPV distribution for a development project.

measure of the project such as resource consumption. Additionally, the symbols can be of different shapes, colors, and textures to represent a range of other properties, including product classification, whether the project is new or a product iteration, and so on. The large amount of information that can be communicated using bubble diagrams makes them an extremely powerful tool that can concisely represent and communicate the various dimensions of NPD projects; they are quite popular. Cooper, Edgett, and Kleinschmidt (1997a, 1997b)also note that some firms allocate resources to development efforts based on the firm's strategy. Using this approach, resources are allocated across different types of projects and into different buckets. Projects are then ranked within the different buckets.

After using one of the methods above, one will have arrived at a rank-ordered list or priority list of the NPD projects under consideration. Once the rank-ordered list is developed, the goal is to determine how fast this set of projects can be moved through the NPD process. Since we assume that there will not be adequate resources to simultaneously pursue all projects with promising returns, the issue of resource constraints must be considered. That is, we must determine a project plan that specifies the order and timing of project tasks so that the resources required to execute the plan do not exceed the available resources. This problem is a resource-constrained planning problem under uncertainty. A significant amount of research has been done to understand and solve this problem, but this work will not be covered here. Interested readers can find out more by consulting Subramanian et al. 2000; Honkomp 1998; Schmidt and Grossmann 1996; and Jain and Grossmann 1999.

In practice, planning issues involved in NPD pipeline management can be addressed with a set decision-making method upon the occurrence of events that may be thought of as trigger events. Decisions could consist of starting a new project or delaying the start of a new project. Trigger events are events that would change the resource requirements for the portfolio. Examples include advancement of a project from one stage of development to the next, termination of a project, or launch of a project. Thus the trigger event/decision pair might be the decision to introduce a new project upon termination of an existing project. It could also be delaying the start of a project until a project is launched.

Consider the following example. We have a set of projects for addition into the NPD process. There are clearly insufficient resources to execute all projects at once. Accordingly, we develop a rank-ordered list of the projects and select the top projects until there are no longer resources available. The decision rule applied is that new projects are added in order from the rank-ordered list as projects currently in the pipeline exit due to termination or launch. Financial performance and resource consumption are monitored as the movement of projects is simulated through the NPD system. The performance of our project ranking system and planning system will be evaluated based on realized NPV and fraction of time that resources were overutilized. Both of these measures provide valuable information. One might be willing to specify a project advancement scenario that overutilizes resources slightly

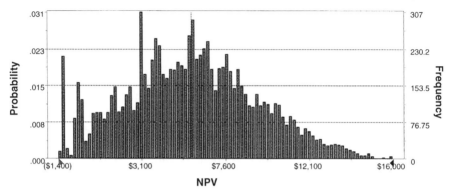

FIGURE 16-12. Simulated NPV distribution for the entire portfolio.

more to dramatically increase NPV, for instance. Ultimately the advancement scenario implemented depends on management's level of comfort with over-committing resources to achieve a financial objective. One might be willing to suggest a plan that would exceed available resources 10 percent of the time if the rewards were significant enough. It is hard to imagine, however, any cir-cumstances under which one would suggest a plan that exceeded available resources 50 percent of the time.

Examples of the NPV distribution and resource requirements distribu-tion are shown in Figure 16-12 and Figure 16-13, respectively. The NPV of the projects with the ranking system chosen and the decision rules selected is $5.4 million. Note in Figure 16-13 that the required resources exceed the available resources 5 percent of the time. Alternative project ranking and advancement scenarios can be evaluated so that the maximum NPV can be achieved while maintaining an acceptable (70-80 percent) level of resource utilization.

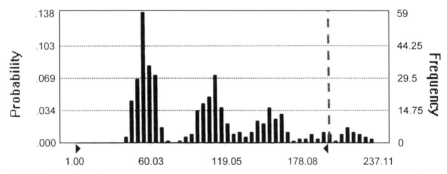

FIGURE 16-13. Distribution of resource requirements for a set of projects and specified project advancement decision rules.
Since available resources are 200, this plan overutilizes in 5 percent of the cases.

SUMMARY

In this chapter, we have provided a framework for considering NPD as a process. Given this framework, simple algebraic models were presented that could be used in R&D planning. Specifically, one model can be used to predict the level of work activity and resources required to achieve a target output level from R&D. The second algebraic model can be used to predict the probabilized resource requirements of a set of projects over a time horizon. These models can be easily implemented in a spreadsheet. Furthermore, if there is uncertainty in the data, Monte Carlo simulations can be applied to determine a range of outputs.

Process simulation was discussed as a way of tracking the movement of individual projects through the NPD process. This method can be used to calculate a range of measures of portfolio performance over time including projects by stage and resource requirements. It was also shown how process simulation could be combined with project selection to predict the performance of a planned set of projects. This method can be used to generate the bimodal NPV distributions for a given project, which can be used as part of the project selection process. Additionally, the method can be used to determine the likelihood that the projects selected will overutilize resources as the projects are executed.

NOTES

1. Monte Carlo simulation is a useful technique to describe the impact of variability in the inputs of a model on the outputs. The technique allows one to estimate the entire range of outputs and the likelihood of achieving each of them. For more detail on this technique, refer to an introductory text on statistical analysis.

BIBLIOGRAPHY

Blau, G. E. 1997. "A Systems Engineering Approach to New Product Development." *CAST Communications* 21, 2.

Blau, G. E., B. Mehta, S. Bose, J. Pekny, G. Sinclair, K. Keunker, and P. R. Bunch. 2000. "Risk Management in the Development of New Products in Highly Regulated Industries." *Computers and Chemical Engineering* 24: 659–64.

Blau, G. E., and K. Rajan. 2000. "Impact of Project Dependencies on New Product Development." AICHE Proceeding, Los Angeles, November.

Blau, G. E., and G. W. Sinclair. 2001. "Dealing with Uncertainty in New Product Development." *Chemical Engineering Progress,* 21, 2.

Bose, S., and G. E. Blau. 2000. "Use of a Network Model Interface to Build Spreadsheet Models of Process Systems: A Productivity Enhancement Tool for Risk Management Studies." Seventh International Symposium on Process Systems Engineering, July, Keystone, CO.

Bunch, P., and A. Schacht. 2002. "Modeling Resource Requirements for Pharmaceutical R&D." *Research Technology Management.* 45, 1: 48–56.

Cooper, R. G., S. J. Edgett, and E. J. Kleinschmidt. 1997a. "Portfolio Management in New Product Development: Lessons from the Leaders—I." *Research Technology Management* 40, 5: 16–28.

Cooper, R. G., S. J. Edgett, and E. J. Kleinschmidt. 1997b. "R&D Portfolio Management Best Practices Study." Michael G. DeGroote School of Business, McMaster University.

Drews, J. 1998. "Innovation Deficit Revisited: Reflections on the Productivity of Pharmaceutical R&D." *Drug Discovery Today* 3, 11: 491–94.

Ford, D. N., and J. D. Sterman. 1998. "Dynamic Modeling of Product Development Processes." *System Dynamic Review* 14, 1: 31–68.

Honkomp, S. J. 1998. "Solving Mathematical Programming Planning Models Subject to Stochastic Task Success." Ph.D. dissertation, Purdue University.

Hopp, W. J., and M. L. Spearman. 1996. *Factory Physics.* Chicago: Irwin.

Jain, V., and I. E. Grossmann. 1999. "Resource-Constrained Scheduling of Tests in New Product Development." *Industrial & Engineering Chemistry Research* 38, 8: 3013–26.

Pisano, G. P. 1996. *The Development Factory.* Boston: Harvard Business School Press.

Pritsker, A. A. B. 1986. *Introduction to Simulation and SLAM II.* New York: Halsted Press.

Rosenau, M. D. 1996. "Choosing A Development Process That's Right For Your Company." *PDMA Handbook of New Product Development.* New York: John Wiley and Sons, 77–92.

Schmidt, C. W., and I. E. Grossmann. 1996. "Optimization Models for the Scheduling of Testing Tasks in New Product Development." *Industrial & Engineering Chemistry Research* 35, 10: 3498–510.

Sterman, J. D. 2000. *Business Dynamics: Systems Thinking and Modeling for a Complex World.* Boston: Irwin McGraw-Hill.

Subramanian, D., J. F. Pekny, and G. V. Reklaitis. 2000. "A Simulation-Optimization Framework for Addressing Combinatorial and Stochastic Aspects of an R&D Pipeline Management Problem." Seventh International Symposium on Process Systems Engineering, July, Keystone, CO.

Wheelwright, S. C., and K. B. Clark. 1992. *Revolutionizing Product Development.* New York: The Free Press.

The PDMA Glossary For New Product Development

Accidental Discovery: New designs, ideas, and developments different from those originally hoped for from research.

Adoption Curve: The phases through which consumers or a market proceed in deciding to adopt a new product or technology. At the individual level, each consumer must move from a cognitive state (becoming aware of and knowledgeable about) to an emotional state (liking and then preferring the product) and into a conative, or behavioral, state (deciding and then purchasing the product). At the market level, the new product is first purchased by the innovators in the marketplace, which are generally thought to constitute about 2.5 percent of the market. Early adopters (13.5 percent of the market) are the next to purchase, followed by the early majority (34 percent), late majority (34 percent), and finally the laggards (16 percent).

Affinity Charting: A "bottom-up" technique for discovering connections between pieces of data. An individual or group starts with one piece of data (say, a customer need). They then look through the rest of the data they have (say, statements of other customer needs) to find other data (needs) similar to the first, and place it in the same group. As they come across pieces of data that differ from those in the first group, they create a new category. The end result is a set of groups where the data contained within a category are similar, and the groups all differ in some way. *See also* Qualitative Cluster Analysis.

Alliance: Formal arrangement with a separate company for purposes of development, and involving exchange of information, hardware, intellectual property, or enabling technology. Alliances involve shared risk and reward (e.g., co-development projects).

Alpha Test: In-house testing of pre-production products to find and eliminate the most obvious design defects or deficiencies, either in a laboratory setting or in some part of the developing firm's regular operations. *See also* Beta Test and Gamma Test.

Analytical Hierarchy Process (AHP): A decision-making tool for complex, multi-criteria problems where both qualitative and quantitative aspects of a problem need to be incorporated. AHP clusters decision elements according to their common characteristics into a hierarchical structure similar to a family tree or affinity chart. The AHP process was designed by T. L. Saaty.

Analyzer: A firm that follows an imitative innovation strategy, where the goal is to get to market with an equivalent or slightly better product very quickly once someone else opens up the market, rather than to be first to market with new products or technologies.

Anticipatory Failure Determination (AFD): A failure analysis method. In this process, developers start from a particular failure of interest as the intended consequence and try to devise ways to assure that the failure always happens reliably. Then the developers use that information to develop ways to better identify steps to avoid the failure.

Applications Development: The iterative process through which software is designed and written to meet the needs and requirements of the user base or the process of enhancing or developing new products.

Architecture: *See* Product Architecture.

Asynchronous Groupware: Software used to help people work as groups, but not requiring those people to work at the same time.

Audit: When applied to new product development, an audit is an appraisal of the effectiveness of the processes by which the new product was developed and brought to market (*see* Chapter 14 of *The PDMA ToolBook*).

Augmented Product: The core product, plus all other sources of product benefits, such as service, warranty, and image.

Autonomous Team: A completely self-sufficient project team with very little, if any, link to the funding organization. Frequently used as an organizational model to bring a radical innovation to the marketplace. Sometimes called a *"tiger"* team.

Awareness: A measure of the percent of target customers who are aware that the new product exists. Awareness is variously defined, including recall of brand, recognition of brand, recall of key features, or positioning.

Backup: A project that moves forward, either in synchrony or with a moderate time lag, and for the same marketplace, as the lead project to provide an alternative asset should the lead project fail in development. A backup has essentially the same mechanism of action performance as the lead project. Normally a company would not advance both the lead and the backup project through to the market place, since they would compete directly with each other.

Balanced Scorecard: A comprehensive performance measurement technique that balances four performance dimensions: (1) customer perceptions of how we are performing; (2) internal perceptions of how we are doing at what we must excel at; (3) innovation and learning performance; (4) financial performance.

Benchmarking: A process of collecting process performance data, generally in a confidential, blinded fashion, from a number of organizations to allow them to assess their performance individually and as a whole.

Benefit: A product attribute expressed in terms of what the user gets from the product rather than its physical characteristics or features. Benefits are often paired with specific features, but they need not be.

Best Practices: Methods, tools, or techniques that are associated with improved performance. In new product development, no one tool or technique assures success; however, a number of them are associated with higher probabilities of achieving success. Best practices likely are at least somewhat context specific.

Best Practices Study: A process of studying successful organizations and selecting the best of their actions or processes for emulation. In new product development it means finding the best process practices, adapting them, and adopting them for internal use. (*See* Chapter 33 in *The PDMA HandBook;* Griffin, "PDMA Research on New Product Development Practices: Updating Trends and Benchmarking Best Practices," *JPIM*, 14: 6, 429–458, November 1997; and "Drivers of NPD Success: The 1997 PDMA Report," PDMA, October 1997.)

Beta Test: An external test of pre-production products. The purpose is to test the product for all functions in a breadth of field situations to find those system faults that are more likely to show in actual use than in the firm's more controlled in-house tests before sale to the general market. *See also* Field Test.

Brainstorming: A group method of creative problem solving frequently used in product concept generation. There are many modifications in format, each variation with its own name. The basis of all of these methods uses a group of people to creatively generate a list of ideas related to a particular topic. As many ideas as possible are listed before any critical evaluation is performed (*see* Chapters 12 and 13 in *The PDMA HandBook*).

Brand: A name, term, design, symbol, or any other feature that identifies one seller's good or service as distinct from those of other sellers. The legal term for brand is *trademark*. A brand may identify one item, a family of items, or all items of that seller.

Brand Development Index (BDI): A measure of the relative strength of a brand's sales in a geographic area. Computationally, BDI is the percent of total national brand sales that occur in an area divided by the percent of U.S. households that reside in that area.

Breadboard: A proof-of-concept modeling technique that represents how a product will work, but not how a product will look.

Breakeven: The interval between the inception of a new product project and the date at which it reaches positive cumulative cash flow.

Business Analysis: An analysis of the business situation surrounding a proposed project. Usually includes financial forecasts in terms of discounted cash flows, net present values, or internal rates of returns.

Business Case: The results of the market, technical and financial analyses, or up-front homework. Ideally defined just prior to the "go to development" decision (gate), the case defines the product and project, including the project justification and the action or business plan.

Business-to-Business: Non-consumer purchasers such as manufacturers, resellers (distributors, wholesalers, jobbers, and retailers, for example), institutional, professional, and governmental organizations. Frequently referred to as "industrial" businesses in the past.

Buyer: The purchaser of a product, whether or not he or she will be the ultimate user. Especially in business-to-business markets, a purchasing agent may contract for the actual purchase of a good or service, yet never benefit from the function(s) purchased.

Buyer Concentration: The degree to which purchasing power is held by a relatively small percentage of the total number of buyers in the market.

Cannibalization: That portion of the demand for a new product that comes from the erosion of the demand for (sales of) a current product the firm markets.

Capacity Planning: A forward-looking activity that monitors the skill sets and effective resource capacity of the organization. For product development, the objective is to manage the flow of projects through development such that none of the functions (skill sets) creates a bottleneck to timely completion. Necessary in optimizing the project portfolio.

Centers of Excellence: A geographic or organizational group with an acknowledged technical, business, or competitive competency.

Certification: A process for formally acknowledging that someone has mastered a body of knowledge on a subject. In new product development, the PDMA has created and manages a certification process to become a New Product Development Professional (NPDP).

Champion: A person who takes a passionate interest in seeing that a particular process or product is fully developed and marketed. This informal role varies from situations calling for little more than stimulating awareness of the opportunity to extreme cases where the champion tries to force a project past the strongly entrenched internal resistance of company policy or that of objecting parties (*see* Chapter 5 in *The PDMA ToolBook*).

Charter: A project team document defining the context, specific details, and plans of a project. It includes the initial business case, problem and goal statements, constraints and assumptions, and preliminary plan and scope. Periodic reviews with the sponsor ensure alignment with business strategies (*see also* Product Innovation Charter).

Checklist: A list of items used to remind an analyst to think of all relevant aspects. It finds frequent use as a tool of creativity in concept generation, as a factor consideration list in concept screening, and to ensure that all appropriate tasks have been completed in any stage of the product development process.

Chunks: The building blocks of product architecture. They are made up of inseparable physical elements. Other terms for chunks may be modules or major subassemblies.

Clockspeed: Industries with a very rapid evolutionary rate.

Cognitive Modeling: A method for producing a computational model for how individuals solve problems and perform tasks, which is based on psychological principles. The modeling process outlines the steps a person goes through in solving a particular problem or completing a task, which allows one to predict the time it will take or the types of errors an individual may make. Cognitive models frequently are used to determine ways to improve a user interface to minimize interaction errors or time by anticipating user behavior.

Cognitive Walkthrough: Once a model of the steps or tasks a person must go through to complete a task is constructed, an expert can role play the part of a user to cognitively "walk through" the user's expected experience. Results from this walkthrough can help make human-product interfaces more intuitive and increase product usability.

Collaborative Product Development: When two firms (frequently one large corporation and one smaller entrepreneurial firm) work together to develop and commercialize a specialized product. The smaller firm may contribute technical or creative expertise, while the larger firm may be more likely to contribute capital, marketing, and distribution capabilities. When two firms of more equal size collaborate, they may each bring some specialized technology capability to the table in developing some highly complex product or system requiring expertise in both technologies.

Co-location: Physically locating project personnel in one area, enabling more rapid and frequent decision making and communication among them.

Commercialization: The process of taking a new product from development to market. It generally includes production launch and ramp-up, marketing materials and program development, supply chain development, sales channel development, training development, training, and service and support development.

Competitive Intelligence: Methods and activities for transforming disaggregated public competitor information into relevant and strategic knowledge about competitors' positions, sizes, efforts, and trends. The term refers to the broad practice of collecting, analyzing, and communicating the best available information on competitive trends occurring outside one's own company.

Computer-Aided Design: A technology that allows designers and engineers to use computers for their design work. Early programs enabled two-dimensional (2-D) design. Current programs allow designers to work in 3-D (three dimensions), and in either wire or solid models.

Computer-Aided Engineering: Using computers in designing, analyzing and manufacturing a product or process. Sometimes refers more narrowly to using computers just at the engineering analysis stage.

Computer-Enhanced Creativity: Using specially designed computer software that aids in the process of recording, recalling, and reconstructing ideas to speed up the new product development process.

Concept: A clearly written and possibly visual description of the new product idea that includes its primary features and consumer benefits, combined with a broad understanding of the technology needed.

Concept Generation: The processes by which new concepts, or product ideas, are generated. Sometimes also called *idea generation* or *ideation* (*see* Chapters 12 and 13 in *The PDMA HandBook*).

Concept Optimization: A research approach that evaluates how specific product benefits or features contribute to a concept's overall appeal to consumers. Results are used to select from the options investigated to construct the most appealing concept from the consumer's perspective.

Concept Statement: A verbal or pictorial statement of a concept that is prepared for presentation to consumers to get their reaction prior to development.

Concept Study Activity: The set of product development tasks in which a concept is given enough examination to determine if there are substantial unknowns about the market, technology, or production process.

Concept Testing: The process by which a concept statement is presented to consumers for their reactions. These reactions can either be used to permit the developer to estimate the sales value of the concept or to make changes to the concept to enhance its potential sales value (*see* Chapters 14 and 15 in *The PDMA HandBook*).

Concurrency: Carrying out separate activities of the product development process at the same time rather than sequentially.

Concurrent Engineering (CE): When product design and manufacturing process development occur concurrently in an integrated fashion, using a cross-functional team, rather than sequentially by separate functions. CE is intended to cause the development team to consider all elements of the product life cycle from conception through disposal, including quality, cost, and maintenance, from the project's outset. Also called *simultaneous engineering* (*see* Chapter 30 of *The PDMA HandBook).*

Conjoint Analysis: A quantitative market research technique that determines how consumers make trade-offs between a small number of different features or benefits.

Consumer: The most generic and all-encompassing term for a firm's targets. The term is used in either the business-to-business or household context and may refer to the firm's current customers, competitors' customers, or current non-purchasers with similar needs or demographic characteristics. The term does not differentiate between whether the person is a buyer or a user target. Only a fraction of consumers will become customers.

Consumer Market: The purchasing of goods and services by individuals and for household use (rather than for use in business settings). Consumer purchases generally are made by individual decision makers, either for themselves or others in the family.

Consumer Need: A problem the consumer would like to have solved. What consumers would like a product to do for them.

Consumer Panels: Specially recruited groups of consumers whose longitudinal category purchases are recorded via the scanner systems at stores.

Contextual Inquiry: A structured qualitative market research method that uses a combination of techniques from anthropology and journalism. Contextual inquiry is a customer needs discovery process that observes and interviews users of products in their actual environment.

Contingency Plan: A plan to cope with events whose occurrence, timing, and severity cannot be predicted.

Continuous Improvement: The review, analysis, and rework directed at incrementally improving practices and processes. Also called *Kaizen*.

Continuous Innovation: A product alteration that allows improved performance and benefits without changing either consumption patterns or behavior. The product's general appearance and basic performance do not functionally change. Examples include fluoride toothpaste and higher computer speeds.

Continuous Learning Activity: The set of activities involving an objective examination of how a product development project is progressing or how it was carried out to permit process changes to simplify its remaining steps or improve the product being developed or its schedule (*see also* Learning Organization).

Contract Developer: An external provider of product development services.

Controlled Store Testing: A method of test marketing where specialized companies are employed to handle product distribution and auditing rather than using the company's normal sales force.

Convergent Thinking: A technique generally performed late in the initial phase of idea generation to help funnel the high volume of ideas created through divergent thinking into a small group or single idea on which more effort and analysis will be focused.

Cooperation (Team Cooperation): The extent to which team members actively work together in reaching team level objectives.

Coordination Matrix: A summary chart that identifies the key stages of a development project, their goals, and key activities within each stage, and who (what function) is responsible for each.

Core Benefit Proposition (CBP): The central benefit or purpose for which a consumer buys a product. The CBP may come either from the physical good or service, or it may come from augmented dimensions of the product (*see also* Value Proposition). (*See* Chapter 3 of *The PDMA ToolBook.*)

Core Competence: That capability at which a company does better than other firms, which provides them with a distinctive competitive advantage and contributes to acquiring and retaining customers. Something that a firm does better than other firms. The purist definition adds "and is also the lowest cost provider."

Corporate Culture: The "feel" of an organization. Culture arises from the belief system through which an organization operates. Corporate cultures are variously described as being authoritative, bureaucratic, and entrepreneurial. The firm's culture frequently impacts which ways to get things are appropriate for the organization.

Cost of Goods Sold (COGS or CGS): The direct costs (labor and materials) associated with producing a product and delivering it to the marketplace.

Creativity: "An arbitrary harmony, an expected astonishment, a habitual revelation, a familiar surprise, a generous selfishness, an unexpected certainty, a formable stubbornness, a vital triviality, a disciplined freedom, an intoxicating steadiness, a repeated initiation, a difficult delight, a predictable gamble, an ephemeral solidity, a unifying difference, a demanding satisfier, a miraculous expectation, and accustomed amazement" (George M. Prince, *The Practice of Creativity*, 1970). Creativity is the ability to produce work that is both novel and appropriate.

Criteria: Statements of standards used by decision makers at each gate. The dimensions of performance necessary to achieve or surpass for product development projects to continue in development. In the aggregate, these criteria reflect a business unit's new product strategy.

Critical Assumption: An explicit or implicit assumption in the new product business case that, if wrong, could undermine the viability of the opportunity.

Critical Path: The set of interrelated activities that must be completed for the project to be finished successfully can be mapped into a chart showing how long each task takes, and which tasks cannot be started before which other tasks are completed. The critical path is the set of linkages through the chart that is the longest. It determines how long a project will take.

Critical Path Scheduling: A project management technique, frequently incorporated into various software programs, that puts all important steps of a given new product project into a sequential network based on task interdependencies.

Critical Success Factors: Those critical few factors that are necessary for, but don't guarantee, commercial success.

Cross-Functional Team: A team consisting of representatives from the various functions involved in product development, usually including members from marketing, engineering, manufacturing, finance, purchasing, and quality. The team is empowered by the departments to represent each function's perspective in the development process (*see* Chapter 9 in *The PDMA HandBook* and Chapter 6 in *The PDMA ToolBook*).

Crossing the Chasm: Making the transition to a mainstream market from an early market dominated by a few visionary customers.

Customer: One who purchases or uses your firm's products or services.

Customer-based Success: The extent to which a new product is accepted by customers and the trade.

Customer Needs: Problems to be solved. These provide new product development opportunities for the firm (*see* Chapter 11 in *The PDMA HandBook*).

Customer Perceived Value (CPV): The result of the customer's evaluation of all the benefits and all the costs of an offering as compared to that customer's perceived alternative. It is the basis on which customers decide to buy things (*see* Chapter 4 of *The PDMA ToolBook*).

Customer Site Visits: A qualitative market research technique for uncovering customer needs. The method involves going to a customer's work site, watching as he or she performs functions associated with the customer needs your firm wants to solve, and then debriefing the customer about what he or she did, why he or she did those things, the problems encountered as he or she was trying to perform the function, and what worked well (*see* Chapter 11 of *The PDMA HandBook*).

Customer Value Added Ratio: The ratio of WWPF (worth what paid for) for your products to WWPF for your competitors' products. A ratio above 100 percent indicates superior value compared to your competitors.

Cycle Time: The length of time for any operation, from start to completion. In the new product development sense, it is the length of time to develop a new product from an early initial idea for a new product to initial market sales. Precise definitions of the start and end point vary from one company to another, and may vary from one project to another within the company.

Dashboard: Colored graphical presentation of a project's status resembling a vehicle's dashboard.

Database: An electronic gathering of information organized in some way to make it easy to search, discover, analyze, and manipulate.

Decision Screens: Sets of criteria that are applied as checklists or screens at new product decision points. The criteria may vary by stage in the process (*see* Chapter 7 in *The PDMA ToolBook*).

Decision Tree: A diagram used for making decisions in business or computer programming. The "branches" of the tree diagram represent choices with associated risks, costs, results, and outcome probabilities. By calculating outcomes (profits) for each of the branches, the best decision for the firm can be determined.

Decline Stage: The fourth and last stage of the product life cycle. Entry into this stage is generally caused by technology advancements, consumer or user preference changes, global competition, or environmental or regulatory changes.

Defenders: Firms that stake out a product turf and protect it by whatever means, not necessarily through developing new products.

Deliverable: The output (such as test reports, regulatory approvals, working prototypes, or marketing research reports) that shows a project has achieved a result. Deliverables may be specified for the commercial launch of the product or at the end of a development stage.

Delphi Processes: A technique that uses iterative rounds of consensus development across a group of experts to arrive at a forecast of the most probable outcome for some future state.

Demographic: The statistical description of a human population. Characteristics included in the description may include gender, age, education level, and marital status, as well as various behavioral and psychological characteristics.

Derivative Product: A new product based on changes to an existing product that modifies, refines, or improves some product features without affecting the basic product architecture or platform.

Design for the Environment (DFE): The systematic consideration of environmental safety and health issues over the product's projected life cycle in the design and development process.

Design for Excellence (DFX): The systematic consideration of *all* relevant life cycle factors, such as manufacturability, reliability, maintainability, affordability, testability, and so forth, in the design and development process.

Design of Experiments (DOE): A statistical method for evaluating multiple product and process design parameters simultaneously rather than one parameter at a time.

Design for Maintainability (DFMt): The systematic consideration of maintainability issues over the product's projected life cycle in the design and development process.

Design for Manufacturability (DFM): The systematic consideration of manufacturing issues in the design and development process, facilitating the fabrication of the product's components and their assembly into the overall product.

Design to Cost: A development methodology that treats costs as an independent design parameter, rather than an outcome. Cost objectives are established based on customer affordability and competitive constraints.

Design Validation: Product tests to ensure that the product or service conforms to defined user needs and requirements. These may be performed on working prototypes or using computer simulations of the finished product.

Development: The functional part of the organization responsible for converting product requirements into a working product.

Development Change Order (DCO): A document used to implement changes during product development. It spells out the desired change, the reason for the change, and the consequences to time to market, development cost, and the cost of producing the final product. It gets attached to the project's charter as an addendum.

Digital Mock-Up: An electronic model of the product created with a solids modeling program. Mockups can be used to check for interface interferences and component incompatibilities. Using a digital mock-up can be less expensive than building physical prototypes.

Discontinuous Innovation: Previously unknown products that establish new consumption patterns and behavior changes. Examples include microwave ovens and the automobile.

Discounted Cash-Flow Analysis: One method for providing an estimate of the current value of future incomes and expenses projected for a project. Future cash flows for a number of years are estimated for the project, and then discounted back to the present using forecast interest rates.

Discrete Choice Experiment: A quantitative market research tool used to model and predict customer buying decisions.

Dispersed Teams: Product development teams that have members working at different locations, perhaps even in different countries.

Distribution: The method and partners used to get the product (or service) from where it is produced to where the end user can buy it.

Divergent Thinking: Technique performed early in the initial phase of idea generation that expands thinking processes to generate, record, and recall a high volume of new or interesting ideas.

Dynamically Continuous Innovation: A new product that changes behavior, but not necessarily consumption patterns. Examples include Palm Pilots, electric toothbrushes, and electric hair curlers.

Early Adopters: For new products, these are customers who, relying on their own intuition and vision, buy into new product concepts very early in the life cycle. For new processes, these are organizational entities that were willing to try out new processes rather than just maintaining the old.

Economic Value Added (EVA): The value added to or subtracted from shareholder value during the life of a project.

Empathic Design: A five-step method for uncovering customer needs and sparking ideas for new concepts. The method involves going to a customer's work site, watching as he or she performs functions associated with the customer needs your firm wants to solve, and then debriefing the customer about what he or she did, why he or she did those things, the problems encountered as he or she was trying to perform the function, and what worked well. By spending time with customers, the developer develops empathy for the problems customers encounter trying to perform their daily tasks. *See also* Customer Site Visits.

Engineering Design: A function in the product creation process where a good or service is configured and specific form is decided.

Engineering Model: The combination of hardware and software intended to demonstrate the simulated functioning of the intended product as currently designed.

Enhanced New Product: A form of derivative product. Enhanced products include additional features not previously found on the base platform, which provide increased value to consumers.

Entrance Requirement: The document(s) and reviews required before any phase of the development process can be started.

Entrepreneur: A person who initiates, organizes, operates, assumes the risk, and reaps the potential reward for a new business venture.

Ethnography: A descriptive, qualitative market research methodology for studying the customer in relation to his or her environment. Researchers spend time in the field observing customers and their environment to acquire a deep understanding of customers' lifestyles or cultures as a basis for better understanding their needs and problems.

Event: Marks the point in time when a task is completed.

Event Map: A chart showing important events in the future that is used to map out potential responses to probable or certain future events.

Excursion: An idea generation technique to force discontinuities into the idea set. Excursions consist of three generic steps: (1) step away from the task; (2) generate disconnected or irrelevant material; (3) force a connection back to the task.

Exit Requirement: The document(s) and reviews required to complete a stage of the development process.

Exit Strategy: A pre-planned process for deleting a product or product line from the firm's portfolio. At a minimum it includes plans for clearing inventory out of the supply chain pipeline at a minimum of losses, continuing to provide for after-sales parts supply and maintenance support, and converting customers of the deleted product line to a different one.

Explicit Customer Requirement: What the customer asks for in a product.

Factory Cost: The cost of producing the product in the production location, including materials, labor, and overhead.

Failure Mode Effects Analysis: A technique used at the development stage to determine the different ways in which a product may fail, and evaluating the consequences of each type of failure.

Failure Rate: The percentage of a firm's new products that make it to full market commercialization but that fail to achieve the objectives set for them.

Feasibility Determination: The set of product development tasks in which major unknowns (technical or market) are examined to produce knowledge about how to resolve or overcome them or to clarify the nature of any limitations. Sometimes called *exploratory investigation.*

Feature: The solution to a consumer need or problem. Features provide benefits to consumers. The handle (feature) allows a laptop computer to be carried easily (benefit). Usually any one of several different features can be chosen to meet a customer need. For example, a carrying case with shoulder straps is another feature that allows a laptop computer to be carried easily.

Feature Creep: The tendency for designers or engineers to add more capability, functions, and features to a product as development proceeds than were originally intended. These additions frequently cause schedule slip, development cost increases, and product cost increases.

Field Testing: Product use testing with users from the target market in the actual context in which the product will be used.

Financial Success: The extent to which a new product meets its profit, margin, and return-on-investment goals.

Firefighting: An unplanned diversion of scarce resources, and the reassignment of some of them to fix problems discovered late in a product's development cycle (*see* Repenning, *JPIM*, September 2001).

Firm-Level Success: The aggregate impact of the firm's proficiency at developing and commercializing new products. Several different specific measures may be used to estimate performance.

First-to-Market: The first product that creates a new product category or a substantial subdivision of a category.

Flexible Gate: A permissive or permeable gate in a Stage-Gate™ process that is less rigid than the traditional "go-stop-recycle" gate. Flexible gates are useful in shortening time-to-market. A permissive gate is one where the next stage is authorized although some work in the almost-completed stage has not yet been finished. A permeable gate is one where some work in a subsequent stage is authorized before a substantial amount of work in the prior stage is completed (R. G. Cooper, *JPIM*, 1994).

Focus Groups: A qualitative market research technique where 8 to 12 market participants are gathered in one room for a discussion under the leadership of a trained moderator. Discussion focuses on a consumer problem, product, or potential solution to a problem. The results of these discussions are not projectable to the general market.

Forecast: A prediction, over some defined time, of the success or failure of implementing with a business plan the decisions derived from an existing strategy.

Function: (1) An abstracted description of work that a product must perform to meet customer needs. A function is something the product or service must do. (2) Term describing an internal group within which resides a basic business capability such as engineering.

Functional Elements: The individual operations that a product performs. These elements are often used to describe a product schematically.

Functional Pipeline Management: Optimizing the flow of projects through all functional areas in the context of the company's priorities.

Functional Schematic: A schematic drawing that is made up of all of the functional elements in a product. It shows the product's functions as well as how material, energy, and signal flow through the product.

Functional Testing: Testing either an element of or the complete product to determine whether it will function as planned and as actually used when sold.

Fuzzy Front End: The messy "getting started" period of product development, which comes before the formal and well-structured product development process, when the product concept is still very fuzzy. It generally consists of the first three tasks (strategic planning, concept generation, and, especially, pre-technical evaluation) of the product development process. These activities are often chaotic, unpredictable, and unstructured. In comparison, the new product development process is typically structured and formal, with a prescribed set of activities, questions to be answered, and decisions to be made.

Fuzzy Gates: Fuzzy gates are conditional or situational, rather than full "go" decisions. Their purpose is to try to balance timely decisions and risk management. Conditional go decisions are "go," subject to a task being successfully completed by a future, but specified, date. Situational gates have some criteria that must be met for all projects, and others that are only required for some projects. For example, a new-to-the world product may have distribution feasibility criteria that a line extension will not have (R. G. Cooper, *JPIM*, 1994).

Gamma Test: A product use test in which the developers measure the extent to which the item meets the needs of the target customers, solves the problems(s) targeted during development, and leaves the customer satisfied.

Gantt Chart: A horizontal bar chart used in project scheduling and management that shows the start date, end date, and duration of tasks within the project.

Gap Analysis: The difference between projected outcomes and desired outcomes. In product development, the gap is frequently measured as the difference between expected and desired revenues or profits from currently planned new products if the corporation is to meet its objectives.

Gate: The point at which a management decision is made to allow the product development project to proceed to the next stage, to recycle back into the current stage to better complete some of the tasks, or to terminate. The number of gates varies by company.

Gatekeepers: The group of managers who serve as advisors, decision makers, and investors in a Stage-Gate™ process. Using established business criteria, this multifunctional group reviews new product opportunities and project progress, and allocates resources accordingly at each gate. This group is also called a Product Approval Committee.

Graceful Degradation: When a product, system, or design slides into defective operation a little at a time, while providing ample opportunity to take corrective preventative action or protect against the worst consequences of failure before it happens. The opposite is catastrophic failure.

Gross Rating Points (GRPs): A measure of the overall media exposure of consumer households (reach times frequency).

Groupware: Software designed to facilitate group efforts such as communication, workflow coordination, and collaborative problem solving. The term generally refers to technologies relying on modern computer networks (external or internal).

Growth Stage: The second stage of the product life cycle. This stage is marked by a rapid surge in sales and market acceptance for the good or service. Products that reach the growth stage have successfully "crossed the chasm."

Heavyweight Team: An empowered project team with adequate resourcing to complete the project. Personnel report to the team leader and are co-located as practical.

Hunting for Hunting Grounds: A structured methodology for completing the Fuzzy Front End of new product development (*see* Chapter 2 of *The PDMA ToolBook*).

Hunting Ground: A discontinuity in technology or the market that opens up a new product development opportunity.

Hurdle Rate: The minimum return on investment or internal rate of return percentage a new product must meet or exceed as it goes through development.

Idea: The most embryonic form of a new product or service. It often consists of a high-level view of the envisioned solution needed to solve the problem identified by a person, team, or firm.

Idea Exchange: A divergent thinking technique that provides a structure for building on different ideas in a quiet, non-judgmental setting that encourages reflection.

Idea Generation (Ideation): All of those activities and processes that lead to creating broad sets of solutions to consumer problems. These techniques may be used in the early stages of product development to generate initial product concepts, in the intermediate stages for overcoming implementation issues, in the later stages for planning launch, and in the post-mortem stage to better understand success and failure in the marketplace (*see* Chapters 12 and 13 in *The PDMA HandBook*).

Idea Merit Index: An internal metric used to impartially rank new product ideas.

Implementation Team: A team that converts the concepts and good intentions of the "should-be" process into practical reality.

Implicit Product Requirement: What the customer expects in a product, but does not ask for, and may not even be able to articulate.

Incremental Improvement: A small change made to an existing product that serves to keep the product fresh in the eyes of customers.

Incremental Innovation: An innovation that improves the conveyance of a currently delivered benefit, but produces neither a behavior change nor a change in consumption.

Industrial Design (ID): The professional service of creating and developing concepts and specifications that optimize the function, value, and appearance of products and systems for the mutual benefit of both user and manufacturer (Industrial Design Society of America) (*see* Chapter 17 of *The PDMA HandBook*).

Information: Knowledge and insight, often gained by examining data.

Information Acceleration: A new concept testing method employing virtual reality. In it, a virtual buying environment is created that simulates the information available (product, societal, political, and technological) in a real purchase situation at some time several years or more into the future.

Informed Intuition: Using the gathered experiences and knowledge of the team in a structured manner.

Initial Screening: The first decision to spend resources (time or money) on a project. The project is born at this point. Sometimes called *idea screening*.

In-licensed: The acquisition from external sources of novel product concepts or technologies for inclusion in the aggregate NPD portfolio.

Innovation: A new idea, method, or device. The act of creating a new product or process. The act includes invention as well as the work required to bring an idea or concept into final form.

Innovation Engine: The creative activities and people that actually think of new ideas. It represents the synthesis phase when someone first recognizes that customer and market opportunities can be translated into new product ideas.

Innovation Strategy: The firm's positioning for developing new technologies and products. One categorization divides firms into Prospectors (those who lead in technology, product and market development, and commercialization, even though an individual product may not lead to profits), Analyzers (fast followers, or imitators, who let the prospectors lead, but have a product development process organized to imitate and commercialize quickly any new product a Prospector has put on the market), Defenders (those who stake out a product turf and protect it by whatever means, not necessarily through developing new products), and Reactors (those who have no coherent innovation strategy).

Innovative Problem Solving: Methods that combine rigorous problem definition, pattern-breaking generation of ideas, and action planning that results in new, unique, and unexpected solutions.

Integrated Architecture: A product architecture in which most or all of the functional elements map into a single or very small number of chunks. It is difficult to subdivide an integrally designed product into partially functioning components.

Integrated Product Development (IPD): A philosophy that systematically employs an integrated team effort from multiple functional disciplines to develop effectively and efficiently new products that satisfy customer needs.

Intellectual Property (IP): Information, including proprietary knowledge, technical competencies, and design information, that provides commercially exploitable competitive benefit to an organization.

Internal Rate of Return (IRR): The discount rate at which the present value of the future cash flows of an investment equals the cost of the investment. The discount rate with a net present value of 0.

Intrapreneur: The large-firm equivalent of an entrepreneur. Someone who develops new enterprises within the confines of a large corporation.

Introduction Stage: The first stage of a product's commercial launch and the product life cycle. This stage is generally seen as the point of market entry, user trial, and product adoption.

ISO-9000: A set of five auditable standards of the International Standards Organization that establishes the role of a quality system in a company and which is used to assess whether the company can be certified as compliant to the standards. ISO-9001 deals specifically with new products.

Issue: A certainty that will affect the outcome of a project, either negatively or positively. Issues require investigation as to their potential impacts, and decisions about how to deal with them. Open issues are those for which the appropriate actions have not been resolved, while closed issues are ones that the team has dealt with successfully.

Journal of Product Innovation Management: The premier academic journal in the field of innovation, new product development, and management of technology. The journal, which is owned by the PDMA, is dedicated to the advancement of management practice in all of the functions involved in the total process of product innovation. Its purpose is to bring to managers and students of product innovation the theoretical structures and the practical techniques that will enable them to operate at the cutting edge of effective management practice.

Kaizen: A Japanese term describing a process or philosophy of continuous improvement.

Launch: The process by which a new product is introduced into the market for initial sale (*see* Chapters 25 and 26 of *The PDMA HandBook*).

Lead Users: Users for whom finding a solution to one of their consumer needs is so important that they have modified a current product or invented a new product to solve the need themselves because they have not found a supplier who can solve it for them. When these consumers' needs are portents of needs that the center of the market will have in the future, their solutions are new product opportunities.

Learning Organization: An organization that continuously tests and updates the experience of those in the organization, and that transforms that experience into improved work processes and knowledge that is accessible to the whole organization and relevant to its core purpose (*see* Continuous Learning Activity).

Life Cycle Cost: The total cost of acquiring, owning, and operating a product over its useful life. Associated costs may include: purchase price, training expenses, maintenance expenses, warrantee costs, support, disposal, and profit loss due to repair downtime.

Lightweight Team: New product team charged with successfully developing a product concept and delivering it to the marketplace. Resources are, for the most part, not dedicated and the team depends on the technical functions for resources necessary to get the work accomplished.

Line Extension: A form of derivative product that adds or modifies features without significantly changing the product functionality.

Long-term Success: The new product's performance in the long run or at some large fraction of the product's life cycle.

"M" Curve: An illustration of the volume of ideas generated over a given amount of time. The illustration often looks like two arches from the letter M.

Maintenance Activity: That set of product development tasks aimed at solving initial market and user problems with the new product or service.

Manufacturability: The extent to which a new product can be easily and effectively manufactured at minimum cost and with maximum reliability.

Manufacturing Assembly Procedure: Procedural documents normally prepared by manufacturing personnel that describe how a component, subassembly, or system will be put together to create a final product.

Manufacturing Design: The process of determining the manufacturing process that will be used to make a new product (*see* Chapter 23 of *The PDMA HandBook*).

Manufacturing Test Specification and Procedure: Documents prepared by development and manufacturing personnel that describe the performance specifications of a component, subassembly, or system that will be met during the manufacturing process, and that describe the procedure by which the specifications will be assessed.

Market Conditions: The characteristics of the market into which a new product will be placed, including the number of competing products, level of competitiveness, and growth rate.

Market Development: Taking current products to new consumers or users. This effort may involve making some product modifications.

Market-Driven: Allowing the marketplace to direct a firm's product innovation efforts.

Market Research: Information about the firm's customers, competitors, or markets. Information may be from secondary sources (already published and publicly available) or primary sources (from customers themselves). Market research may be qualitative in nature, or quantitative (*see* Qualitative Market Research and Quantitative Market Research).

Market Segmentation: The act of dividing an overall market into groups of consumers with similar needs, where each of the groups differs from others in the market in some way that is important to the design or marketing of the product.

Market Share: A company's sales in a product area as a percent of the total market sales in that area.

Market Testing: The product development stage when the new product and its marketing plan are tested together. A market test simulates the eventual marketing mix and takes many different forms, only one of which bears the name *test market*.

Maturity Stage: The third stage of the product life cycle. This is the stage where sales begin to level off due to market saturation. It is a time when heavy competition, alternative product options, and (possibly) changing buyer or user preferences start to make it difficult to achieve profitability.

Metrics: A set of measurements to track product development and allow a firm to measure the impact of process improvements over time. These measures generally vary by firm but may include measures characterizing both aspects of the process, such as time to market, and duration of particular process stages, as well as outcomes from product development such as the number of products commercialized per year and percentage of sales due to new products.

Modular Architecture: A product architecture in which each functional element maps into its own physical chunk. Different chunks perform different functions, the interactions between the chunks are minimal, and they are generally well defined.

Monitoring Frequency: The frequency with which performance indicators are measured.

Morphological Analysis: A matrix tool that breaks a product down by needs met and technology components, allowing for targeted analysis and idea creation.

Multifunctional Team: A group of individuals brought together from the different functional areas of a business to work on a problem or process that requires the knowledge, training, and capabilities across the areas to successfully complete the work (*see* Chapter 9 in *The PDMA HandBook* and Chapter 6 in *The PDMA ToolBook*).

Needs Statement: Summary of consumer needs and wants, described in customer terms, to be addressed by a new product.

Net Present Value (NPV): Method to evaluate comparable investments in very dissimilar projects by discounting the current and projected future cash inflows and outflows back to the present value based on the discount rate, or cost of capital, of the firm.

Network Diagram: A graphical diagram with boxes connected by lines that shows the sequence of development activities and the interrelationship of one task with another. Often used in conjunction with a Gantt Chart.

New Concept Development Model: A theoretical construct that provides for a common terminology and vocabulary for the Fuzzy Front End. The model consists of three parts: the uncontrollable influencing factors, the controllable engine that drives the activities in the Fuzzy Front End, and five activity elements: Opportunity Identification, Opportunity Analysis, Idea Generation and Enrichment, Idea Selection, and Concept Definition (*see* Chapter 1 of *The PDMA ToolBook*).

New Product: A term of many opinions and practices, but most generally defined as a product (either a good or service) new to the firm marketing it. Excludes products that are only changed in promotion.

New Product Development (NPD): The overall process of strategy, organization, concept generation, product and marketing plan creation and evaluation, and commercialization of a new product. Also frequently referred to as just *product development*.

New Product Development Process (NPD Process): A disciplined and defined set of tasks and steps that describe the normal means by which a company repetitively converts embryonic ideas into salable products or services (*see* Chapters 6 and 7 of *The PDMA HandBook*).

New Product Development Professional (NPDP): A New Product Development Professional is certified by the PDMA as having mastered the body of knowledge in new product development, as proven by performance on the certification test. To qualify for the NPDP certification examination, a candidate must hold a bachelor's or higher university degree (or an equivalent degree) from an accredited institution and have spent a minimum of two years working in the new product development field.

New Product Idea: A preliminary plan or purpose of action for formulating new products or services.

New Product Introduction (NPI): The launch or commercialization of a new product into the marketplace. Takes place at the end of a successful product development project (*see* Chapters 25 and 26 of *The PDMA HandBook*).

New-to-the-World Product: A good or service that has never before been available to either consumers or producers. The automobile was new-to-the-world when it was introduced, as were microwave ovens and pet rocks.

Nominal Group Process: A brainstorming process in which members of a group first write their ideas out individually, and then participate in group discussion about each idea.

Non-Destructive Test: A test of the product that retains the product's physical and operational integrity.

Non-Product Advantage: Elements of the marketing mix that create competitive advantage other than the product itself. These elements can include marketing communications, distribution, company reputation, technical support, and associated services.

Operations: A term that includes manufacturing but is much broader, usually including procurement, physical distribution, and, for services, management of the offices or other areas where the services are provided.

Operator's Manual: The written instructions to the users of a product or process. These may be intended for the ultimate customer or for the use of the manufacturing operation.

Opportunity: A business or technology gap that a company or individual realizes, by design or accident, that exists between the current situation and an envisioned future in order to capture competitive advantage, respond to a threat, solve a problem or ameliorate a difficulty.

Outsourcing: The process of procuring a good or service from someone else, rather than the firm producing it themselves.

Outstanding Corporate Innovator: Firms acknowledged through a formal vetting process as being outstanding innovators. The basic requirements for receiving this award, which is given yearly by the PDMA, are: (1) sustained success in launching new products over a five-year time frame; (2) significant company growth from new product success; (3) a defined new product development process that can be described to others; (4) distinctive innovative characteristics and intangibles.

Pareto Chart: A bar graph with the bars sorted in descending order used to identify the largest opportunity for improvement. Pareto charts distinguish the "vital few" from the "useful many."

Participatory Design: A democratic approach to design that does not simply make potential users the subjects of user testing, but empowers them to be a part of the design and decision-making process.

Payback: The time, usually in years, from some point in the development process until the commercialized product or service has recovered its costs of development and marketing. While some firms take the point of full-scale market introduction of a new product as the starting point, others begin the clock at the start of development expense.

Payout: The amount of profits and their timing expected from commercializing a new product.

Perceptual Mapping: A quantitative market research tool used to understand how customers think of current and future products. Perceptual maps are visual representations of the positions that sets of products hold in consumers' minds.

Performance Indicators: Criteria on which the performance of a new product in the market are evaluated.

Performance Measurement System: The system that enables the firm to monitor the relevant performance indicators of new products in the appropriate time frame.

PERT (Program Evaluation and Review Technique): An event-oriented network analysis technique used to estimate project duration when there is a high degree of uncertainty in estimates of duration times for individual activities.

Phase Review Process: A staged product development process in which first one function completes a set of tasks, then passes the information generated sequentially to another function, which in turn completes the next set of tasks and then passes everything along to the next function. Multifunctional teamwork is largely absent in these types of product development processes, which may also be called *phase review* or *baton-passing processes.* Most firms have moved from these processes to Stage-Gate™ processes using multifunctional teams.

Physical Elements: The components that make up a product. These can be both components (or individual parts) in addition to minor subassemblies of components.

Pilot Gate Meeting: A trial, informal gate meeting usually held at the launch of a Stage-Gate™ process to test the design of the process and familiarize participants with the Stage-Gate™ process.

Pipeline (Product Pipeline): The scheduled stream of products in development for release to the market.

Pipeline Alignment: The balancing of project demand with resource supply.

Pipeline Inventory: Production of a new product that has not yet been sold to end consumers, but which exists within the distribution chain.

Pipeline Management: A process that integrates product strategy, project management, and functional management to continually optimize the cross-project management of all development-related activities.

Pipeline Management Enabling Tools: The decision-assistance and data-handling tools that aid managing the pipeline. The decision-assistance tools allow the pipeline team to systematically perform trade-offs without losing sight of priorities. The data-handling tools deal with the vast amount of information needed to analyze project priorities, understand resource and skill-set loads, and perform pipeline analysis.

Pipeline Management Process: Consists of three elements; pipeline management teams, a structured methodology, and enabling tools.

Pipeline Management Teams: The teams of people at the strategic, project, and functional levels responsible for resolving pipeline issues.

Platform Product: The design and components that are shared by a set of products in a product family. From this platform, numerous derivative products can be designed.

Portfolio: A set of items. *See also* Project Portfolio and Product Portfolio.

Portfolio Criteria: The set of criteria against which the business judges proposed product development projects to create a balanced and diverse mix of ongoing efforts.

Portfolio Management: A business process by which a business unit decides on the mix of active projects, staffing, and dollar budget allocated to each project currently being undertaken. *See also* Pipeline Management (*see* Chapter 13 of *The PDMA ToolBook*).

Pre-Production Unit: A product that looks like and acts like the intended final product but is made either by hand or in pilot facilities rather than by the final production process.

Process Champion: The person responsible for the daily promotion of and encouragement to use a formal business process throughout the organization. They are also responsible for the ongoing training, innovation input, and continuous improvement of the process.

Process Managers: The operational managers responsible for ensuring the orderly and timely flow of projects through the process.

Process Map: A workflow diagram that uses an x-axis for process time and a y-axis that shows participants and tasks.

Process Mapping: The act of identifying, defining, and sequencing all of the steps associated with completing any particular process.

Process Maturity Level: The amount of movement of a reengineered process from the "as-is" map, which describes how the process operated initially, to the "should-be" map of the desired future state of the operation.

Process Owner: The executive manager responsible for the strategic results of the process. This includes process throughput, quality of output, and participation within the organization (*see* Section 3 of *The PDMA ToolBook* for four tools that process owners might find useful, and see Chapter 29 of *The PDMA HandBook*).

Process Re-engineering: A discipline to measure and modify organizational effectiveness by documenting, analyzing, and comparing an existing process to "best-in-class" practice, and then implementing significant process improvements or installing a whole new process.

Product: Term used to describe all goods, services, and knowledge sold. Products are bundles of attributes (features, functions, benefits, and uses) and can be either tangible, as in the case of physical goods, or intangible, as in the case of those associated with service benefits, or a combination of the two.

Product Advisory Committee: The multifunctional group of managers who serve as advisors, decision-makers, and resource allocators at the decision gates in a Stage-Gate process. They use established criteria to review product development projects at each gate (*see* Chapter 7 of *The PDMA ToolBook*).

Product Approval Committee: The group of managers who serve as advisors, decision makers, and investors in a Stage-Gate™ process: a company's NPD executive committee. Using established business criteria, this multifunctional group reviews new product opportunities and project progress, and allocates resources accordingly at each gate (*see* Chapter 7 of *The PDMA ToolBook*).

Product Architecture: The way in which the functional elements are assigned to the physical chunks of a product and the way in which those physical chunks interact to perform the overall function of the product (*see* Chapter 16 of *The PDMA HandBook*).

Product and Process Performance Success: The extent to which a new product meets its technical performance and product development process performance criteria.

Product Definition: Defines the product, including the target market, product concept, benefits to be delivered, positioning strategy, price point, and product requirements and design specifications.

Product Development: The overall process of strategy, organization, concept generation, product and marketing plan creation, execution, and evaluation, and commercialization of a new product (*see* Chapters 19–22 of *The PDMA HandBook*).

Product Development Check List: A pre-determined list of activities and disciplines responsible for completing those activities used as a guideline to ensure that all the tasks of product development are considered prior to commercialization.

Product Development & Management Association (PDMA): A not-for-profit professional organization whose purpose is to seek out, develop, organize, and disseminate leading edge information on the theory and practice of product development and product development processes. The PDMA uses local, national, and international meetings and conferences, educational workshops, a quarterly newsletter (*Visions*), a bi-monthly scholarly journal (*Journal of Product Innovation Management*), research proposal and dissertation proposal competitions, *The PDMA HandBook of New Product Development*, and *The PDMA ToolBook for New Product Development* to achieve its purposes. The association also manages the certification process for New Product Development Professionals (www.pdma.org).

Product Development Portfolio: The collection of new product concepts that are within the firm's ability to develop, are most attractive to the firm's customers, and deliver short- and long-term corporate objectives, spreading risk and diversifying investments (*see* Chapter 13 of *The PDMA ToolBook*).

Product Development Process: A disciplined and defined set of tasks and steps that describe the normal means by which a company repetitively converts embryonic ideas into salable products or services (*see* Chapters 6 and 7 of *The PDMA HandBook*).

Product Development Strategy: The strategy that guides the product innovation program.

Product Development Team: A multifunctional group of individuals chartered to plan and execute a new product development project.

Product Discontinuation: A product or service withdrawn or removed from the market because it no longer provides an economic, strategic, or competitive advantage in the firm's portfolio of offerings (*see* Chapter 28 of *The PDMA HandBook*).

Product Discontinuation Timeline: The process and timeframe in which a product is carefully withdrawn from the marketplace. The product may be discontinued immediately after the decision is made, or it may take a year or more to implement the discontinuation timeline, depending on the nature and conditions of the market and product.

Product Failure: A product development project that does not meet the objective of its charter or the marketplace.

Product Family: The set of products that have been derived from a common product platform. Members of a product family normally have many common parts and assemblies.

Product Innovation Charter (PIC): The summary statement of strategy that will guide a department or project team in their efforts to generate a new product. The PIC specifies the arena within which the project will operate, its goals and objectives, and the general approaches to be used. It may apply to a single project or to a program of projects.

Product Interfaces: Internal and external interfaces impacting the product development effort, including the nature of the interface, action required, and timing.

Product Life Cycle: The four stages that a new product is thought to go through from birth to death: introduction, growth, maturity, and decline. Controversy surrounds whether products go through this cycle in any predictable way.

Product Life-Cycle Management: Changing the features and benefits of the product, elements of the marketing mix, and manufacturing operations over time to maximize the profits obtainable from the product over its life cycle.

Product Line: A group of products marketed by an organization to one general market. The products have some characteristics, customers, and/or uses in common and may also share technologies, distribution channels, prices, services, and other elements of the marketing mix.

Product Management: Ensuring over time that a product or service profitably meets the needs of customers by continually monitoring and modifying the elements of the marketing mix, including: the product and its features, the communications strategy, distribution channels, and price.

Product Manager: The person responsible for overseeing all of the various activities that concern a particular product. Sometimes called a *brand manager* in consumer packaged goods firms.

Product Plan: Detailed summary of all the key elements involved in a new product development effort, such as product description, schedule, resources, financial estimations, and interface management plan.

Product Platforms: Underlying structures or basic architectures that are common across a group of products or that will be the basis of a series of products commercialized over a number of years.

Product Portfolio: The current set of products and product lines the firm has in the market (*see* Chapter 13 of *The PDMA ToolBook*).

Product Rejuvenation: The process by which a mature or declining product is altered, updated, repackaged, or redesigned to lengthen the product life cycle and in turn extend or increase sales demand.

Product Requirements Document: The contract between, at a minimum, marketing and development, describing completely and unambiguously the necessary attributes (functional performance requirements) of the product to be developed, as well as information about how achievement of the attributes will be verified (i.e., through testing).

Product Superiority: A product differentiated from those offered by competitors by offering consumers benefits and value for money beyond what similar products offer. This is one of the critical success factors in commercializing new products.

Program Manager: The organizational leader charged with responsibility of executing a portfolio of NPD projects (*see* Section 4 of *The PDMA ToolBook* for four product development tools a program manager may find helpful).

Project Leader: The person responsible for managing an individual new product development project to market or stage completion. He or she is responsible for ensuring that milestones and deliverables are achieved and that resources are utilized effectively. *See also* Team Leader (*see* Sections 1 and 2 of *The PDMA ToolBook* for eight product development tools for project leaders).

Project Management: The set of people, tools, techniques, and processes used to define the project's goal, plan all the work necessary to reach that goal, lead the project and support teams, monitor progress, and ensure that the project is completed in a satisfactory way.

Project Pipeline Management: Fine-tuning resource deployment smoothly for projects during ramp-up, ramp-down, and mid-course adjustments.

Project Plan: A formal, approved document used to guide both project execution and control. Documents planning assumptions and decisions, facilitates communication among stakeholders, and documents approved scope, cost, and schedule.

Project Portfolio: The set of projects in development at any point in time. These will vary in the extent of newness or innovativeness (*see* Chapter 13 in *The PDMA ToolBook*).

Project Sponsor: The authorization and funding source of the project. The person who defines the project goals and to whom the final results are presented. Typically a senior manager.

Project Strategy: The goals and objectives for an individual product development project. It includes how that project fits into the firm's product portfolio, who the target market is, and what problems the product will solve for those customers (*see* Chapter 10 in *The PDMA HandBook*).

Project Team: A multifunctional group of individuals chartered to plan and execute a new product development project.

Prospectors: Firms that lead in technology, product and market development, and commercialization, even though an individual product may not lead to profits. Their general goal is to be first to market with innovation.

Protocol: A statement of the attributes (mainly benefits; features only when required) that a new product is expected to have. A protocol is prepared prior to assigning the project to the technical development team. The benefits statement is agreed to by all parties involved in the project.

Prototype: A physical model of the new product concept. Depending upon the purpose, prototypes may be non-working, functionally working, or both functionally and aesthetically complete.

Psychographics: Characteristics of consumers that, rather than being purely demographic, measure their attitudes, interests, opinions, and lifestyles.

Pull-Through: The revenue created when a new product or service positively impacts the sales of other products or services (the obverse of cannibalization).

Qualitative Cluster Analysis: An individual- or group-based process using Informed Intuition for clustering and connecting data points.

Qualitative Marketing Research: Research conducted with a small number of respondents, either in groups or individually, to gain an impression of their beliefs, motivations, perceptions, and opinions. Frequently used to gather initial consumer needs and obtain initial reactions to ideas and concepts. Results are not representative of the market in general or projectable. Qualitative marketing research is used to show why people buy a particular product, whereas quantitative marketing research reveals how many people buy it (*see* Chapter 11 of *The PDMA HandBook*).

Quality: The collection of attributes, which when present in a product, means a product has conformed to or exceeded customer expectations.

Quality Assurance/Compliance: Function responsible for monitoring and evaluating development policies and practices, to ensure they meet company and applicable regulatory standards.

Quality Control Specification and Procedure: Documents that describe the specifications and the procedures by which they will be measured which a finished subassembly or system must meet before judged ready for shipment.

Quality-by-Design: The process used to design quality into the product, service, or process from the inception of product development.

Quality Function Deployment (QFD): A structured method employing matrix analysis for linking what the market requires to how it will be accomplished in the development effort. This method is most frequently used during the stage of development when a multifunctional team agrees on how customer needs relate to product specifications and the features that deliver those needs. By explicitly linking these aspects of product design, QFD minimizes the possibility of omitting important design characteristics or interactions across design characteristics. QFD is also an important mechanism in promoting multifunctional teamwork.

Quantitative Market Research: Consumer research, often surveys, conducted with a large enough sample of consumers to produce statistically reliable results that can be used in predicting outcomes to the general consumer population. Used to determine importance levels of different customer needs, performance ratings of and satisfaction with current products, probability of trial, repurchase rate, and product preferences. These techniques are used to reduce the uncertainty associated with many other aspects of product development (*see* Chapter 18 of *The PDMA HandBook*).

Radical Innovation: A new product, generally containing new technologies, that significantly changes behaviors and consumption patterns in the marketplace.

Rapid Prototyping: A variety of processes that avoid tooling time in producing prototypes or prototype parts and therefore allow (generally nonfunctioning) prototypes to be produced in hours or days rather than weeks. These prototypes are frequently used to test quickly the product's technical feasibility or consumer interest.

Reactors: Firms that have no coherent innovation strategy. They only develop new products when absolutely forced to by the competitive situation.

Realization Gap: The time between first perception of a need and the launch of a product that satisfies that need.

Render: Process that industrial designers use to visualize their ideas by putting their thoughts on paper with any number of combinations of color markers, pencils, and highlighters, or computer visualization software.

Reposition: To change the position of the product in the minds of customers, either on failure of the original positioning or to react to changes in the marketplace. Most frequently accomplished through changing the marketing mix rather than redeveloping the product.

Resource Matrix: An array that shows the percentage of each person's time that is to be devoted to each of the current projects in the firm's portfolio.

Resource Plan: Detailed summary of all forms of resources required to complete product development, including personnel, equipment, time, and finances.

Responsibility Matrix: Matrix indicating the specific involvement of each functional department or individual in each task or activity in each stage.

Return on Investment (ROI): A standard measure of project profitability, this is the discounted profits over the life of the project expressed as a percentage of development investment.

Rigid Gate: A review point in a Stage-Gate™ process at which all the prior stage's work and deliverables must be complete before the next stage can commence.

Risk: An event or condition that may or may not occur, but if it does occur will impact the ability to achieve a project's objectives. In new product development, risks may take the form of market, technical, or organizational issues. For more on managing product development risks, *see* Chapters 8 and 15 in the *PDMA ToolBook*.

Risk Acceptance: An uncertain event or condition for which the project team has decided not to change the project plan. A team may be forced to accept an identified risk when it is unable to identify any suitable response to the risk.

Risk Avoidance: Project teams who change the project plan to eliminate a risk or to protect the project objectives from any potential impact due to the risk are practicing risk avoidance.

Risk Management: The process of identifying, measuring, and mitigating the business risk in a product development project.

Risk Mitigation: Actions taken to reduce the probability and/or impact of a risk to below some threshold of acceptability.

Risk Tolerance: The level of risk that a project stakeholder is willing to accept. Tolerance levels are context specific. That is, stakeholders may be willing to accept different levels of risk for different types of risk, such as risks of project delay, price realization, and technical potential.

Risk Transference: Actions taken to shift the impact of a risk and the ownership of the risk response actions to a third party.

Roadmapping: A multi-step process to forecast future market and/or technology changes, and then plan the products to address these changes.

Robust Design: Products that are designed to be less sensitive to variations, including manufacturing variation and consumer misuse, increasing the probability that they will perform as intended.

"Rugby" Process: A product development process in which stages are partially or heavily overlapped rather than sequential with crisp demarcations between one stage and its successor.

S-Curve (Technology S-Curve): Technology performance improvements tend to progress over time in the form of an "S" curve. When first invented, technology performance improves slowly and incrementally. Then, as experience with a new technology accrues, the rate of performance increase grows and technology performance increases by leaps and bounds. Finally, some of the performance limits of a new technology start to be reached and performance growth slows. At some point, the limits of the technology may be reached and further improvements are not made. Frequently, the technology then becomes vulnerable to a substitute technology that is capable of making additional performance improvements.

Scanner Test Markets: Special test markets that provide supermarket scanner data from panels of consumers to help assess the product's performance.

Scenario Analysis: A tool for envisioning alternate futures so that a strategy can be formulated to respond to future opportunities and challenges (*see* Chapter 16 of the *PDMA ToolBook*).

Screening: The process of evaluating and selecting new ideas or concepts to put into the project portfolio. Most firms now use a formal screening process with evaluation criteria that span customer, strategy, market, profitability, and feasibility dimensions.

Segmentation: The process of dividing a large and heterogeneous market into more homogeneous subgroups. Each subgroup, or segment, holds similar views about the product, and values, purchases, and uses the product in similar ways (*see* Chapters 3 and 4 of *The PDMA HandBook*).

Senior Management: That level of executive or operational management above the product development team that has approval authority or controls resources important to the development effort.

Sensitivity Analysis: A calculation of the impact that an uncertainty might have on the new product business case. It is conducted by setting upper and lower ranges on the assumptions involved and calculating the expected outcomes (*see* Chapter 16 of *The PDMA ToolBook*).

Services: Products, such as an airline flight or insurance policy, that are intangible or at least substantially so. If totally intangible, they are exchanged directly from producer to user, cannot be transported or stored, and are instantly perishable. Service delivery usually involves customer participation in some important way. Services cannot be sold in the sense of ownership transfer, and they have no title of ownership.

Short-term Success: The new product's performance immediately after launch, well within the first year of commercial sales.

Should-Be Map: A version of a process map depicting how a process will work in the future. A revised "as-is" process map. The result of the team's re-engineering work.

Simulated Test Market: A form of quantitative market research and pre-test marketing in which consumers are exposed to new products and to their claims in a staged advertising and purchase situation. Output of the test is an early forecast of expected sales or market share, based on mathematical forecasting models, management assumptions, and input of specific measurements from the simulation.

Six Sigma: A level of process performance that produces only 3.4 defects for every one million operations.

Slip Rate: Measures the accuracy of the planned project schedule according to the formula:

Slip Rate = ([actual schedule/planned schedule] − 1) * 100%.

Specification: A detailed description of the features and performance characteristics of a product. For example, a laptop computer's specification may read as a 90 megahertz Pentium, with 16 megabytes of RAM and 720 megabytes of hard disk space, 3.5 hours of battery life, a weight of 4.5 pounds, with an active matrix 256 color screen.

Speed to Market: The length of time it takes to develop a new product from an early initial idea to initial market sales. Precise definitions of the start and end point vary from one company to another, and may vary from one project to another within a company (*see* Chapter 24 of *The PDMA HandBook*).

Sponsor: An informal role in the product development project, usually a higher-ranking person in the firm who is not personally involved in the project, but who is ready to extend a helping hand if needed, or provide a barrier to interference by others.

Stage: One group of concurrently accomplished tasks, with specified outcomes and deliverables, of the overall product development process.

Stage-Gate™ Process: A widely employed product development process that divides the effort into distinct time-sequenced stages separated by management decision gates. Multifunctional teams must successfully complete a prescribed set of related cross-functional tasks in each stage prior to obtaining management approval to proceed to the next stage of product development. The framework of the Stage-Gate™ process includes workflow and decision-flow paths and defines the supporting systems and practices necessary to ensure the process's ongoing smooth operation.

Staged Product Development Activity: The set of product development tasks commencing when it is believed there are no major unknowns and that result in initial production of salable product, carried out in stages.

Standard Cost: *See* Factory Cost.

Stop Light Voting: A convergent thinking technique by which participants weight their idea preferences using colored adhesive dots. Also called *preference voting*.

Strategic Balance: Balancing the portfolio of development projects along one or more of many dimensions such as focus versus diversification, short versus long term, high versus low risk, extending platforms versus development of new platforms.

Strategic New Product Development (SNPD): The process that ties new product strategy to new product portfolio planning (*see* Chapter 2 of *The PDMA HandBook*).

Strategic Partnering: An alliance or partnership between two firms (frequently one large corporation and one smaller, entrepreneurial firm) to create a specific new product. Typically, the large firm supplies capital, and the necessary marketing and distribution capabilities, while the small firm supplies specialized technical or creative expertise.

Strategic Pipeline Management: Focuses on achieving strategic balance, which entails allocating resources among the numerous opportunities and adjusting the organization's skill sets to deliver products.

Strategic Plan: Establishes the vision, mission, values, objectives, goals, and strategies of the organization's future state.

Strategy: The organization's vision, mission, and values. One subset of the firm's overall strategy is its Innovation Strategy.

Subassembly: A collection of components that can be put together as a single assembly to be inserted into a larger assembly or final product. Often the subassembly is tested for its ability to meet some set of explicit specifications before inclusion in the larger product.

Success: A product that meet's its goals and performance expectations. Product development success has four dimensions. At the project level, there are three dimensions: financial, customer-based, and product technical performance. The fourth dimension is new product contribution to overall firm success (*see* Chapters 1, 31, and 32 of *The PDMA HandBook)*.

Support Service: Any organizational function whose primary purpose is not product development but whose input is necessary to the successful completion of product development projects.

System Hierarchy Diagram: The diagram used to represent product architectures. This diagram illustrates how the product is broken into its chunks.

Systems and Practices: Established methods, procedures, and activities that either drive or hinder product development. These may relate to the firm's day-to-day business or may be specific to product development.

Systems and Practices Team: Senior managers representing all functions who work together to identify and change those systems and practices hindering product development and who establish new tools, systems, and practices for improving product development.

Target Cost: A cost objective established for a new product based on consideration of customer affordability. Target cost is treated as an independent variable that must be satisfied along with other customer requirements.

Target Market: The group of consumers or potential customers selected for marketing. The market segment of consumers likely to buy the products within a given category. These are sometimes called the *prime prospects*.

Task: The smallest describable unit of accomplishment in completing a deliverable.

Team: That group of persons who participate in the product development project. Frequently each team member represents a function, department, or specialty. Together they should represent the full set of capabilities needed to complete the project (*see* Chapter 9 in *The PDMA HandBook* and Chapter 6 in *The PDMA ToolBook*).

Team Leader: The person leading the new product team. Responsible for ensuring that milestones and deliverables are achieved, even though they may not have any authority over project participants (*see* Sections 1 and 2 of *The PDMA ToolBook* for eight product development tools for team leaders).

Team Spotter's Guide: A questionnaire used by a team leader (or team members) to assess and diagnose the quality of the team's functioning (*see* Chapter 6 in the *PDMA ToolBook*).

Technology-Driven: A new product or new product strategy based on the strength of a technical capability. If overemphasized, sometimes called *solutions in search of problems*.

Technology Road Map: A graphic representation of technology evolution or technology plans mapped against time. It is used to guide new technology development for or technology selection in developing new products.

Technology Stage-Gate™ (TSG): A process for managing the technology development efforts when there is high uncertainty and risk. The process brings a structured methodology for managing new technology development without thwarting the creativity needed in this early stage of product development. It is specifically intended to manage high-risk technology development projects when there is uncertainty and risk that the technology discovery may never occur and therefore the ultimate desired product characteristics might never be achieved (*see* Chapter 11 in *The PDMA ToolBook*).

Technology Transfer: The process of converting scientific findings from research laboratories into useful products.

Test Markets: The launching of a new product into one or more limited geographic regions in a very controlled manner, and measuring consumer response to the product and its launch. When multiple geographies are used in the test, different advertising or pricing policies may be tested and the results compared.

Think Links: Stimuli used in divergent thinking to help participants make new connections using seemingly unrelated concepts from a list of people, places, or things.

Think-Tank: Environments, frequently isolated from normal organizational activities, created by management to generate new ideas or approaches to solving organizational problems.

Thought Organizers: Tools that help categorize information associated with ideas such that the ideas can be placed into groups that can be more easily compared or evaluated.

Three Rs: The fundamental steps of Record, Recall, and Reconstruct, which most creative minds go through when generating new product ideas.

Threshold Criteria: The minimum acceptable performance targets for any product development project being proposed or considered.

Thumbnail: The most minimal form of sketching, usually using pencils, to represent a product idea.

Time to Market: The length of time it takes to develop a new product from an early initial idea to initial market sales. Precise definitions of the start and end point vary from one company to another, and may vary from one project to another within a company.

Tone: The feeling, emotion, or attitude most associated with using a product. The appropriate tone is important to include in consumer new product concepts and advertising.

Total Quality Management (TQM): A business improvement philosophy that comprehensively and continuously involves all of an organization's functions in improvement activities.

Tracking Studies: Surveys of consumers (usually conducted by telephone) following the product's launch to measure consumer awareness, attitudes, trial, adoption, and repurchase rates.

TRIZ: A Russian acronym for the Theory of Inventive Problem Solving, which is a systematic method of solving problems and creating multiple-alternative solutions. It is based on an analysis and codification of technology solutions from millions of patents. The method enhances creativity by getting individuals to think beyond their own experience and to reach across disciplines to solve problems using solutions from other areas of science.

Uncertainty Range: The spread between the high (best case) and low (worst case) values in a business assumption.

User: Any person who uses a product or service to solve a problem or obtain a benefit, whether or not they purchase it. Users may consume a product, as in the case of a person using shampoo to clean their hair or eating a potato chip to assuage hunger. Users may not directly consume a product, but may interact with it over a longer period of time, like a family owning a car, with multiple family members using it for many purposes over a number of years. Products also are employed in the production of other products or services, where the users may be the manufacturing personnel who operate the equipment.

Utilities: The weights derived from conjoint analysis that measure how much a product feature contributes to purchase interest or preference.

Value: Any principle to which a person or company adheres with some degree of emotion. It is one of the elements that enter into formulating a strategy.

Value-added: The act or process by which tangible product features or intangible service attributes are bundled, combined, or packaged with other features and attributes to create a competitive advantage, reposition a product, or increase sales.

Value Analysis: A technique for analyzing systems and designs. Its purpose is to help develop a design that satisfies users by providing the needed user requirements in sufficient quality at an optimum (minimum) cost.

Value Chain: As a product moves from raw material to finished good delivered to the customer, value is added at each step in the manufacturing and delivery process. The value chain indicates the relative amount of value added at each of these steps.

Value Proposition: A short, clear, and simple statement of how and on what dimensions a product concept will deliver value to prospective customers. The essence of "value" is embedded in the trade-off between the benefits a customer receives from a new product and the price a customer pays for it (*see* Chapter 3 of the *PDMA ToolBook*).

Vertical Integration: A company that operates across multiple levels of the value chain is vertically integrated. In the early 1900s, Ford Motor Company was extremely vertically integrated, as it owned forests and operated logging and wood finishing and glass-making businesses. They made all of the components that went into automobiles, as well as most of the raw materials used in those components. This is called "backward integration." Today, Ford is much less vertically integrated in total, and the direction of their vertical integration has changed. They are now more "forward integrated," providing services closer to the customer such as financing for automobile purchases.

Virtual Customer: A set of web-based customer methods for gathering voice-of-the-customer data in all phases of product development (*see* Dahan and Hauser, *JPIM*, July 2002).

Virtual Product Development: Paperless product development. All design and analysis is computer-based.

Virtual Reality: Technology that enables a designer or user to "enter" and navigate a computer-generated 3-D environment. Users can change their viewpoint and interact with the objects in the scene in a way that mimics real-world experiences.

Virtual Team: Geographically dispersed teams that communicate and work primarily electronically may be called *virtual teams*.

Vision: An act of imagining, guided by both foresight and informed discernment, that reveals both possibilities and practical limits in new product development. The most desirable, ultimate state of a product or organization.

Visionary Companies: Leading innovators in their industries, they rank first or second in market share, profitability, growth, and shareholder performance. A substantial number (e.g., 30 percent or more) of their sales are from products introduced in the last three years, and everyone wants to benchmark them.

Visions: The new product development practitioner-oriented magazine of the PDMA.

Voice of the Customer (VOC): A process for eliciting needs from consumers that uses structured in-depth interviews to lead interviewees through a series of situations in which they have experienced and found solutions to the set of problems being investigated. Needs are obtained through indirect questioning, thereby coming to understand how the consumers found ways to meet their needs, and, more important, why they chose the particular solutions they found (*see* Chapter 11 of *The PDMA ToolBook*).

Workflow Design Team: Functional contributors who work together to create and execute the workflow component of a Stage-Gate™ system. They decide how the firm's Stage-Gate™ process will be structured, what tasks it will include, what decision points will be included, and who is involved at all points.

Worth What Paid For (WWPF): The quantitative evaluation by a person in your customer segment of the question: "Considering the products and services that your vendor offers, are they worth what you paid for them?"

Index

Abbott Laboratories, 204
Acterna, 195
Agreed-to-kill points, 279–280
AirNet Systems, 99
Albright, Rich, 252
Alcoa, concept selection process, 28
Algebraic models, process modeling, 412–414
American Demographics, 43
Anthropological research approach, 97
AnyStream, elevator test, 65
Approval of product. *See* Product approval
Arm & Hammer, lead user research, 244
Arm function, affecting factors, 309–310
Assumptions analysis
    assumption generation, 397, 402
    risk identification method, 196, 397
    uncertainty ranges, 404–406
Attributes
    definition of, 98
    life cycle of, 101–102
    See also Customer-perceived value (CPV)
Awards program, employee ideas programs, 237–238

Bank One Corporation, One Great Idea program, 224–227
Basic attributes, elements of, 102
Benefit and cost attributes, customer-perceived value (CPV) attributes, 98–100
Black & Decker Appliances, 39, 44, 56
Body function, affecting factors, 307–308
Bottom-up approach, strategic fit method, 347
Brainstorming
    advantages of, 46
    risk identification method, 196
Breakeven analysis, in business case, 131–132
Bubble diagrams, 339–345, 351
    risk-reward diagram, 341, 342–344
    scoring model combination, 345
    types of charts, 341
    variations of, 343–345
Business case, 126–132
    champion, writing of case, 138
    credibility of, 133
    financial information, 126, 131–132
    market opportunity worksheet, 130–131
    outline for, 126–127
    product market matrix, 128–129
    and resources for project, 132–134
    risk management, 389, 396–397
    strategy, statement of, 132
    tone and content of, 127–128
    See also Valley of Death model
Business periodicals, as information source, 104
Business strategy, and new concept development (NCD), 12–14
Buying needs, 95

Capacity analysis, 350–353
    purposes of, 350–351
    resource demand and active projects, 352–353
    resource demand and new product goals, 353
Case development, 53–54
    components of case, 54
    and scenario analysis, 54
Champions
    and project approval. *See* Valley of Death model
    skills of, 27–28, 137–138
    traits of, 119

Charters
    authors of, 273, 274
    and concept screening, 56
    contents of, 39
    minicharter building, 49, 51–53
    outline for, 274
    purpose of, 40, 273
    in technology stage-gate (TechSG) process, 273–275
Charts
    for customer-perceived value (CPV) relative performance ratings, 108–111
    types of, 339, 346
Checklists
    portfolio management tool, 351, 354
    risk identification method, 196
Choice modeling
    cost of, 107
    method of, 107
Church & Dwight Company, lead user research, 244
Cluster analysis
    cluster, nature of, 49, 51
    cluster testing, 49
    purpose of, 49–50
    qualitative cluster analysis, 49–50
Cognitive capabilities, variations in, 302–303
Cold calls, difficulties of, 69–70
Coleman, 51
Commercial risk, sources of, 198
Commercialization stage, 5, 6
Companies, information sources on, 104
Competition, versus complementors, 10
Competitive attributes, nature of, 102
Competitive intelligence analysis
    functions of, 16
    for large-scale opportunity, 18–19
    SWOT analysis, 103
Complementors, and industry building, 10
Concept definition, 26–29
    gatekeeper decision-making, 26
    goal deliberation process, 27
    and technology Stage-gate™ (TSG) process, 26, 28–29
Concept screening process, 56–57
    multistage process, 56
    risk management, 388, 394–396
Concept statement, 56
    elements of, 395
    example of, 400–401
    purpose of, 294
    risk discussion, 396
Confidentiality, lead user research, 262
Confirmatory research, 93–94
    goals of, 93
    pros/cons of, 93–94
    survey research, 106–107
Conjoint analysis, 107
Contextual inquiry, customer needs research, 97
Contingency planning, 393, 406–407
    trade-offs in, 406
Control Data Corporation (CDC), off-site facility, 153
Copco, Inc., 300
Corning, catalytic converter development, 11–12
Cousins, Morison, 324
Customer-based research
    advisory panels, 97
    customer events, 97
    customer shadowing, 97